PHOTOCHEMISTRY AND
REACTION KINETICS

PHOTOCHEMISTRY AND REACTION KINETICS

EDITED BY

P. G. ASHMORE
F. S. DAINTON
T. M. SUGDEN

CAMBRIDGE
AT THE UNIVERSITY PRESS
1967

CAMBRIDGE UNIVERSITY PRESS
Cambridge, New York, Melbourne, Madrid, Cape Town, Singapore,
São Paulo, Delhi, Dubai, Tokyo, Mexico City

Cambridge University Press
The Edinburgh Building, Cambridge CB2 8RU, UK

Published in the United States of America by Cambridge University Press, New York

www.cambridge.org
Information on this title: www.cambridge.org/9780521147477

First published 1967
First paperback printing 2010

A catalogue record for this publication is available from the British Library

Library of Congress Catalogue Card Number: 67–10157

ISBN 978-0-521-04065-5 Hardback
ISBN 978-0-521-14747-7 Paperback

Dedicated to

RONALD GEORGE WREYFORD NORRISH

*to commemorate his lasting contributions to
photochemistry and reaction kinetics*

CONTENTS

List of authors   *page* xiii

Preface by F. S. Dainton   xv

CHAPTER 1

The Contributions of R. G. W. Norrish to Photochemistry
by W. A. Noyes, Jr.
page 1

CHAPTER 2

The Contributions of R. G. W. Norrish to Combustion
by Bernard Lewis
page 22

CHAPTER 3

Photochemistry in the Liquid Phase
by C. H. Bamford and R. P. Wayne

1   The influence of phase   26

2   Photochemistry of inorganic systems   28

  2.1   Water   28

  2.2   Photovoltaic phenomena in water   30

  2.3   Hydrogen peroxide   33

  2.4   Ozone   35

  2.5   Chlorine dioxide   39

  2.6   Insertion reactions of oxygen atoms and of methylene   41

3   Photochemistry of organic systems   42

  3.1   Aldehydes and ketones   42

  3.2   Amines and amides   47

  3.3   Organic peroxides   49

  3.4   Photosensitization of polymerization   56

CHAPTER 4

Gaseous photochlorination
by F. S. Dainton and P. B. Ayscough

1   Introduction   *page* 64

2   Classical studies on photochlorination   66

    2.1   The hydrogen–chlorine reaction   66

    2.2   Classical studies of the photochlorination of olefins   70

3   Measurement of absolute rate constants   71

    3.1   Reactions of Cl atoms with CO, NOCl and saturated   72
        hydrocarbons

4   General mechanisms for the photochlorination of hydrocarbons   74

    4.1   Limiting conditions   75

    4.2   Combination of chlorine atoms   79

    4.3   Excited chloroalkyl radicals   81

5   Competitive photochlorination reactions   84

    5.1   Addition–addition   84

    5.2   Abstraction–abstraction   85

    5.3   Addition–abstraction   87

    5.4   Addition–dehydrochlorination   88

6   The theoretical interpretation of reaction rates   88

CHAPTER 5

Flash Photolysis
by George Porter

1   Introduction   93

2   Flash photolysis and other fast reaction techniques   93

3   The experimental method   95

    3.1   Flash spectroscopy   95

    3.2   Kinetic spectrophotometry   96

4 Applications of flash photolysis   *page* 98

   4.1 General   98

   4.2 Free radical spectroscopy   100

   4.3 Excited state absorption spectra   101

   4.4 Gas kinetics   102

   4.5 Solution kinetics   105

   4.6 Photobiology   107

5 Recent and future developments   108

CHAPTER 6

Flash Photolytic Studies of Free Radicals
In the Gas Phase
by B. A. Thrush

1 Introduction   112

2 Kinetic studies of free radicals   114

   2.1 The halogens   114

   2.2 Other kinetic studies   118

3 Spectroscopic studies of free radicals   121

4 The study of combustion by flash photolysis   126

CHAPTER 7

Energy Transfer in Molecular Collisions
by A. B. Callear

1 Introduction   133

2 Theory   135

3 Rotation–translation energy transfer   145

4 Vibration–translation energy transfer   149

5 Vibration–vibration energy transfer   157

6 Rotation–vibration energy transfer   *page* 164

7 Electronic–vibration and electronic–translation energy transfer with $\Delta E < 1\ eV$ 165

8 Electronic–vibration and electronic–translation energy transfer with $\Delta E > 1\ eV$ 175

9 Electronic–electronic energy transfer   180

10 General applications and discussion   189

CHAPTER 8

Polymer Chemistry
by J. C. Bevington

1 Introduction   198

2 Initiation of radical polymerization   199

3 Acceleration during polymerization   204

4 Polymerization at high pressures   210

5 Co-polymerization   212

6 Cross-linking   219

7 Polymerization of aldehydes   220

8 Cationic and Ziegler–Natta polymerizations   224

CHAPTER 9

Modern Concepts of the Mechanism of Hydrocarbon Oxidation in the Gas-Phase
by N. N. Semenov

1 A summary of modern concepts   229

2 Dehydrogenation scheme   237

3 Possible participation of excited species in branching   241

CHAPTER 10

The Interpretation of Cool Flame and Low-Temperature Combustion Phenomena
by John H. Knox

1 Introduction   page 250

2 Experimental features of cool flame combustion   259

3 Recent work on the slow oxidation mechanism of lower
  hydrocarbons   269

4 The mechanism of the slow oxidation of lower hydrocarbons   276

  4.1 Chain-propagating reactions   276

  4.2 Reactions forming minor products   280

  4.3 Branching and termination reactions   281

5 An interpretation of the negative temperature coefficient   281

CHAPTER 11

The Sensitization and Inhibition of Ignitions
by P. G. Ashmore

1 Introduction   287

2 Spontaneous ignition boundaries   288

  2.1 Thermal ignitions   290

  2.2 Isothermal ignitions   292

3 The branched-chain thermal theory of ignitions   294

  3.1 Norrish's early work   294

  3.2 Recent work on the system $H_2 + O_2 + NO_2 + NO$ 298

  3.3 Hydrocarbons and ignitions in $H_2 + O_2$ 302

  3.4 Self-inhibition in $H_2 + O_2$ ignitions   306

  3.5 Halogenated compounds and ignitions   308

  3.6 Other sensitizers and inhibitors in $H_2 + O_2$ ignitions   310

4 Sensitized ignitions of methane and oxygen   310

5 Sensitized ignitions of carbon monoxide and oxygen   *page* 316

6 Sensitized ignitions of hydrogen and chlorine   320

7 Conclusions   326

CHAPTER 12

The Pyrolyses of Paraffins
by J. H. Purnell and C. P. Quinn

1 Introduction   330

2 Inhibition of paraffin pyrolyses   331

   2.1 The mechanism of inhibition   333

3 The uninhibited pyrolyses   336

   3.1 Surface effects   336

   3.2 Oxygen effects   338

4 Mechanisms of uninhibited pyrolyses   339

   4.1 *n*-Butane pyrolysis   339

   4.2 Ethane pyrolysis   340

   4.3 Propane pyrolysis   344

5 The derivation of fundamental rate constants   346

6 Conclusion   348

Index   353

LIST OF AUTHORS

PROFESSOR P. G. ASHMORE. Department of Chemistry, University of Manchester Institute of Science and Technology

DR P. B. AYSCOUGH. School of Chemistry, University of Leeds

PROFESSOR C. H. BAMFORD. Department of Inorganic, Physical and Industrial Chemistry, University of Liverpool

PROFESSOR J. C. BEVINGTON. Department of Chemistry, University of Lancaster

DR A. B. CALLEAR. Department of Physical Chemistry, University of Cambridge

DR F. S. DAINTON. Vice-Chancellor, University of Nottingham

DR J. H. KNOX. Department of Chemistry, University of Edinburgh

DR BERNARD LEWIS. Honorary President, The Combustion Institute, Pittsburgh

PROFESSOR W. A. NOYES, JR. Department of Chemistry, University of Texas

PROFESSOR G. PORTER. The Royal Institution, London

PROFESSOR J. H. PURNELL. Department of Chemistry, University College of South Wales, Swansea

ACADEMICIAN N. N. SEMENOV. Academy of Sciences, Institute of Chemical Physics, Moscow

DR B. A. THRUSH. Department of Physical Chemistry, University of Cambridge

DR R. P. WAYNE. Physical Chemistry Laboratory, University of Oxford

PREFACE

Professor R. G. W. Norrish retired from the Chair of Physical Chemistry in the University of Cambridge on 30 September 1965, almost fifty years after he entered Emmanuel College as a Scholar. Apart from his student days and service in World War I, the whole of this period was spent as a researcher and teacher in the Cambridge University Chemical Laboratories, either in the old buildings in Pembroke Street and in Free School Lane, or latterly in the palatial premises in Lensfield Road. Physical chemistry penetrated English universities much more slowly than German universities, and its recognition in England as a separate discipline is essentially a phenomenon of the twentieth century. When Norrish began his researches, the quantitative approach to chemistry was dominated by the thermodynamics of systems in equilibrium. In contrast, theory and experiment concerned with the *rate* of chemical change were both in a rudimentary state and in importance were regarded as subsidiary to thermodynamics and electrochemistry. Einstein's Law dominated the infant photochemistry; polymers were often considered as unwelcome, insoluble, non-crystalline manifestations of the experimental ineptitude of organic chemists; the pursuit of collisional mechanisms for first-order homogeneous reactions was thought to be self-defeating; flame, explosions and combustion, whilst admirably suited to dramatic lecture bench demonstrations, were largely the preserve of the mines department, the fuel chemist, and the gas engineer; chemiluminescence was similarly seen to be well adapted to a popular evening discourse in a dimly lit lecture theatre, but otherwise was hardly worthy of serious investigation. Now, often largely as a result of Norrish's own studies, our insights into the mechanisms of photochemistry, combustion, luminescence, polymerization, chain reactions, and the flow of energy between molecules and between different modes within molecules are incomparably deeper. Reaction kinetics has grown in stature to stand beside thermodynamics, quantum mechanics, and statistics (which links all three) as one of the four great divisions of physical chemistry. It therefore seemed appropriate to some of us that on the occasion of Norrish's retirement his achievement should be marked by a book comprising a series of articles dealing in a contemporary way with themes which, at one time or another, have been his major preoccupation. We hope that each article will, in its different way, be of use to undergraduates and graduate students alike as well as affording them a glimpse of the way in which the attitudes and thinking of a chemist in a given field develop over several decades.

This is not the place to describe Norrish the man, but I cannot forbear to comment that these pages will illustrate the contribution which can be made by an obsessive experimentalist, for whom a sufficient reward is that the description in quantitative as well as in qualitative terms of an observed phenomenon will stand the test of time. Norrish has made additions to theory, but these are largely as by-products of his elegant experiments, and were not generated *a priori*. His arguments about theory will be recalled by many of his collaborators as having a unique personal character, so that one sensed his deep sympathy with Berzelius, who is reputed to have remarked, 'when I see a wrong theoretical interpretation I feel it even if I do not know the right one'.

For many years Norrish gave a lecture course to undergraduates called 'Photochemistry and Reaction Kinetics'. We have therefore chosen this title for this book, which is offered to him in affection and admiration by some of his former research students and three of his distinguished near-contemporaries from the U.S.A. and the U.S.S.R., whose work has affinities with his.

F. S. D.

December, 1965

NOTE

Superior figures in brackets refer to the numbered
reference lists at the end of chapters

1

THE CONTRIBUTIONS OF
R. G. W. NORRISH
TO PHOTOCHEMISTRY

W. A. NOYES, JR.

In 1923 there appeared in the *Proceedings of the Royal Society* two papers [1] by E. K. Rideal and R. G. W. Norrish on the photochemistry of potassium permanganate. These descriptions of research performed by a young man working in the laboratory of his preceptor signalled the beginning of a career which has had a lasting impact on a branch of chemistry previously neglected and all too little understood. Few people live to see a field grow from infancy to maturity during their lifetimes and have the right to feel that without their own contributions this growth could not have occurred so well and so rapidly.

To realize the significance of the changes in the field of photochemistry since 1923, it is necessary to recall the situation as it was then. Chlorination reactions had been studied for nearly a hundred years, and these studies had culminated in the proof, through photochemistry, that many of the reactions are chain reactions and that in the hydrogen–chlorine reaction the chain propagators are atoms. This proof of the existence and nature of chains could not have been realized without the fundamental law of photochemical equivalence based on the quantum theory and enunciated by Einstein. Nevertheless, the detailed proof of all that occurs in photochlorinations was difficult to obtain because of the reactivity of chlorine and the effect of impurities. If, perhaps, undue attention had been paid to photochlorinations, this may have been because they are chain reactions, so that problems of analysis of products are minimized, and because chlorine absorbs in the visible and near ultraviolet ranges, so that relatively inexpensive reaction vessels and optical equipment could be used. The essential outline of the hydrogen–chlorine reaction was known, even though details were still ambiguous.

In 1923 James Franck had not yet published his epoch-making paper [2] which correlated spectrum types with dissociation. Predissociation had not been discovered by Henri and Teves. [3] The classification of the spectra of diatomic molecules had hardly been started and the symbolism was in a state of utter confusion.

Thus the photochemist in the early 1920's was engaged in determining quantum yields and in discussing mechanisms without having been provided by the physicist with the fundamental tools of the trade. Reactions were sometimes classified by their temperature coefficients; actually this had some merit because chain-type reactions often have rates which increase rapidly with temperature while reactions in which products are formed in single steps do not.

It was into this embryonic and confused state of knowledge that Norrish stepped as a young man. In the early 1920's perhaps only three or four laboratories in the entire world were doing serious research in photochemistry. During the period up to World War II the number increased only slowly. In some ways these were halcyon times for the photochemist, for, by personal correspondence and a few trips to maintain personal contacts with half a dozen institutions in North America and an equivalent number in Europe, one could keep abreast of the field. There was a spirit of comradeship and co-operation among the photochemists of the time which has inevitably been lost as numbers have increased. The field has grown immeasurably, and when one reads articles today one often cannot picture the face, the voice, the manner of thinking of the writer as one could formerly. No wonder there is more competition and more unnecessary duplication of work.

The Faraday Society organized a conference on Photochemistry in October 1925, the first important postwar scientific meeting to which scientists of the Central Powers had been invited. It proved, in more ways than one, to be the starting-point of the modern era, for the officers of the Faraday Society had the foresight to invite physicists as well as chemists. Norrish attended and presented a paper [4] on 'The role of water in the synthesis of hydrogen chloride'.

Photochemistry, therefore, was born at about the time Norrish started his scientific career. His contributions have been of the greatest importance in making the subject what it is today.

The years 1923 to 1926 were very active ones for Norrish, who published several papers with his preceptor and several over his own name only. They covered reactions of hydrogen with sulphur [5] and the catalytic effect of oxygen, the chlorine-sensitized reaction of oxygen with hydrogen, [6] the induction period in the hydrogen–chlorine reaction, [7] the role of water in the hydrogen–chlorine reaction [2, 8] and the effect of surface polarity on chlorination reactions such as the chlorination of ethylene. [9] Norrish showed a desire to deal with fundamentals and his technique was a distinct improvement over those currently in use in this difficult field. The broad outlines of the conclusions were suggestive, interesting, and essentially correct.

There followed during the years 1927 to 1929 several papers [10] on the photochemistry of nitrogen dioxide. Norrish showed that the photochemically active species is NO_2 rather than N_2O_4 and that there is a photochemical threshold which lies not far from 4,000 Å. At longer wavelengths little reaction occurs and at shorter wavelengths a photochemical stationary state is reached in which the products nitric oxide and oxygen react to reform nitrogen dioxide. He also published a very interesting paper [11] on the photochemical reaction between nitric oxide and cyanogen. Although the details of the mechanism could not be stated, the chemistry of the overall process was clearly elucidated.

Studies of the inhibiting effect of ammonia on the photochemical hydrogen–chlorine reaction led Norrish to work on nitrogen trichloride. [12] This undergoes a chlorine-sensitized decomposition and the inhibiting effect of nitrogen trichloride (derived from ammonia if initially present) on the hydrogen–chlorine reaction lasts until the nitrogen trichloride is decomposed. This study was characteristic of the thoroughness with which Norrish approached problems of mechanism.

Interspersed among these and other [13, 14] interesting papers on chlorine reactions and related subjects, there appeared in 1928 a paper by Norrish and Griffiths [15] entitled 'The photochemical decomposition of glyoxal'. This molecule undergoes complicated reactions, and Norrish and Griffiths discussed the spectrum and the possible behaviour of CHO (formyl) radicals. Quite possibly Norrish himself did not realize that this paper would be the first of a long series dealing with carbonyl compounds.

While this paper on glyoxal could not be called startling because of the great difficulties in dealing with this substance and the consequent uncertainty in some of the conclusions, it nevertheless seems to show a departure from the early work of Norrish and hence to be, in a very real sense, the forerunner of his later work. It would be interesting—but perhaps fruitless—to speculate on whether Norrish turned away from chlorinations and related reactions because he thought that field had become essentially worked out, or whether he really saw great possibilities in the photochemical study of the carbonyl compounds. Quite possibly he was forced, as were many other photochemists, to choose substances which absorbed in easily accessible regions of the spectrum. Thus simplicity of optics rather than simplicity of chemistry was the guiding point.

After this first paper about glyoxal there was a lapse of three years before the appearance of the next one devoted to this class of compounds. In 1931 Kirkbride and Norrish discussed [16] 'The photochemical properties of the carbonyl group'. The data on carbonyl compounds were meagre at that time and yet this paper shows a very clear insight into some of the basic problems of photochemistry. A single step elimination of carbon monoxide is now generally accepted for aldehydes, as proposed in this article, and it is now known not to be a radical process. For aliphatic ketones (other than certain cyclic ketones) a similar proposal has not stood the test of time but the data were not then available to disprove it. Much of the speculation about photochemical thresholds and bond energies was interesting, but the data were too inexact at that time to permit the drawing of many quantitative conclusions.

Early work, either in photochemistry or in any other field, must not be judged on its 'rightness' or 'wrongness'. Certainly data improve and ideas change. Work must be judged to be of real value if it stimulates further good work. Beyond any question this early paper on the mechanisms of photochemical dissociation of carbonyl compounds had this much-to-be-desired result.

The classical work [17] of Paneth and Hofeditz in 1929, as well as work with scavengers, had already indicated that radicals are formed both in photochemical and in thermal reactions. Thus there were means available for ascertaining whether or not a photo-chemical dissociation into radicals occurred and the determination of thresholds for the onset of dissociation was quite popular. Attempts to uncover relationships between bond energies and spectral appearance led to much valuable work; in this field Norrish's laboratory was a pioneer. These attempts involved a long controversy, to which Norrish contributed in a paper[18] with Long in 1946, about the heat of sublimation of carbon.

The papers with Griffiths and with Kirkbride were followed in 1932 by another,[19] the very title of which is suggestive of a programme which must have been forming in Norrish's mind: 'Primary photochemical processes. I. The decomposition of formaldehyde'. An attempt was made to determine the threshold for the dissociation

$$HCHO + h\nu \rightarrow H + HCO \qquad (1)$$

A split into complete molecules

$$HCHO + h\nu \rightarrow H_2 + CO$$

would require much less energy than (1) and had been proposed in the first paper[16] by Kirkbride and Norrish.

From estimates of bond energies, then very uncertain because of the uncertainty about the heat of sublimation of carbon, it was believed that (1) could only take place at wavelengths below about 3,000 Å. Norrish and Kirkbride showed[19] that the yield was approximately unity at 3,650 Å and also at shorter wavelengths around 3,130 and 2,540 Å. They concluded that the primary process is mainly (1). This was a very fundamental conclusion of far-reaching importance.

A still later paper[20] on formaldehyde showed that there was no evidence that the carbon–oxygen bond in formaldehyde was ruptured by absorption of radiation in the fluorite region. Wavelengths below about 1,800 Å provide enough energy for dissociation of that bond.

While a careful study of the absorption spectrum of the reacting system is an obvious prerequisite to the study of its photo-chemistry, Norrish realized that a study of emission might also provide valuable information about the behaviour of the absorbing species. In recent years it has become recognized that emission may sometimes be from molecules other than those which initially absorb the radiation.

Norrish brought good spectroscopists to the laboratory in Free School Lane and by the middle 1930's it became one of the centres which clearly recognized the continuity between atomic and molecular physics and photochemistry. Even for diatomic molecules, the detailed history of absorption is often hard to elucidate; for polyatomic molecules the situation is far worse. Structure in the spectrum is so abundant in many cases that any simple interpretation would be of dubious value. Lifetimes following absorption, and prior to dissociation, may be so long that the fact of dissociation would have little bearing on the appearance of the spectrum. Photochemistry may aid the spectroscopist about as much as the reverse.

The acetone molecule was one of those to receive early attention during the modern era of photochemistry. By 1965 so much has been published from so many laboratories about the photochemistry of acetone that the early stages tend to be forgotten. The classical work of Berthelot and Gaudechon [21] and of Ciamician [22] had indicated the general ways in which ketones must dissociate, but modern quantitative work came along much later. It started in several places almost simultaneously. Norrish studied [23] the emission from acetone and so did Damon and Daniels. [24] While the correct explanation of the main characteristics of this emission came years later, the paper [23] by Crone and Norrish showed a sound desire to relate what might be called the physics to what might be called the chemistry of the process.

Norrish published in 1934 a paper [25] entitled 'The primary photochemical production of some free radicals'. In this paper he gave a preview of ideas of the utmost importance, including statements about work published later with Appleyard [26] on the nonfree-radical decomposition of methyl butyl ketone. Work on other primary processes [27-30] also appeared about this time.

Norrish was eventually led to the conclusion that part of the primary process in acetone could be written

$$CH_3COCH_3 + h\nu \rightarrow 2CH_3 + CO \qquad (2)$$

Norrish, in his 1934 paper, [25] does not say that (2) is a primary process but that, under some experimental conditions, the acetyl radical dissociates sufficiently rapidly to give (2) as the main result within a very short time interval. The primary process, as originally given by Norrish, is now believed to be the correct one

$$CH_3COCH_3 + h\nu \rightarrow CH_3 + COCH_3$$

The pioneering work of Norrish led rapidly to the widespread use of acetone as a source of methyl radicals. Thus the many studies of reactions of the type

$$CH_3 + RH \rightarrow CH_4 + R$$

almost invariably used acetone as the source of methyl radicals, and the foresight of Norrish in pointing out that the primary process in ketones must be the formation of radicals has paid rich dividends. In this respect the simple ketones differ markedly from the aldehydes, as Norrish pointed out.

The study of methyl ethyl ketone by Norrish and Appleyard [26] confirmed beyond any question the primary split of ketones into radicals since ethane, propane and butane were found in the products. Nevertheless, the photochemistry of methyl butyl ketone first mentioned [25] by Norrish and presented in more detail by Norrish and Appleyard [26] is one of the most significant pieces of work to come from Norrish's laboratory. This was followed by beautiful work, mainly with Bamford [31-33] and published between 1936 and 1938, in which three types of decomposition of ketones and aldehydes are mentioned. These are often referred to as Norrish Type I, Norrish Type II, and Norrish Type III. They are distinguished by the following general equations for the primary processes:

$$\text{Type I} \qquad R_1COR_2 + h\nu \rightarrow R_1 + R_2CO \qquad (3)$$
$$\rightarrow R_1CO + R_2 \qquad (4)$$

This is the customary dissociation into free radicals found for the very simple ketones which have no more than two carbon atoms in any radical attached to the carbonyl group.

$$\text{Type II} \quad R_1COCH_2CH_2CH_2R_2 + h\nu \rightarrow R_1COCH_3 + CH_2{=}CHR_2$$

Methyl n-butyl ketone was the first example of this to be studied but the reaction is a general one for ketones which have hydrogen atoms on gamma carbon atoms.

Type III $R_1COCH_2CH_2R_2 + h\nu \rightarrow R_1CHO + CH_2{=}CHR_2$ (5)

This reaction, while less common and less fully substantiated than either Type I and Type II, might occur whenever there are hydrogen atoms on β-carbon atoms.

The work of Norrish and Bamford extended the study of the photochemistry of ketones and aldehydes to condensed phases, particularly to paraffinoid solutions. [33] In such solutions, the Type I reaction may be considerably suppressed or, if it occurs, the products may be modified by reaction of radicals with the solvent. However, the Type II reaction may have yields closely approximating those in the gas phase. Thus one is led to the inevitable conclusion that Type II reactions occur by some sort of intramolecular step. This exciting postulate has given rise to a long series of researches in many laboratories and today, after much speculation and many false starts, a coherent explanation can be given.

The evidence for the Type III reaction is less conclusive, but time has shown that the original postulate made by Norrish is correct and that a few reactions analogous to (5) may occur.

The work [33] of Bamford and Norrish with paraffinoid solutions was based on the fact that radical reactions lead to unsaturation in the solvent whereas a non-free-radical reaction does not. The evidence that Type II reactions occur rapidly after the absorption of radiation and with direct dissociation into complete molecules is conclusive. This situation may be contrasted with the reaction so often postulated for carbon monoxide formation

$$R_1COR_2 + h\nu \rightarrow R_1R_2 + CO \tag{6}$$

This reaction has not been verified for simple ketones and Norrish came to the conclusion that for such molecules it does not occur. It does occur for aldehydes. It is evident that in the papers of this period Norrish was referring to Type I processes as those in which the carbon–carbon bond adjacent to the carbonyl group breaks, while the Type II process was one in which a carbon–carbon bond in the paraffinic chain is ruptured. A detailed discussion of the way in which this could occur was not really given. Thus (6) together with (3) and (4) were considered to be Type I reactions.

In 1935 there appeared a paper[28] by Saltmarsh and Norrish entitled 'Primary photochemical reactions. Part VI. The photochemical decomposition of certain cyclic ketones'. Once again this article stimulated much research throughout the world and today this research might be said to be more important than ever. It was followed in 1938 by Part X, published with Bamford,[31] on the photolysis of cyclic ketones in the gas phase.

In the first of these papers it was concluded that the simple cyclic ketones such as *cyclo*pentanone, *cyclo*hexanone, and *cyclo*heptanone undergo Type I photochemical reactions, in the sense that carbon monoxide is eliminated. The rest of the molecule either closes to form a cyclic hydrocarbon or splits further to give two olefins. The 1938 paper showed clearly that if radicals are attached to the ring in the proper positions there may be a Type II reaction in which an olefin is formed from part of the side chain and a simpler cyclic ketone is produced. Thus menthone gives propylene and 3-methyl *cyclo*hexanone as the principal products. The Type I and Type II reactions proceed concurrently for these ring compounds, just as they do for the aliphatic straight-chain ketones. Thus Norrish was able to show that these two types of reaction are of general significance. It would require volumes to discuss the literature which has arisen as a result of these generalizations.

It might be inferred from the preceding paragraphs that Norrish's entire attention was devoted to carbonyl compounds after 1930 and until the war inevitably diverted much of his time to other matters. However, this is far from being the case and we must review hastily some of the other work in photochemistry which appeared during those productive years.

The work with chlorine was by no means terminated by 1930, and we find several later papers on the formation of hydrogen peroxide and explosions in the chlorine-sensitized hydrogen–oxygen reaction,[13] the photosensitized decomposition of nitrogen trichloride,[12] and the decomposition of ozone photosensitized by chlorine.[34] Papers also appeared[14, 35] on the photochemical hydrogen–chlorine reaction since Norrish continued to be interested in chain reactions. The effects of impurities, of surfaces, and of inhibitors, including hydrogen chloride itself, were all studied. The effect of oxygen was very carefully studied and successfully elucidated. Both the mechanism based on reactions of atoms with oxygen and the rate expressions derived therefrom have stood the test of time. Thus Norrish's work put the final touches on the Nernst atom chain mechanism for the hydrogen–chlorine reaction. This was a very difficult field of study from an experimental standpoint, and the work published by Norrish and his co-workers stands as a monument to progress in reaction kinetics.

Other types of reaction were also being studied in the laboratory in Free School Lane. The first paper on nitrogen peroxide (nitrogen dioxide) appeared in 1927 and this was followed by many other papers on work related in one way or another to this molecule.

As already mentioned, Norrish showed earlier [10] that the active photochemical species is the monomer NO_2. He had also shown that there is a threshold near 4,000 Å above which photochemical effects are negligible. Thus the photochemistry of this molecule was well established by the early work, and later work has served mainly to confirm and to extend the basic conclusions.

Norrish's interest in photosensitization naturally extended to the use of nitrogen dioxide since this molecule, when exposed to radiation of the proper wavelength, dissociates into nitric oxide and atomic oxygen. While the reaction of oxygen atoms with nitrogen dioxide is quite rapid, at low pressures of nitrogen dioxide they may react primarily with other molecules. The studies of reactions sensitized by nitrogen dioxide covered both the explosive and the non-explosive reactions of hydrogen and oxygen, the combustion of methane and of other hydrocarbons. While these reactions were often initiated photochemically, thermal initiation was also used and the entire problem of sensitized oxidations was very carefully and thoroughly investigated.

These sensitized oxidations, including explosions, made a very large contribution to the general field of kinetics and they will be discussed in Chapter 11. Apart from studies of the primary process in the photolysis of nitrogen dioxide [36, 37] and of nitrosyl chloride, [38] the work made less impact on photochemistry than did the work on carbonyl compounds.

There was, however, one aspect of the oxidation work which concerned photochemistry more directly. It was the work on the photochemical oxidation of formaldehyde and of acetaldehyde published with Carruthers.[39] These reactions were found to proceed by short chains. No peroxides were detected in the oxidation of formaldehyde but they were found in the oxidation of acetaldehyde. Once more there was initiated a type of work which has extended over the years to many different laboratories.

In 1936 there appeared a paper with Hirschlaff[40] entitled 'Primary photochemical processes. Part IX. A preliminary study of the decomposition of nitromethane and nitroethane'. This described their very interesting conclusions that the main primary processes are (7) and (8):

$$CH_3NO_2 + h\nu \rightarrow CH_2O + NOH \tag{7}$$
$$C_2H_5NO_2 + h\nu \rightarrow CH_3CHO + NOH \tag{8}$$

Secondary reactions of decomposition and of oxidation of oximino-radicals accounted for the main products other than the aldehydes.

Primary processes (7) and (8) are indeed interesting, and Hirschlaff and Norrish speculated on the details of these processes. Thus the migration of a hydrogen atom to one of the oxygen atoms of the nitro group was visualized. This is a process often suggested by organic chemists and is now known to be of far-reaching importance in photochemical reactions.

We have already cited one example of work done in Norrish's laboratory in the fluorite region, namely the attempt to dissociate the carbon–oxygen bond in formaldehyde.[29] One other paper[41] in this series appeared before the war, 'Photochemical reactions in the fluorite regions. I. Photochemical decomposition of ethylene'. Even though ethylene is a simple molecule its reactions are often very difficult to disentangle and this bit of photochemistry proved to be no exception. The authors concluded that the ethylene molecule probably loses two hydrogen atoms simultaneously, either as H_2 or as free atoms. Acetylene, ethane and butane were found among the products. The correctness of the conclusion about the main primary process has certainly been substantiated by subsequent work.

The difficulties of work in the Schumann region were very great in the 1930's and most workers who started to study this region soon abandoned it. However, the importance of this part of the spectral region is unquestioned today and Norrish deserves great credit for sensing its importance.

The last pre-war paper on photochemistry from Norrish's laboratory appeared[42] in 1939, although a paper with W. MacF. Smith[43] on the quenching of the resonance radiation of sodium appeared in 1940, and four papers on kinetics not related to photochemistry appeared in 1941 and 1942. Then the war had its inevitable effect and the next papers did not appear until 1946. Besides the papers on carbon, already mentioned, an important paper with Porter entitled 'The structure of methylene' was read at the first postwar Faraday Discussion.[44] CH_2 fragments were obtained by the photolysis of ketene and it was concluded that they react much as methyl radicals do, i.e. they behave like real free radicals and not like quasi-molecules. This statement has since been shown to be essentially correct, although much work had to be done to make certain of this point.

Norrish's prewar interest in oxidations was also resumed, and papers describing work on the role of aldehydes in hydrocarbon oxidation included one on the effect of light on intermediates in the combustion of hydrocarbons.[45] As someone has said, the addition of oxygen to any system causes unbelievable complications and usually no two authors agree completely about mechanisms. Norrish's use of photolysis was of the pioneering variety and served very clearly to place in relief the detailed problems which must be solved in studying these mechanisms.

In every field of research there occurs a piece of work of such grandeur that one can speak of events before and events after it, just as one can detect changes in the trend of history, say at the time of the Thirty Years' War or of the Napoleonic Wars. Photochemistry was marked by such an event in 1949. In that year there appeared in *Nature* a brief note by Norrish and Porter[46] entitled 'Chemical reactions produced by very high light intensities'. Thus the scientific world was introduced to 'flash photolysis', although that term does not appear in the text of the note. The term does appear in an article[47] by Porter in 1950 entitled 'Flash photolysis and spectroscopy. A new method for the study of free radical reactions'.

This technique made use of batteries of condensers (available from war surplus stores) to pass large currents of short duration through flash lamps containing noble gases. In the first note it was estimated that 10^{24} quanta, of wavelengths absorbed by the uranyl oxalate actinometer, were emitted by a single flash of about 2 ms duration. Extraordinary effects were observed. Much acetone was changed to elemental carbon which appeared in the form of fine filaments. The same effect was found for ketene. Chlorine and carbon monoxide gave no permanent amount of phosgene, presumably because the latter when formed was rapidly decomposed.

Single flashes of about 4,000 joules energy were employed. The ordinary photochemist is happy to have light sources which emit a few watts of effective radiation, so that perhaps an hour would normally be required to obtain the same number of quanta as the flash gives in a millisecond or less.

A complete review of published work based on flash techniques would now require several volumes. Not only are flashes used to produce large photochemical effects in short periods of time but, by absorption spectroscopy after the flash, it is possible to identify radicals, atoms, and molecules in excited states. If proper timing of the absorption spectra can be arranged, it is also possible to measure lifetimes of these intermediate species. It would be interesting to list the institutions which now have flash equipment and to describe the various modifications which have been introduced.

Flash photolysis brought to photochemistry a new and exciting tool. By increasing the intensity, one can obtain measurable amounts of products in short periods of time. Many new methods of analysis have been developed, such as gas chromatography, nuclear magnetic resonance, electron spin resonance, and polarography. Combinations of these new methods give photochemists the opportunity to use experiments of finite length and still obtain measurable amounts not only of the major but also of many of the minor products. Thus a complete picture of a photochemical reaction became possible to an extent hardly visualized before 1939.

Norrish's own laboratory quite naturally continued to use the flash technique for a variety of problems: 'Spectroscopic studies of the hydrogen–oxygen explosion initiated by the flash photolysis of nitrogen dioxide';[48] 'The photochemical decomposition of ketene by means of light of very high intensity';[49] 'The recombination of atoms. I. Iodine atoms in the rare gases';[50] 'The photolysis of acetaldehyde, diacetyl and acetone at high intensity';[51] 'The combustion of hydrogen sulphide studied by flash photolysis and kinetic spectroscopy';[52] 'Studies of the ClO radical by flash photolysis'.[53] These are the titles of only a few of the papers but they show clearly the wide range of applications of the flash technique in just one laboratory. In the hands of others the triplet states of many molecules, particularly in the aromatic series of compounds, have been identified. The state of the methylene radical has been established and its behaviour investigated.

It is not the purpose of this chapter to review the flash technique in all its aspects. Separate accounts appear in Chapters 5 and 6. It is sufficient here to say that the technique is certainly one of the most powerful tools ever made available to photochemistry.

The work in Norrish's laboratory became even more diverse after than before the war. As early as 1936 Norrish and Carruthers had published a paper[54] entitled 'The polymerization of gaseous formaldehyde and acetaldehyde' and since then work on polymerization has continued. Part of this work was photochemical and part was based on kinetic studies of other processes. From 1942 to 1947 no papers on polymerizations appeared, but since 1947 there has been a steady stream of them and this must, therefore, be regarded as one of Norrish's major interests. Since the main emphasis was rarely photochemical, we leave to others the discussion of this field (Chapter 8), mentioning only an investigation[55] of the photolysis of methyl vinyl ketone polymers. The decrease in the molecular weight of this polymer may be described in terms of a Type I process.

Norrish's interest in initiators of polymerization led him to study the photolysis of t-butyl hydroperoxide, both in carbon tetrachloride and in hydrocarbon solutions.[56] The main rupture is at the oxygen–oxygen bond. The discussion of the mechanism was imaginative and thorough, although the explanations were not really novel. In the same field Norrish and Searby[57] made a very careful study of the photochemistry of dicumyl peroxide and of cumene hydroperoxide.

A paper with Booth on the photochemistry of some organic nitrogen derivatives appeared[58] in 1952. This work was conducted in solution in hydrocarbons. Comparison was made between the decomposition of ammonia and the decomposition of primary and secondary amines. The primary processes were considered to be very similar.

$$NH_3 + h\nu \to NH_2 + H$$
$$RNH_2 + h\nu \to RNH + H$$

Hydrogen was almost the only gaseous product and unsaturation in the solvent could be accounted for by hydrogen abstraction by the hydrogen atoms formed in the primary process.

With acid amides, Type I and Type II processes were found. For example

$$CH_3CH_2CH_2CONH_2 + h\nu \to CH_3CH_2CH_2NH_2 + CO \qquad \text{Type I}$$
$$CH_3CH_2CH_2CONH_2 + h\nu \to CH_2{=}CH_2 + CH_3CONH_2 \qquad \text{Type II}$$

The discussion went on to cover possibilities for the dissociation of proteins. Thus the generalizations found earlier for the dissociation of ketones could be extended to molecules containing atoms other than carbon, hydrogen and oxygen. It should be noted that in the reactions studied the NH_2 group remained intact.

The use of photochemistry as a tool to study some of the fundamental problems of kinetics is almost certainly Norrish's finest work of the past few years. This type of problem had already been under attack in the work on iodine atom recombination,[50] and a most significant paper[59] with Lipscomb and Thrush entitled 'The study of energy transfer by kinetic spectroscopy. I. The production of vibrationally excited oxygen' appeared in 1956. The photolysis of chlorine dioxide proceeds by the primary process

$$ClO_2 + h\nu \to ClO + O \qquad (9)$$

This is followed by several secondary processes of which one is

$$O + ClO_2 \to O_2 + ClO \qquad (10)$$

The striking feature is that the oxygen molecule, although formed in the ground electronic state, possesses several quanta of vibrational energy. This could be verified by absorption spectroscopy in the Schumann–Runge band region. Reactions (9) and (10) provide, therefore, an unparalleled means for studying the loss of vibrational energy by collision, uncomplicated by problems of electronic energy transfer.

The paper by Lipscomb, Norrish and Thrush was followed by several others. Norrish discussed the general problem of vibrational energy transfer at Liége [60] and at Brussels. [61] Edgecombe, Norrish and Thrush [53] studied the behaviour of the ClO radical by flash photolysis.

The really significant thing about this work is that for the first time it became possible to assign definite numbers to the effectiveness of various molecules in removing vibration energy from oxygen molecules. In one sense the results were as expected, that is fairly complex molecules with low vibration frequencies were more effective than the noble gases and simple rigid molecules such as nitrogen. This work was, however, brilliant in its conception and of fundamental importance in kinetics.

Other molecules were studied by methods similar to those used for chlorine dioxide: ozone, [62] chlorine monoxide, [53, 63] sulphur dioxide and sulphur trioxide. [64]

The study of ozone was particularly interesting. In visible light the ozone must dissociate to give an oxygen atom in the ground (^3P) state. In dry ozone the chains can only be very short. In ozone containing water vapour, the chains are also short unless radiation of short wavelengths is absorbed. In that event longer chains occur which must in some way be associated with reaction of water molecules. An energy chain is indicated. Below 3,130 Å ozone may dissociate to give O_2 ($^1\Delta_g$) and O (^1D). Oxygen atoms may react with ozone to give some form of excited oxygen molecule which could then start an energy chain. McGrath and Norrish [62] showed that these oxygen molecules are in the ground electronic state but possess up to seventeen quanta of vibration energy. This fascinating work opened vistas long since forgotten. In the early days of photochemistry, energy chains were often postulated but evidence for them was very slight, almost non-existent. However, this work of Norrish brought them back into the picture and provided a sound basis for their consideration.

Several papers with Callear and with Basco (and some with both these men) extended studies of energy transfer. Of particular significance is the work with nitric oxide.[65, 66] Vibrationally excited nitric oxide molecules could be formed either by photolysis of nitrosyl chloride

$$NOCl + h\nu \rightarrow NO^* + Cl$$

where NO* has several quanta of vibrational energy, or by electronic excitation of nitric oxide followed by collisional deactivation of the electronic energy. The latter method proved to be most useful for quantitative work and many rate constants were determined for collisional loss of vibrational energy by nitric oxide molecules in the ground electronic state.

The hydroxyl radical in a vibrationally excited state was formed by the flash photolysis of ozone in the presence of several hydrogen-containing molecules.[67] The radicals were rotationally cold in spite of having many quanta of vibrational energy. At times oxygen molecules with up to 16 quanta of vibrational energy were produced, presumably by the reaction

$$O\,(^1D) + O_3 \rightarrow O_2^* + O_2$$

where O_2^* indicates a vibrationally excited oxygen molecule in the ground electronic state.

Thus in these and related studies flash photolysis provided a powerful tool for the study of many kinetic problems of fundamental importance.

Since the classical work of Cario and Franck[68] on mercury sensitization of several reactions, much work has been done on the reactivities of various metastable mercury atoms which may be present. Callear and Norrish,[69] using pressure broadening of the absorption line of mercury at 2,537 Å, were able to produce sufficient concentrations of 6^3P_1 and of 6^3P_0 mercury atoms by flash photolysis to study their relative reactivities. Another fundamental problem had been approached in a totally new way. In all instances 6^3P_0 mercury atoms were found to be less reactive than 6^3P_1 atoms, the difference being particularly great with carbon dioxide.

Recent publications from the Laboratory of Physical Chemistry, now located on Lensfield Road, show both occasional reversions to earlier loves and branching into totally new fields. Thus Briggs and Norrish [70] have returned to the chlorine-sensitized decomposition of nitrogen trichloride. Transient spectra were found by flash photolysis, but only if chlorine and nitrogen trichloride were both present. Nitrogen trichloride alone did not show such spectra, which were ascribed to the radicals NCl_2 and NCl. A discussion of the inhibition of the hydrogen–chlorine reaction by nitrogen trichloride was given at the conclusion of this article and the quantitative explanation of this inhibition given by Griffiths and Norrish [12] was supported by new evidence.

The use of flash photolysis to study photovoltaic effects [71] gave further proof, if any is really needed, of Norrish's versatility.

Thus a career in photochemistry which began in 1923 is still continuing in 1965. We have not mentioned every publication, nor, perhaps, have we singled out the most important for emphasis. Norrish's career has been so long, so fruitful, and so varied, that choices had necessarily to be made. Perhaps no two photochemists would agree completely on where the greatest significance in this vast body of work will be found. Certainly history will provide the verdict far better than can be done by any person living today.

When one looks at the scientific attainments of Professor Norrish one realizes that they are truly monumental. Perhaps above everything else they are noted for their originality. At no time was Norrish bound by tradition. Each paper shows careful thought and each one shows that the conclusions were reached without prejudice being exerted by previously published work. This is the mark of a great as distinguished from merely a good scientist.

To survey the work of a productive scholar who has had an active life of over forty years is indeed a difficult task. Norrish's work has not been limited to photochemistry. In that sense our task has been somewhat simplified. Nevertheless, a high fraction of his bibliography is found in this area so that our duties have been heavy.

Final judgement about the permanent value of any single publication or group of publications must be made by posterity.

Professor Norrish would probably not fully agree with opinions of his colleagues and former students about the relative merits of his various contributions to science. But whatever may be the belief about any single article, there can be no doubt about the overall appraisal: the sum total of Norrish's contributions is extremely impressive.

Norrish returned to Cambridge after military service in World War I. To many persons the return to civilian life meant a difficult period of adjustment, but his intense interest in, and devotion to, science were soon apparent. His first published contribution came in 1922. From that year to the present (1965) time there has been an unending stream of outstanding articles.

There are many tests of greatness of a scientific career. The first of these is originality. By this standard Norrish's career has few equals. The second is in starting a series of events carried on by others. Again in this respect Norrish's work deserves the highest of praise. A third relates to men trained in his laboratory. They are now scattered all over the world, many of them eminent in their own rights and many in high positions in university life.

One can rarely say that a single person has mainly by his own efforts determined the trend in a scientific field, but this has been true of Professor Norrish. Photochemistry has been shaped largely in his image. What more can one say in praise of a scientist?

REFERENCES

1   Rideal, E. K. and Norrish, R. G. W. *Proc. Roy. Soc.* A, 1923, **103**, 342, 366.
2   Franck, J. *Trans. Faraday Soc.* 1925, **21**, 536.
3   Henri, V. and Teves, M. C. *Nature, Lond.*, 1924, **114**, 894.
4   Norrish, R. G. W. *Trans. Faraday Soc.* 1925, **21**, 575.
5   Norrish, R. G. W. and Rideal, E. K. *J. chem. Soc.* 1923, **123**, 696, 1689, 3202; 1924, **125**, 2070.
6   Norrish, R. G. W. and Rideal, E. K. *J. chem. Soc.* 1925, **127**, 788.
7   Norrish, R. G. W. *J. chem. Soc.* 1925, **127**, 2316.
8   Norrish, R. G. W. *Z. phys. Chem.* 1926, **120**, 205.
9   Norrish, R. G. W. *J. chem. Soc.* 1926, **128**, 55.
10 Norrish, R. G. W. *J. chem. Soc.* 1927, p. 761; 1929, pp. 1158, 1604, 1611; *J. Chim. Phys.* 1928, p. 133.
11 Norrish, R. G. W. and Smith, F. F. P. *Trans. Faraday Soc.* 1928, **24**, 620.

12 Griffiths, J. G. A. and Norrish, R. G. W. *Trans. Faraday Soc.* 1931, **27**, 451; *Proc. Roy. Soc.* A, 1931, **130**, 592; 1932, **135**, 69.

13 Norrish, R. G. W. *Trans. Faraday Soc.* 1931, **27**, 461; *Proc. Roy. Soc.* A, 1932, **135**, 334.

14 Ritchie, M. and Norrish, R. G. W. *Proc. Roy. Soc.* A, 1933, **139**, 147; 1933, **140**, 112, 713.

15 Norrish, R. G. W. and Griffiths, J. G. A. *J. chem. Soc.* 1928, p. 2829.

16 Kirkbride, F. W. and Norrish, R. G. W. *Trans. Faraday Soc.* 1931, **27**, 404.

17 Paneth, F. A. and Hofeditz, W. *Ber. dt. chem. Ges.* 1929, **62**B, 1335.

18 Norrish, R. G. W. and Long, L. H. *Proc. Roy. Soc.* A, 1946, **187**, 337.

19 Norrish, R. G. W. and Kirkbride, F. W. *J. chem. Soc.* 1932, p. 1518.

20 Norrish, R. G. W. and Noyes, W. A., Jr. *Proc. Roy. Soc.* A, 1937, **163**, 221.

21 Berthelot, D. and Gaudechon, H. *C. r. Séanc. Soc. Biol.* 1910, **150**, 1169; **151**, 478.

22 Ciamician, G. and Silber, P. *Ber. dt. chem. Ges.* 1901, **34**, 1530 (and succeeding volumes). Also *Atti Accad. naz. Lincei Rc.*, 1961, **10**, 92.

23 Crone, H. G. and Norrish, R. G. W. *Nature, Lond.*, 1933, **132**, 241.

24 Damon, G. H. and Daniels, F. *J. Amer. chem. Soc.* 1933, **55**, 2363.

25 Norrish, R. G. W. *Trans. Faraday Soc.* 1934, **30**, 103.

26 Norrish, R. G. W. and Appleyard, M. E. S. *J. chem. Soc.* 1934, p. 874.

27 Norrish, R. G. W., Crone, H. G. and Saltmarsh, O. D. *J. chem. Soc.* 1933, p. 1533; 1934, p. 1456.

28 Saltmarsh, O. D. and Norrish, R. G. W. *J. chem. Soc.* 1935, p. 455.

29 Bamford, C. H. and Norrish, R. G. W. *J. chem. Soc.* 1935, p. 1504.

30 Bloch, B. M. and Norrish, R. G. W. *J. chem. Soc.* 1935, p. 1638.

31 Norrish, R. G. W. and Bamford, C. H. *Nature, Lond.*, 1936, **138**, 1016; 1937, **140**, 195.

32 Bamford, C. H. and Norrish, R. G. W. *J. chem. Soc.* 1938, p. 1521.

33 Bamford, C. H. and Norrish, R. G. W. *J. chem. Soc.* 1938, pp. 1531, 1544.

34 Norrish, R. G. W. and Neville, G. H. J. *Nature, Lond.*, 1933, **131**, 544; *J. chem. Soc.* 1934, p. 1864.

35 Griffiths, J. G. A. and Norrish, R. G. W. *Proc. Roy. Soc.* A, 1934, **147**, 140.

36 Norrish, R. G. W. and Griffiths, J. G. A. *Proc. Roy. Soc.* A, 1933, **139**, 147.

37 Dainton, F. S. and Norrish, R. G. W. *Proc. Roy. Soc.* A, 1941, **177**, 393.

38 Dainton, F. S. and Norrish, R. G. W. *Proc. Roy. Soc.* A, 1941, **177**, 411.

39 Carruthers, J. E. and Norrish, R. G. W. *J. chem. Soc.* 1936, p. 1036.

40 Hirschlaff, E. and Norrish, R. G. W. *J. chem. Soc.* 1936, p. 1580.

41 McDonald, R. D. and Norrish, R. G. W. *Proc. Roy. Soc.* A, 1936, **57**, 480.

42 Norrish, R. G. W. *Trans. Faraday Soc.* 139, **35**, 21.

43 Norrish, R. G. W. and Smith, W. MacF. *Proc. Roy. Soc.* A, 1940, **176**, 295.

44 Norrish, R. G. W. and Porter, G. *Disc. Faraday Soc.* 1947, **2**, 97.
45 Norrish, R. G. W. and Patnaik, D. *Nature, Lond.*, 1949, **163**, 883.
46 Norrish, R. G. W. and Porter, G. *Nature, Lond.*, 1949, **164**, 658.
47 Porter, G. *Proc. Roy. Soc.* A, 1950, **200**, 284.
48 Norrish, R. G. W. and Porter, G. *Proc. Roy. Soc.* A, 1952, **210**, 439.
49 Knox, J., Norrish, R. G. W. and Porter, G. *J. chem. Soc.* 1952, pp. 270, 1477.
50 Christie, M., Norrish, R. G. W. and Porter, G. *Proc. Roy. Soc.* A, 1952, **216**, 152.
51 Khan, M. A., Norrish, R. G. W. and Porter, G. *Proc. Roy. Soc.* A, 1953, **219**, 312.
52 Norrish, R. G. W. and Zeelenberg, A. P. *Proc. Roy. Soc.* A, 1957, **240**, 293.
53 Edgecombe, F. H. C., Norrish, R. G. W. and Thrush, B. A. *Chem. Soc. Special Publication*, no. 9, 1958, 121.
54 Carruthers, J. E. and Norrish, R. G. W. *Trans. Faraday Soc.* 1936, **32**, 195.
55 Guillet, J. E. and Norrish, R. G. W. *Proc. Roy. Soc.* A, 1955, **233**, 153, 172; *Nature, Lond.*, 1954, **173**, 625; *Symposio Internazionale di Chemica Macromolecolare*, Milan, 1955.
56 Martin, J. T. and Norrish, R. G. W. *Proc. Roy. Soc.* A, 1953, **220**, 322.
57 Norrish, R. G. W. and Searby, M. H. *Proc. Roy. Soc.* A, 1956, **237**, 464.
58 Booth, G. H. and Norrish, R. G. W. *J. chem. Soc.* 1952, p. 188.
59 Lipscomb, F. J., Norrish, R. G. W. and Thrush, B. A. *Proc. Roy. Soc.* A, 1956, **233**, 455.
60 Norrish, R. G. W. *Mem. Soc. R. Liége*, 1957, **18**, 4.
61 Norrish, R. G. W. The Solvay Institute Twelfth Discussion, Brussels, 1962. *The Transference of Energy in Gases*, 1964, p. 99. New York: Interscience.
62 McGrath, W. D. and Norrish, R. G. W. *Proc. Roy. Soc.* A, 1957, **242**, 265; *Nature, Lond.*, 1957, **180**, 1272; Bonhoeffer Memorial Issue, *Z. phys. Chem.* 1958, **15**, 246.
63 Edgecombe, F. H. C., Norrish, R. G. W. and Thrush, B. A. *Proc. Roy. Soc.* A, 1957, **243**, 24.
64 Norrish, R. G. W. and Oldershaw, G. A. *Proc. Roy. Soc.* A, 1959, **249**, 490.
65 Basco, N. and Norrish, R. G. W. *Nature, Lond.*, 1961, **189**, 455; *Proc. Roy. Soc.* A, 1962, **268**, 291; *Disc. Faraday Soc.* 1962, **33**, 99.
66 Basco, N., Callear, A. B. and Norrish, R. G. W. *Proc. Roy. Soc.* A, 1961, **260**, 459; *Proc. Roy. Soc.* A, 1962, **269**, 180.
67 Basco, N. and Norrish, R. G. W. *Proc. Roy. Soc.* A, 1961, **260**, 293.
68 Cario, G. and Franck, J. *Z. Phys.* 1922, **11**, 161.
69 Callear, A. B. and Norrish, R. G. W. *Proc. Roy. Soc.* A, 1962, **266**, 299.
70 Briggs, A. G. and Norrish, R. G. W. *Proc. Roy. Soc.* A, 1964, **278**, 27.
71 Norrish, R. G. W. and Paszyc, S. *Roczn. Chem.* 1963, **37**, 1305.

2

THE CONTRIBUTIONS OF
R. G. W. NORRISH
TO COMBUSTION

BERNARD LEWIS

I am happy to be invited to contribute to a book honouring my old friend and colleague, R. G. W. Norrish, for his lasting contributions to reaction kinetics. Indeed, it is probably true that much of our knowledge of the mechanism of gaseous oxidation reactions stems from the detailed and painstaking experimental work of Norrish and his students over the past forty years. Over this period some 200 research papers have enriched the literature. Not only has this work clarified the complex mechanism and the kinetics of the chemical interactions but it has pointed the way, in many instances, to the solution of some of the outstanding problems, both experimental and theoretical, with which future students of the subject will be involved.

My own discussion of Norrish's work is limited to his contribution to combustion. Combustion, as it is known today, after over a quarter of a century of almost breathless activity on the part of thousands, covers a very broad field. Its extension from concern with chemical interactions which utilize essentially the disciplines of chemistry, physical chemistry and spectroscopy, to studies of combustion waves (deflagration and detonation) and the interaction of flow with combustion waves, has enormously complicated the subject; so that today one is unable to solve contemporary combustion problems without invoking the use of numerous additional disciplines such as physics, fluid dynamics, and thermodynamics, as well as mathematical methods for the evaluation and correlation of data by high speed computer techniques. Not very long ago some aerodynamicists who had first experienced the fascination of combustion phenomena believed that problems of combustion were really problems in fluid dynamics. While it is true that many of the practical combustion problems of today lean heavily for their solution on considerations of fluid dynamics and aerodynamics, the older workers continue to find enlightenment for the deeper understanding of combustion phenomena in physical chemistry as well.

The studies of Norrish and his school were not concerned with the properties of combustion waves or the theory of propagation of such waves. They made no studies of flames as such. Rather, they were concerned, in the broadest sense, with the chemical processes that lead to ignition or flame—that is, with the atomic and molecular details of the developing reaction starting with an initial stimulus, be it thermal or photochemical. By and large the methodology consists in making observations, in particularly designed experiments, of the reaction rate and the effect of temperature, pressure, concentration, additives, geometry, amplitude of stimulus and other useful kinetic devices, on such rate. From this, the chemical kineticist is often able to demonstrate the uniqueness of a complex chemical mechanism that will account for the kinetic data. The method is powerful and it frequently allows an insight into the kinetics and dynamics of elementary reactions (the single steps, generally of an atomic or radical nature, which make up the overall complex reaction) as it has in the hands of Norrish and his school. Such then is the nature of Norrish's interest and contributions to combustion.

The combustion studies have concentrated on a limited number of fuels, chiefly hydrogen, carbon monoxide and methane, although more complex higher hydrocarbons and other compounds have been studied with less concentration and thus with less finality to the results.

An interesting phase of Norrish's earlier work in combustion, aside from a few studies on the reaction between hydrogen and sulphur, was with the system hydrogen–chlorine–oxygen. The explosive oxidation of hydrogen by chlorine had interested many before him (Bodenstein, Weigert, Christiansen, Polanyi, and others) and although the chain character of the reaction had been postulated and, in a sense, demonstrated, Norrish's studies can be said to have really elucidated the mechanism and the special inhibitive role of elements such as oxygen that had baffled earlier workers. These investigators had found that the purer they prepared the reactants hydrogen and chlorine, the more prone the mixture was to reaction with explosive violence even by relatively mild initiating sources like sunlight. Norrish demonstrated in a beautifully simple way the powerful role of traces of oxygen in removing chain carriers and thus stopping the otherwise long chains. There followed many studies of explosions chiefly of hydrogen and oxygen sensitized by such additives as nitrogen dioxide, hydrogen peroxide, chlorine, nitrosyl chloride, chloropicrin, and of carbon monoxide and oxygen sensitized by hydrogen and hydrogen peroxide, which are models in themselves of the application of the kinetic method to the elucidation of chemical interaction and which as a group established a firm basis for understanding the complexities of sensitized chain reactions.

It is certainly outside the scope of this brief sketch to discuss details of Norrish's combustion studies. These will surely come to light through the papers of some of his eminent students who are contributing to this volume. All the contributions are in areas in which Norrish pioneered and they show the extraordinary breadth of his interests. As one reads through his many papers, as no student can afford not to do if he wishes to appreciate the basis of our present-day understanding of combustion kinetics, one is impressed with the continuity of the developing theme and relationship of the previous studies to succeeding ones. To have maintained for about four decades this persistent effort to solve the kinetics of combustion and other reactions is a tribute to Norrish's scholarly approach to research.

Perhaps his outstanding contribution to the understanding of combustion reactions is the experimental demonstration of degenerate chain branching in the oxidative combustion of hydrocarbons. The conception of degeneracy (delay) in branching was due to N. N. Semenov in 1930. Semenov envisaged the slow combustion of hydrocarbons taking place by a kinetic process, which he described as a chain reaction exhibiting degenerate branching, wherein the hydrocarbon is first oxidized to an intermediate, possessing moderate stability, which is itself oxidized to the final products, carbon monoxide, water, etc. A stationary concentration of the intermediate is built up during the reaction and reaches a maximum when the reaction velocity is a maximum. The intermediate is oxidized by an alternative reaction with oxygen and is responsible for starting new chains. In this way delayed or 'degenerate' chain branching results. In a series of papers with several students Norrish showed how, in the combustion of methane, the intermediate, formaldehyde, builds up to a steady state concentration when its rate of removal by chain oxidation equals the rate at which it is formed by chain oxidation of methane, and that the attainment of this steady state concentration coincides in time with the time at which the overall reaction rate reaches a maximum. Other kinetic experiments, concerned chiefly with the effect on the induction period (the period of build-up of intermediate) of adding formaldehyde to the mixture initially, concluded a brilliant proof of this complicated process.

Other topics in combustion to which Norrish and his students have made important contributions include the mechanism of anti-knock and cool flame phenomena, associated with the low temperature phase of the oxidation of ethane and higher hydrocarbons, in which excited formaldehyde intermediate plays the role of emitter. In other respects Norrish's work established a firm basis for the chain character of combustion reactions.

The kinetics of chemical transformation in a propagating combustion wave is very complex. Other facets of the flame problem have only in recent times been developed to the point where advantage can now be taken of the pioneering work of Norrish and his students, with the prospect of a more complete understanding of flame processes.

3

PHOTOCHEMISTRY IN THE LIQUID PHASE

C. H. BAMFORD AND R. P. WAYNE

It is not our intention in the present chapter to attempt a survey of all known photochemical reactions which occur in the liquid phase. We wish, rather, to concentrate attention on certain reactions typical of their class. Many of these have been chosen because they originate from Professor R. G. W. Norrish's pioneering investigations, or because they have been of special interest to him. Wherever appropriate we shall compare the liquid-phase reactions with the gas-phase photochemistry. Consideration of the reactions in the gas phase is, indeed, frequently a profitable way in which to approach the solution chemistry; it may, therefore, be desirable to begin by indicating the general ways in which photochemistry can differ in the two phases.

1. The influence of phase

The effects peculiar to photochemistry may be divided into those concerning the absorption of radiation and those affecting the dissipation of energy from excited species. Since no well-defined rotational energy levels exist in the liquid phase, rotational structure of electronic absorption bands does not appear. Thus it may be difficult to demonstrate the existence of a predissociation in condensed phase systems by other than circumstantial evidence. Such evidence suggests [50] that the wavelength of the onset of predissociation, as of absorption maxima, may be different in the gas and liquid phases. It is evident that in the presence of solvent the probability of quenching electronically (or vibrationally) excited molecules is very high, so that primary quantum efficiencies may be correspondingly small in solution. Further, excited molecules can themselves react with the solvent, thus altering the whole course of the reaction. For example, the photolysis of an aqueous solution of monochloracetic acid proceeds with a quantum yield close to one, hydroxyacetic acid [1] being the product. It is suggested that excited monochloracetic acid molecules may react with water in the reaction

$$CH_2ClCOOH^* + H_2O \rightarrow CH_2OHCOOH + HCl \qquad (1)$$

In addition to the phenomena already described, there are certain formal ways in which reactions may differ in solution and in the gas phase. Both the absolute rate theory [2] (using experimental data for heats and entropies of solution) and the kinetic theory of collisions [3] lead to the conclusion that collision frequencies in solution are higher than those in the gas phase by a factor of two or three. A further complication is introduced by the way in which collisions in solution are distributed in time. The early work of Rabinowitch and Wood [3] and many subsequent investigations have shown that in the liquid phase collisions are not uniformly distributed in time, but occur in sets or 'encounters'. This arises because colliding species are enclosed in a solvent 'cage' which tends to prevent their separation by diffusion. The two particles may, therefore, make several mutual collisions before leaving each other's sphere of influence; the number of these collisions will be strongly dependent on the viscosity of the solvent. For reactions between atoms and radicals, in which the activation energy is very small, the first collision of a set may lead to reaction so that the frequency factor for the overall reaction will be reduced by a factor dependent on the number of collisions in an encounter. On the other hand, if the activation energy be large, then the number of collisions required for reaction is much greater than the number within a set, and the reaction rate is then independent of whether or not the collisions occur in sets.

A related phenomenon is the 'cage-effect', first postulated [4] by Franck and Rabinowitch. The recombination of atoms or radicals necessitates the presence of a third body to remove excess energy, and, in the absence of a surface reaction, occurs relatively slowly at normal pressures in the gas phase. In solution, however, recombination may occur at every collision since solvent molecules can act as third bodies, and two atoms or radicals produced by the dissociation of an excited species may recombine before they have an opportunity to escape from the cage. Such a process is referred to as a 'primary' recombination and occurs within 10^{-11} s of the generation of the radicals, before they have separated by a molecular diameter. 'Secondary' recombination takes place within 10^{-9} s of radical formation; in this case the radicals may have separated somewhat, but there is still a high probability of further mutual collisions. Noyes[5] has more recently discussed these phenomena quantitatively, and has introduced the term 'geminate' recombination to cover both primary and secondary recombination. For our purposes we shall refer to geminate recombination as the cage effect. The importance of the effect is related to the kinetic energy of the species formed, since if they have sufficient energy they may force their way out of the cage. For this reason the primary quantum yield of some photochemical reactions in solution increases as the wavelength of the photolysing radiation is decreased. Recombination in the bulk of the solution occurs between atoms or radicals which have diffused out of the solvent cage, and takes place randomly. It occurs about 10^3 to 10^4 times more readily than in the gas phase because of the high concentration of third-body molecules. A useful review[6] of these diffusion-controlled processes has been given by Noyes.

As a result of the factors mentioned above, photochemical reactions are frequently simpler in solution than in the gas phase. In the first place the reactions of excited species may be neglected except in special circumstances. Secondly, the radicals formed often recombine, or in suitable cases react with the solvent. Except for condensed-phase systems in which the 'solvent' is the reactant, reaction of radicals with the reactant is of much less importance than in the gas phase. Finally, surface reactions are usually negligible in liquid systems.

2. Photochemistry of inorganic systems

2.1 *Water.* The discussion of individual photochemical reactions in solution or in the liquid phase may conveniently begin with some inorganic reactions, and the photochemistry of water is an obvious starting-point. Water absorbs radiation to a significant extent at wavelengths below 2,000 Å and the spectrum of the vapour is continuous to a first peak at 1,650 Å. There is sufficient energy in this wavelength region for either primary reaction (2) or (3) to take place:

$$H_2O + h\nu \rightarrow H_2 + O \quad -117 \text{ kcal/mole} \tag{2}$$

or

$$H_2O + h\nu \rightarrow H + OH - 118 \text{ kcal/mole} \tag{3}$$

Reaction (2) was postulated [7] as the primary photolytic step by Goodeve and Stein, although Herzberg [8] subsequently showed that it would require an excitation energy considerably in excess of 117 kcal/mole. Barrett and Baxendale [9] have confirmed that reaction (3) is the important primary process; they have further shown, from measurements of the rate of hydrogen evolution when methanol or ethyl acetate is present, that the quantum efficiency of (3) is 0·6 at 1,849 Å. Again, flash photolysis [10] with flash times of about 2 μs, during which no significant reaction occurs, suggests that only hydrogen atoms and hydroxyl radicals are produced in the photolysis of water at a wavelength of 1,800 Å. It is of interest that while radiolysis of water produces hydrated electrons as the dominant reducing species, yet in the photolysis their yield must be less than one-tenth of the hydrogen atom yield. Another difference between radiolysis and photolysis is that 50 % of the hydrogen atoms produced in the former process, and only one-sixth of the hydrogen atoms in the latter process, give hydrogen gas by

$$H + H \rightarrow H_2 \tag{4}$$

There is some uncertainty about the formation of hydrogen peroxide in the photolysis of aerated water. Some work [11] suggests that hydrogen peroxide appears only in the presence of trace impurities. On the other hand, Allen and Holroyd [12] report that a stationary concentration of 3–4 μmole l.$^{-1}$ of hydrogen peroxide is set up on irradiation at 1,849 Å of pure water. Although Barrett and Baxendale [9] are unable to detect hydrogen peroxide when they use pure water in their system, they point out that their intensity of radiation at 2,537 Å, which would decompose any peroxide formed, is several times that at 1,849 Å. Transient absorbing species have recently been observed [13] in the flash photolysis of water; the absorption at 254 mμ produced by the photolysis of dilute aerated methanolic solutions of water shows second-order decay kinetics, and the observations are consistent with the absorption of light by the HO_2 radical. On this basis the rate constant for the reaction

$$HO_2 + HO_2 \rightarrow H_2O_2 + O_2 \qquad (5)$$

is about 2×10^8 l. mole^{-1} s^{-1}.

A considerable number of photosensitized decompositions of water have been reported, of which one of the more interesting[14] is the decomposition sensitized by zinc oxide. In the absence of molecular oxygen no reaction occurs, but hydrogen peroxide is formed when oxygen is present, and one mole of oxygen is consumed per mole of hydrogen peroxide formed. It is suggested that the primary process is an electron transfer to a water molecule

$$ZnO . H_2O + h\nu \rightarrow ZnO^+ . H_2O^- \qquad (6)$$

Immediate dissociation of H_2O^- then occurs, OH^- becoming attached to the zinc oxide and hydrogen atoms reacting with oxygen.

$$H + O_2 \rightarrow HO_2 \qquad (7)$$

Hydrogen peroxide is then formed by the recombination reaction (5).

2.2. *Photovoltaic phenomena in water.* The decomposition of water photosensitized at metal surfaces has been suggested to account for the changes of potential observed when an electrode of a voltaic cell is illuminated. The heat of absorption of H and OH on the metal surface is greater than that of H_2O, thus explaining why light of wavelength longer than 2,422 Å (equivalent to 118 kcal/mole) is active. Such effects have been observed in aqueous solutions of sulphuric acid, potassium hydroxide and potassium sulphate. Heyrovsky and Norrish[15] have recently combined the techniques of photochemistry and polarography to confirm the mechanism. By irradiation of a dropping mercury electrode in sulphuric acid they were able to alter the polarographic curves in a way that could be attributed to the competitive action of oxidizing and reducing species formed in water under the incident light. Qualitatively similar results were observed for solutions of perchloric acid, hydrochloric acid and acetic acid. Although the polarographic curves of pure solutions of potassium hydroxide and potassium sulphate do not normally exhibit any change on irradiation, they do so in the presence of nitrous oxide. The latter gas also strengthens the effect in sulphuric acid, while ethanol suppresses it. It is proposed that in acid solution H^+ ions at the negative electrode combine with hydrogen atoms to give H_2^+ and that both the H_2^+ and hydroxyl radicals can be reduced at the electrode to an extent sufficient to produce a polarographic current. Nitrous oxide reacts with hydrogen atoms to produce hydroxyl radicals, while ethanol acts as a radical scavenger and suppresses the current.

It is appropriate at this stage to mention some other examples of photovoltaic phenomena. In 1923, Norrish and Rideal [16] studied the photolysis of aqueous solutions of potassium permanganate by following the change in potential of an electromotive cell on irradiation by ultraviolet light. A very rapid change occurred in 30 min, followed by a slow recovery; if the light were turned off the electrode potential returned to its initial value. The observations were consistent with the establishment of a photostationary state, the initial drop of potential corresponding to a change in pH from about 6 to 10, followed by a continuous decomposition of the salt. The electrode potential changes result from the formation of potassium hydroxide by the photodecomposition of the permanganate; the subsequent removal of hydroxide by combination with hydrated manganese dioxide leads to the stationary state. It occurred to Norrish to devise a photolytic cell, of the type described earlier [17] for uranyl nitrate by Titlestadt. Two water-cooled solutions of $N/100$ potassium permanganate were connected: both had platinum electrodes, but one was in glass, the other in quartz. On irradiation by the light from a mercury arc, an e.m.f., as high as 0·2 V, was created which increased to a maximum, and decayed on cutting off the light. Forty years later, Norrish and Paszyc [18] reinvestigated this photovoltaic system by the new technique of flash photolysis. A reaction cell with two platinum electrodes was constructed so that only one electrode was illuminated by the light from the flash-lamp, and potential changes resulting from the flash were displayed on an oscilloscope. Experiments were performed with aqueous solutions of potassium permanganate, potassium ferric oxalate, uranyl oxalate and potassium halides. Fig. 1 (a) and

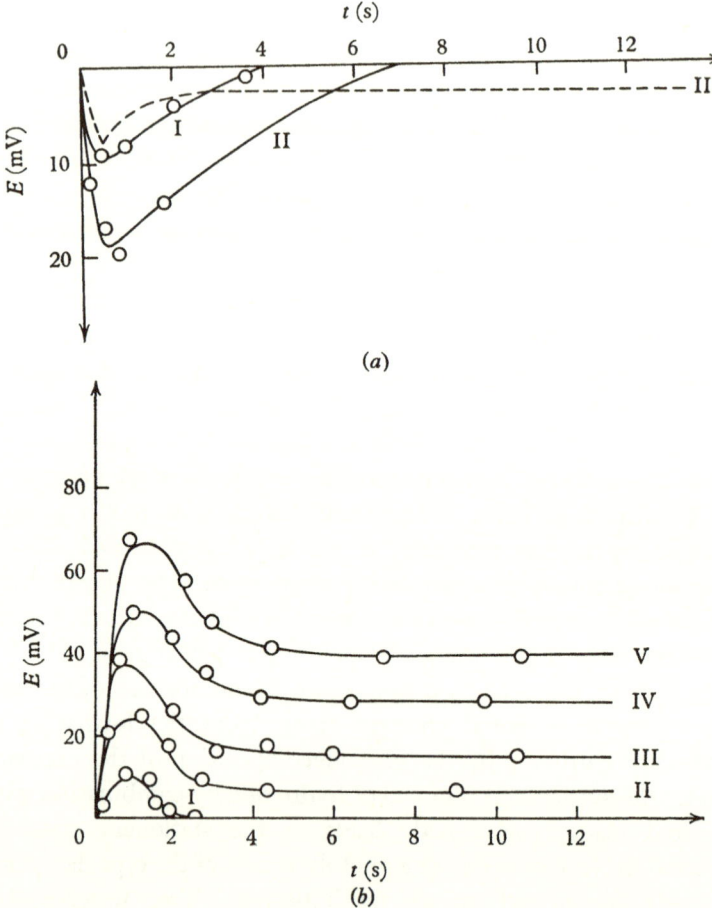

Fig. 1. (a) The photovoltaic effect of aqueous 0·001 N solution of potassium permanganate (quartz cell): I, aqueous solution of potassium permanganate, flash energy 500 J; II, aqueous solution of potassium permanganate in the presence of 0·001 N sulphuric acid, flash energy 500 J; III, aqueous solution of potassium permanganate in the presence of 0·001 N sulphuric acid repeatedly flashed, flash energy 500 J. (b) The photovoltaic effect of uranyl oxalate solution (0·01 M) in the presence of 0·01 M oxalic acid (pyrex cell). Flash energy: I, 20; II, 80; III, 180; IV, 320; V, 500 J.

(b) show the flash photovoltaic effect obtained for potassium permanganate and uranyl oxalate–oxalic acid solutions. In each case investigated the potential rises to a maximum after the flash, and then falls to a limiting value where it remains practically constant during the following period. These phenomena are explained by relatively high conversions in the solution surrounding the highly reflecting electrode: after a short time the resulting large concentration gradients around this electrode are removed by diffusion, and the final potential corresponds to the ratio of oxidized to reduced species in the bulk of the solution. Curve II of Fig. 1(a) shows that the photochemical change in potassium permanganate is more rapid in the presence of sulphuric acid than in a neutral solution, an observation also made[14] in the classical system. Although the bulk concentration of tetravalent manganese is not sufficient to cause a measurable potential change after one flash (curve I), repeated flashing results in a permanent potential change (curve III). In the decomposition of oxalic acid sensitized by the uranyl ion, the valency of the uranium is unaltered[19, 20] so that the change in potential must result from a decrease in oxalic acid concentration: the number of uranyl ions complexed is reduced. An extensive review of photovoltaic phenomena is given in reference 21.

2.3. Hydrogen peroxide. The photolysis of hydrogen peroxide, as of water, has attracted much attention, and the reactions appear to be more easily interpreted when the photolysis is carried out[22] at high light intensities. A considerable weight of evidence exists in favour of a primary process in which hydroxyl radicals are formed. Thus the quantum yield for the decomposition of aqueous solutions of hydrogen peroxide[23, 24] at $\lambda = 2,537$ Å (at high intensity and low peroxide concentration) is 0.94 ± 0.06 while in the presence of allyl alcohol[23] the yield drops to 0.54 ± 0.05. Evidence has been obtained for the formation of tetrahydroxyhexane in this system; it must be formed by the addition of hydroxyl radicals to allyl alcohol followed by the association of the intermediate. It is shown[23] that under conditions of high light intensity and low peroxide concentration, the quantum yield for peroxide decomposition expected, if (8)

$$H_2O_2 + h\nu \rightarrow 2OH \tag{8}$$

is the primary process, would be $2 k_8$ in the absence of allyl alcohol. If the radicals are removed by reaction with allyl alcohol, the quantum yield should drop to half this value, which is the experimental result. It has also been shown by E.S.R. spectroscopy[25] that the irradiation of hydrogen peroxide in a glass containing allyl alcohol leads to abstraction of a hydrogen atom from the alcohol. In a similar way, Daniels has shown [26] that at $\lambda = 2,537$ Å the quantum yield of decomposition of oxygen-free hydrogen peroxide drops to 0·55 in the presence of the arsenite ion, and that the rate of consumption of peroxide is exactly equivalent to the rate of formation of arsenate. Confirmation that hydroxyl radicals are a photolytic product is adduced from the initiation of the chain photo-oxidation of carbon monoxide by hydrogen peroxide [27] at $\lambda = 2,537$ Å. Investigations of this system lead to the further conclusions that hydroxyl radicals react 13 times more rapidly with carbon monoxide than they do with hydrogen peroxide and that the reaction of carbon monoxide with HO_2 is considerably slower than the bimolecular reaction of HO_2, (5). Hochanadel [28] has suggested the scheme (9)–(12) for the secondary reactions of hydroxyl radicals in a system containing hydrogen peroxide and molecular hydrogen and oxygen:

$$H_2O_2 + OH \rightarrow H_2O + HO_2 \qquad (9)$$

$$2HO_2 \rightarrow H_2O_2 + O_2 \qquad (5)$$

$$H_2 + OH \rightarrow H_2O + H \qquad (10)$$

$$H_2O_2 + H \rightarrow H_2O + OH \qquad (11)$$

$$O_2 + H \rightarrow HO_2 \qquad (12)$$

His measurements indicate that $k_{10}/k_9 = 0·93$ and $k_{11}/k_{12} = 2·2 \times 10^{-5}$; the ratios are not sensitive to the acidity. It is of interest that in the photolysis of water the ratio k_{10}/k_9 is the same, while k_{11}/k_{12} is 0·54; the observation is interpreted in terms of differing degrees of association of hydrogen atoms with H_3O^+ ions. A rather complete investigation of both the photochemistry and the radio-chemistry of hydrogen peroxide has recently been made [29] by Currie and Dainton. In the photochemical investigations at 10° and 25 °C, and at natural pH and pH = 1, they find that the curves of decomposition rate against peroxide concentration always have a maximum. The rate is proportional to the square root of the light intensity; and the kinetic chain lifetime (measured by the rotating sector technique) is independent of peroxide concentration, although the lifetime is shorter at natural pH than at pH = 1. The results can be explained on the basis that in acid solution the chain carriers are OH and HO_2 radicals, as formerly believed, but that at natural pH the carriers are OH radicals and O_2^- ions. It is also necessary to postulate that the form of hydrogen peroxide which reacts with HO_2 or O_2^- is the dihydrate. Currie and Dainton calculate [29] from their data that the rate constant of the reaction (5) is about 2×10^6 l. mole^{-1} s^{-1}, which conflicts somewhat with the value [11] of 2×10^8 l. mole^{-1} s^{-1} obtained by the flash photolysis of water. It is of interest that in strongly alkaline solutions of hydrogen peroxide, ozone appears [30] as a product, and the yield is increased by an increase in either base or molecular oxygen concentration. The photochemical formation of the species H_2O_4 has been postulated [31] to account for the difference between rates of peroxide decomposition measured by gas-volumetric and permanganate-titration methods: if H_2O_4 diffuses to the walls to decompose to peroxide and oxygen the observations could be reconciled. It appears, [32] however, that while oxygen is evolved slowly for several hours after irradiation of aqueous solutions of hydrogen peroxide, no concomitant change in peroxide concentration takes place. In fact, a similar oxygen evolution occurs if peroxide solutions are saturated with oxygen and the pressure then reduced, so that it seems unnecessary to retain the species H_2O_4.

2.4. *Ozone*. The formation [33] of hydrogen peroxide from the photolysis of aqueous solutions of ozone is also known. In the presence of dilute hydrochloric or acetic acid (which inhibit chain reactions of ozone with hydrogen peroxide), the reaction

$$O_3 + H_2O \rightarrow O_2 + H_2O_2 \qquad (13)$$

takes place with almost exact stoichiometry, and tracer experiments have shown that the hydrogen peroxide formed derives half its oxygen from the ozone and half from the water. The quantum yields for ozone decomposition at 254, 310 and 600 mμ are 0·62, 0·23 and 0·002–0·005 respectively, and at 600 mμ no hydrogen peroxide is formed. These observations are in complete accord with predictions made on the basis of the gas-phase photolysis, and a short description will now be given of the gas-phase photochemistry of ozone. Ozone has two absorption regions, one in the red part of the visible spectrum and the other in the ultraviolet, and the photochemical behaviour is distinct in the two regions. The quantum yield for the decomposition of ozone by red light[34, 35] is two, and is unaffected by the presence of water vapour. On the other hand, in the ultraviolet[36-39] the quantum yield of dry ozone photolysis may be considerably in excess of two and addition of water vapour increases the quantum yield yet further. Flash photolysis of ozone–water vapour mixtures shows the spectrum of the hydroxyl radical, and McGrath and Norrish[40] suggest that the hydroxyl radical takes part in a chain decomposition of ozone to give the high quantum yields observed:

$$OH + O_3 \rightarrow HO_2 + O_2 \tag{14}$$

$$HO_2 + O_3 \rightarrow OH + 2O_2 \tag{15}$$

The difference in behaviour in the two absorption regions is now readily understood. The primary photochemical process is the production of oxygen atoms

$$O_3 + h\nu \rightarrow O + O_2 \tag{16}$$

and these must react with water to produce hydroxyl radicals

$$O + H_2O \rightarrow 2OH \tag{17}$$

Reaction (17) is 17 kcal/mole endothermic if the O atom produced in (16) is in the ground electronic state (^3P), and hydroxyl radicals cannot be formed at room temperature. In the ultraviolet, however, there is sufficient energy available for the formation of a ^1D oxygen atom in (16), and for this species (17) is 29 kcal/mole exothermic. It remains to be seen why the quantum yield for the ultraviolet photolysis of dry ozone should be greater than the value of two predicted by the simple scheme

$$O_3 + h\nu \rightarrow O + O_2 \tag{16}$$

$$O + O_3 \rightarrow 2O_2 \tag{18}$$

An energy chain has long been proposed for the ultraviolet photolysis of ozone and it is now clear that this is initiated by ^1D oxygen atoms.

$$O(^1D) + O_3 \rightarrow O_2{}^* + O_2 \tag{19}$$

$$O_2{}^* + O_3 \rightarrow 2O_2 + O(^1D) \tag{20}$$

where $O_2{}^*$ represents an energy-rich molecule of ozone. Norrish and co-workers[41, 42] have observed the spectrum of vibrationally excited oxygen in the flash photolysis of dry ozone, and identify

O_2^* in (19) and (20) with this species. A direct diagnostic test for the presence of $O(^1D)$ in the ultraviolet photolysis of ozone should be the formation of nitrous oxide in the association reaction

$$O(^1D) + N_2(^1\Sigma_g^+) + M \rightarrow N_2O(^1\Sigma) + M \qquad (21)$$

a reaction which would be spin-forbidden for ground state oxygen atoms. Norrish and Wayne [39] and DeMore and Raper [43] in fact detect only small yields of nitrous oxide in the gas phase. The latter workers suggest that the low yield could be attributed to a pre-dissociation of the nitrous oxide (22) which results in the formation of a ground state oxygen atom:

$$N_2O(^1\Sigma_g^+) \rightarrow N_2O(^3\Pi \text{ or } ^3\Sigma) \rightarrow N_2(^1\Sigma_g^+) + O(^3P) \qquad (22)$$

They therefore extended their investigations to solutions of ozone in liquid nitrogen, and under these conditions they were able to determine the quantum yield for the formation of nitrous oxide at a number of different wavelengths. In the ultraviolet photolysis there is sufficient energy available for any of the following primary steps to occur:

$$O_3(^1A) + h\nu \rightarrow O(^1D) + O_2(^3\Sigma_g^-) \quad -70 \text{ kcal/mole} \qquad (23)$$

$$O_3(^1A) + h\nu \rightarrow O(^1D) + O_2(^1\Delta_g) \quad -91 \text{ kcal/mole} \qquad (24)$$

$$O_3(^1A) + h\nu \rightarrow O(^1D) + O_2(^1\Sigma_g^+) \quad -106 \text{ kcal/mole} \qquad (25)$$

The process requiring least energy, and also permitted by the law of spin conservation, is (24). The energy required for this reaction is equivalent to a wavelength of 3,100 Å, and it is significant that the quantum yield for nitrous oxide formation shows a marked increase at 3,020 Å. In this connection it is noteworthy that Norrish and Wayne [39] postulate the formation of a $^1\Delta_g$ or a $^1\Sigma_g^+$ oxygen molecule in the primary photolytic step on purely kinetic grounds.

The photolysis of oxygen at short wavelengths should produce one ^1D oxygen atom as well as a ground state atom. Attempts [44] to detect nitrous oxide on irradiation of solutions of liquid oxygen in liquid nitrogen have been only partially successful. Nitrous oxide is formed, but arises from the secondary photolysis of ozone formed. DeMore and Raper have extended [45] their investigations of the low temperature photolysis of ozone to solutions in liquid carbon monoxide. At a wavelength of 2,537 Å the quantum yield for ozone decomposition is 0·07, and there is a 1 : 1 equivalence of ozone decomposition and carbon dioxide formation. The quantum yield for carbon dioxide formation is decreased by the addition of nitrogen, argon, or oxygen, but increased by a decrease in temperature. The carbon dioxide is produced by the reaction

$$O(^1D) + CO(^1\Sigma) + M \rightarrow CO_2(^1\Sigma) + M \qquad (26)$$

but the low quantum yield arises by deactivation in a manner similar to that proposed for nitrous oxide

$$O(^1D) + CO + M \rightarrow O(^3P) + CO + M \qquad (27)$$

Both reaction (22) and reaction (27) are examples of association processes in which the excited association complex may undergo unimolecular decomposition to products of lower electronic energy in the course of stepwise vibrational deactivation. The interpretation of the photochemistry of aqueous solutions of ozone is now apparent. At 600 mμ the oxygen atom produced in the primary step is in its ground electronic state, and cannot react with water. A ^1D oxygen atom may be produced at the two shorter wavelengths, of 310 and 254 mμ (although with a lower efficiency at 310 mμ as described above). This atom reacts with water to give two hydroxyl radicals of which the main reaction in solution will be recombination. The hydrogen peroxide thus formed will have one oxygen atom derived from the ozone and the other from the water.

The importance of metastable molecules such as $O_2(^1\Sigma_g^+)$ in gas-phase reactions has only been touched upon in the preceding discussion. It might be mentioned in passing that such species may also be produced in solution. The red chemiluminescence which arises from the oxidation of polyhydric phenols has long been known. Bowen and Lloyd [46] have recently investigated the chemiluminescence from the oxidation of alkaline pyrogallol with 'formalin' and hydrogen peroxide. They find that the emission consists of a narrow band at 630 mμ, identical with the similar emission obtained from the reaction of hypochlorite with hydrogen peroxide. Khan and Kasha [47] identify the emission as the (0, 0) band of the forbidden $^1\Sigma_g^+ \rightarrow {}^3\Sigma_u^-$ transition of oxygen (the 'atmospheric band'). If this identification is correct, then there must be considerable solvent interaction with the upper excited level, since the (o, o) band lies at 760 mμ in the gas phase. Some recent evidence suggests [48] that the emitting species is, in fact, a pair of $^1\Delta_g$ oxygen molecules. Other reactions in which this emission is observed will be described in our discussion of the photochemistry of organic peroxides (cf. refs. [102–104]).

2.5. *Chlorine dioxide.* In many respects, the photochemistry of chlorine dioxide is similar to that of ozone. The primary photochemical process gives an oxygen atom

$$ClO_2 + h\nu \rightarrow ClO + O \tag{28}$$

and this atom may react in the gas phase to give vibrationally excited oxygen molecules in the ground electronic state [49]

$$O + ClO_2 \rightarrow O_2^* + ClO \tag{29}$$

The solution photochemistry of chlorine dioxide has been studied with both water and carbon tetrachloride as solvents. In the latter solvent, as will appear, the reaction is relatively simple, but in aqueous solution the photodecomposition is complicated by reactions with the solvent. Bowen and Cheung [50] find that the main products from the photolysis of aqueous chlorine dioxide solutions are chloric acid and hydrochloric acid in equimolecular proportion, together with molecular oxygen, and they suggest that the secondary reactions responsible for these products are

$$ClO + H_2O \rightarrow H_2ClO_2 \tag{30}$$
$$H_2ClO_2 + ClO \rightarrow HClO_3 + HCl \tag{31}$$

The quantum efficiency rises from o·2 at $\lambda = 4,360$ Å to 1·0 at $\lambda = 3,000$ Å. Taube [33] extended the observations to shorter wavelengths ($\lambda = 2,537$ Å), but did not obtain the exact stoichiometry observed by Bowen and Cheung. In some experiments the net change could be represented by the equation

$$\tfrac{3}{2}H_2O + 3ClO_2 = 2ClO_3^- + Cl^- + \tfrac{3}{4}O_2 \tag{32}$$

In other cases no oxygen was formed although the quantum yield remained constant. It is possible that, at short wavelengths, the oxygen atom produced in (28) might be electronically excited; hydrogen peroxide would then be formed by the recombination of hydroxyl radicals derived from reaction (17) in a manner exactly similar to that proposed for its formation in the photolysis of ozone solutions. The direct observation of hydrogen peroxide is not to be expected in the presence of chlorine dioxide, but the appearance of oxygen derived from the water would be an indication that hydrogen peroxide is formed. Tracer experiments show that in fact most of the oxygen liberated in the photodecomposition

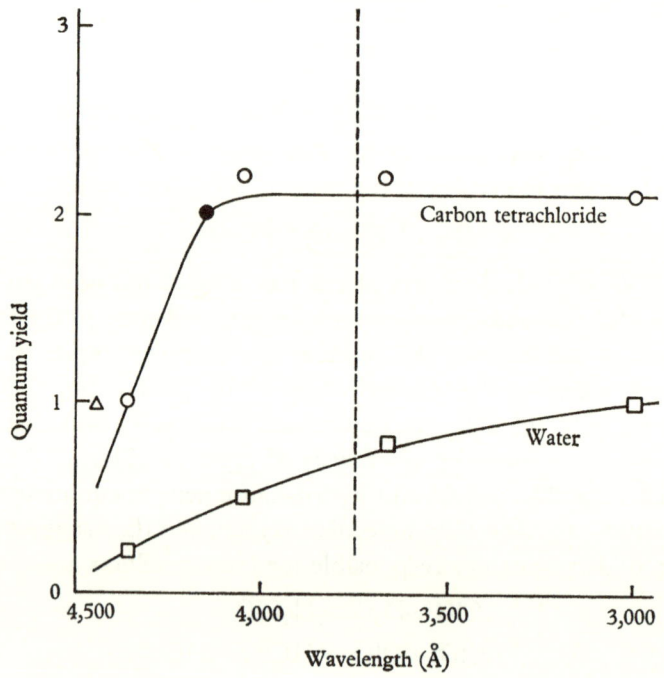

Fig. 2. The photolysis of solutions of chlorine dioxide: quantum yields as a function of wavelength. The broken line shows the wavelength of the onset of predissociation in the gas phase. (Bowen and Cheung.[50])

originates in the chlorine dioxide, and imply that only 30% of the oxygen atoms formed at $\lambda = 254$ mμ, and 10% at longer wavelengths, react as in (17). In carbon tetrachloride the photolysis[51,52] follows a more straightforward route. The products are chlorine and oxygen; chlorine hexoxide, formed in the gas-phase photolysis,[53] does not appear.[54] The dependence of the quantum yield for the photolysis upon wavelength seems rather sharp: at $\lambda = 4{,}450$ Å the quantum yield[51] is one, and at $\lambda = 4{,}150$ Å the quantum yield[50] is two. Bowen and Cheung[50] believe that this dependence results from the onset of predissociation in the wavelength region investigated. Predissociation occurs at about 3,753 Å in gaseous chlorine dioxide, so there is clearly some interaction of the solvent with chlorine dioxide which allows the molecule to predissociate at lower energy values in solution. Fig. 2 summarizes the quantum yields obtained for the decomposition of chlorine dioxide in solution. If, indeed, the rise in quantum yield of the reaction is evidence for the onset of predissociation, then it seems probable that the polar solvent water distorts the molecules even more than does carbon tetrachloride, and the onset of predissociation is less sharp.

2.6. *Insertion reactions of oxygen atoms and of methylene.* Before leaving the reactions of oxygen atoms in solution, we should perhaps note that certain 'insertion' processes[53] have been proposed for this species. This process is suggested by Dainton[55] to be responsible for the formation of hydrogen peroxide in the photolysis[33] of aqueous solutions of ozone, and experiments have been performed in which $O^{18}(^1D)$, derived from the photolysis of N_2O^{18} at a wavelength of 1,849 Å, enters the water with about 100% efficiency. It is concluded that

$$H{-}OH + O \rightarrow HO{-}OH* \xrightarrow{M} H_2O_2 \qquad (33)$$

must account for the formation of hydrogen peroxide, since it seems that if free hydroxyl radicals were formed, then some might escape from the cage and lower the efficiency of the process. Similar reactions are proposed for the liquid-phase reaction of $O(^1D)$ with CH, OH and NH bonds although in the gas phase hydroxyl radicals[39] are certainly formed. There is clearly some conflict between insertion and radical abstraction ideas, and Dainton points out the need for experiments in concentration regions intermediate between those in gases and liquids. Analogous behaviour of the isoelectronic species methylene, formed by the photolysis of diazomethane, ketene or diazirine solutions has been reported, for insertion into both C—H bonds[56] and C—Cl bonds.[57] It might again be argued, however, that the effect arises from a radical abstraction reaction, followed by recombination within the cage: this process leads immediately to the 'insertion' products. Investigations of the gas-phase photolysis of mixtures of ketene or diazomethane with molecules containing C—H or C—Cl bonds [58, 59] does show that in these systems all the cross-combination products expected from a radical abstraction mechanism are in fact observed, although some workers [60] interpret the evidence in the light of true insertion. Bamford, Casson and Wayne [58] have made the observation that in the gas-phase photolysis of ketene–methyl chloride mixtures, although the product of insertion into either C—H or C—Cl bonds (ethyl chloride) always appears more slowly than products such as ethane, yet the addition of about an atmosphere pressure of nitrogen does enhance its formation. DeMore and Benson [61] have suggested in order to account for the observed high rates of 'insertion' that the reaction path must be an initial attack of methylene on a hydrogen atom,

$$
\begin{array}{c}
\text{H} \\
\diagdown \\
\quad\quad \text{C} \cdots \text{H—C} \diagdown \\
\diagup \\
\text{H}
\end{array}
\rightarrow
\left[
\begin{array}{c}
\text{H} \\
\diagdown \\
\quad\quad \text{C—H} \cdots \text{C} \diagdown \\
\diagup \\
\text{H}
\end{array}
\right]
\tag{34}
$$

$$(\text{I})$$

followed by rotation of the methyl group to allow π-orbital overlap. The question now arises whether (I) in fact dissociates and, in solution, immediately recombines, or whether in the liquid phase, but not in the gas, (I) is highly stabilized. It would be more conventional to attribute Bamford's effect of nitrogen in the gas phase to stabilization of (I), but it seems possible that two free alkyl fragments, for which the recombination efficiency is high, could be held together in a gas-phase cage long enough to show a slight enhancement of the primary recombination product. A suggestion that such gas-phase cage effects should not be overlooked has recently been made by R. K. Lyon. [62]

3. Photochemistry of organic systems

3.1. *Aldehydes and ketones.* The use of ketene as a photochemical source of methylene radicals leads us to a general discussion of the photochemical behaviour of the organic carbonyl group. Although the photochemistry of aldehydes and ketones had aroused attention previously, it was not until the now classical studies of Norrish and his co-workers in the 1930's that the processes began to be understood. Norrish studied at this time the photolysis of formaldehyde[63, 64], ketene[65], methyl ethyl ketone[66], methyl butyl ketone[66,67], i-valeraldehyde[68], di-n-propyl ketone[69,70], and also certain cyclic ketones. Finally, from a comparison of reactions in the gas phase and in solution, carried out[70,71] by Norrish and Bamford, a general formulation[72] grew of the photochemistry of aldehydes and ketones. It was concluded that the absorption of light could bring about two distinct primary effects. There could be scission of bonds joined to the carbonyl group giving rise to carbon monoxide and radical fragments which in the gas phase could undergo combination and disproportionation reactions; this process was designated as 'Type I'. In addition, a second process (first noted explicitly by Norrish and Appleyard[66])—'Type II'— could result in fission of the C—C bond α—β to the carbonyl group. Thus for methyl propyl ketone the steps may be written

$$\text{CH}_3\text{COCH}_2\text{CH}_2\text{CH}_3 + h\nu \xrightarrow{\text{Type I}} \text{CH}_3 + \text{C}_3\text{H}_7 + \text{CO} \qquad (35)$$

$$\text{CH}_3\text{COCH}_2\text{CH}_2\text{CH}_3 + h\nu \xrightarrow{\text{Type II}} \text{CH}_3\text{COCH}_3 + \text{C}_2\text{H}_4 \qquad (36)$$

Sometimes, in fact, very little of the Type I process occurred; for example, the photolysis of isovaleraldehyde yields little carbon monoxide, and must, therefore, be mainly of Type II

$$\begin{array}{c}\text{CH}_3 \\ \diagdown \\ \diagup \\ \text{CH}_3\end{array}\!\!\text{CHCH}_2\text{CH}_2\text{CHO} \xrightarrow{h\nu} \begin{array}{c}\text{CH}_3 \\ \diagdown \\ \diagup \\ \text{CH}_3\end{array}\!\!\text{C=CH}_2 \;+\; \text{CH}_3\text{CHO}$$

$$(37)$$

Apparently aldehydes which lead to Type I products do so without the alkyl radical becoming free: acetaldehyde yields no ethane, although it does give methane and carbon monoxide. The quantum yields for these photochemical decompositions are generally less than unity in the gas phase, although a chain reaction does occur at elevated temperatures (not with acetone). In paraffinoid solution at room temperature the nature of the reaction products from the photolysis of aldehydes and ketones alters considerably. While the

Type II reaction is completely unchanged in solution, the radical fragments from the Type I photolysis undergo primary recombination to a large extent, although a small proportion hydrogenate themselves at the expense of the solvent. Thus photolysis of di-i-propyl ketone in solution yields propane, carbon monoxide and butyraldehyde by Type I reaction, but only with very low efficiency.

$$C_3H_7COC_3H_7 + h\nu \rightarrow C_3H_7 + C_3H_7CO \qquad (38)$$

$$C_3H_7CO + \text{solvent} \rightarrow C_3H_7CHO \qquad (39)$$

$$C_3H_7CO \rightarrow C_3H_7 + CO \qquad (40)$$

$$C_3H_7 + \text{solvent} \rightarrow C_3H_8 \qquad (41)$$

At higher temperatures, hydrogenation by (39) and (41) can become relatively less inefficient and the quantum yield of decomposition by the Type I process rises from effectively zero at room temperature to about 0·3 at 96 °C. The Type II reaction, giving ethylene and methyl propyl ketone, is independent of the solvent cage and is not affected by changes in temperature. Thus, considered as a whole, the evidence seems conclusive in favour of two independent processes, one involving a free radical mechanism and the other being intramolecular. It was proposed by Davies and Noyes [73] that the intramolecular Type II process proceeds via a 6-membered ring, formed by internal hydrogen bonding, which can dissociate directly into a lower aldehyde or ketone and an olefin; subsequent experiments have confirmed [74, 75] this view. Another intramolecular process, known as Norrish Type III, has been postulated to account for the formation of olefinic products in the photolysis of compounds where fission of the α—β bond cannot yield them; for example, the photolysis of methyl iso-propyl ketone forms acetaldehyde and propylene amongst its products. It has recently been shown by Zahra and Noyes [76] that this step occurs via a 4-membered ring formed by hydrogen bonding between the carbonyl carbon and the β-carbon of the alkyl chain.

The carbonyl group in acids and esters behaves photochemically in a manner directly analogous to that in aldehydes and ketones, and both Types I and II processes may occur. Thus n-butyric acid and its methyl ester form ethylene intramolecularly [77] (as does ethyl acetate [78]). Sigal [79] has presented evidence for the occurrence of a novel Type II process. The liquid-phase photolysis of N-methyl-β-alanine methyl ester gives, in addition to products from normal ester and amine (see later) cleavage, an imine:

$$CH_3OCCH_2CH_2NH(CH_3) + h\nu \rightarrow CH_3OCCH_3 + CH_2{=}NCH_3$$
$$\overset{\|}{O} \qquad\qquad\qquad\qquad \overset{\|}{O}$$

$$(42)$$

Even in polymeric compounds such as poly(methylvinylketone) a Type II reaction may occur. Guillett and Norrish [80] photolysed dioxan solutions of poly(methylvinylketone) in an attempt to prepare a graft copolymer by the scission of CH_3CO and subsequent addition of a new monomer: the experiment failed in this respect as a result of Type II processes leading to the breakage of the original chain.

Compounds in which there are two interacting carbonyl groups are particularly light-sensitive. Studies have been made of the photolysis of α-ketoacids [81] both in aqueous and organic solutions, and of α-ketoesters [82] in benzene. The photolysis of pyruvic acid or of benzoylformic acid in aqueous solution liberates carbon dioxide, and acetoin or benzaldehyde remains in solution: the precise mechanism is not clear. Photolysis in organic solvents follows a more explicable course. For example, photodecarboxylation occurs with several α-ketoesters, leaving caged radicals which disproportionate to an aldehyde (from the 'acid' fragment) and an aldehyde or ketone (from the 'alcohol' fragment).

Cyclic ketones decompose under the influence of light in a manner which suggests [68, 70, 83] that a biradical is formed. In the gas phase, for example, cyclopentanone [83] decomposes in three ways:

$$\text{\Large $\diagdown$}C{=}O + h\nu \rightarrow 2C_2H_4 + CO \qquad\qquad (43)$$

$$\rightarrow \boxed{} + CO \qquad\qquad (44)$$

$$\rightarrow CH_2{=}CHCH_2CH_2CHO \qquad (45)$$

and these routes may certainly be explained on the basis of a primary production of $\cdot CH_2CH_2CH_2CH_2CO\cdot$. It appears that ethylene is not formed by the decomposition of 'hot' cyclobutane produced by closure of $\cdot CH_2CH_2CH_2CH_2\cdot$ (itself derived from the primary radical). Nevertheless, there is considerable evidence that vibrationally hot molecules are of importance in the gas-phase photolysis. The only reaction common to cyclic ketones with four, five, six and seven carbon atoms in the ring is

$$\left[\begin{array}{c}\text{—CO—}\\ \text{—(CH}_2)_{n-1}\text{—}\end{array}\right] + h\nu \rightarrow \left[\begin{array}{c}\\ \text{—(CH}_2)_{n-1}\text{—}\end{array}\right] + CO \quad (46)$$

although reaction (47) has been observed for $n = 4$, 6, or 7.

$$\left[\begin{array}{c}\text{—CO—}\\ \text{—(CH}_2)_{n-1}\text{—}\end{array}\right] + h\nu \rightarrow CH_2{=}CH(CH_2)_{n-4}CH_3 + CO \quad (47)$$

The importance of the isomerization reaction exemplified by (45) increases with an increase in ring size. In the condensed phase, reactions (46) and (47), which are favoured by high degrees of vibrational excitation, become unimportant. The isomerization reaction then dominates, together with a reaction involving interaction of the photochemically excited ketone with the solvent.

$$\left[\begin{array}{c}\text{—CO—}\\ \text{—(CH}_2)_{n-1}\text{—}\end{array}\right] + RH \xrightarrow{h\nu} \left[\begin{array}{c}\overset{\text{OH} \quad \text{R}}{\diagdown\diagup}\\ \text{—C—}\\ \text{—(CH}_2)_{n-1}\text{—}\end{array}\right] \quad (48)$$

A rather different primary process has recently been described [84] for the photolysis of 6-, 7- and 8-membered cyclic ketones in the liquid phase. A ring contraction occurs, with an isomeric migration of a methyl group:

$$\begin{array}{c}\overset{O}{\overset{\|}{C}}\\ H_2C \diagup \diagdown CH_2 \\ H_2C \diagdown_C\diagup (CH_2)_{n-5} \\ \overset{|}{H_2}\end{array} + h\nu \rightarrow \begin{array}{c}\overset{O}{\overset{\|}{C}}\\ H_2C \diagup \diagdown CH{-}CH_3 \\ H_2C{-}(CH_2)_{n-5}\end{array} \quad (49)$$

A rather complete review of the photochemistry of cyclic ketones in both gas and liquid phases is given [85] by Srinivasan

The nature of the excited state of the carbonyl compound which decomposes by the Type II mechanism has aroused some interest. Borrell and Norrish [77] studied the mercury-sensitized photolysis of a number of compounds, and observed no Type II products. They then inferred that since for reasons of spin conservation the excited state of the reactant must be triplet in the sensitized reaction, Type II products arise from a singlet excited molecule. Borkowski and Ausloos [86] questioned this contention, and it was subsequently shown [87] that in the gas-phase system the absence of olefinic products could be attributed to mercury-sensitized hydrogenations. In solution, however, Borrell [88] has found a similar correlation between non-appearance of Type II products and the use of benzophenone as a sensitizer for decomposition. Benzophenone is believed to transfer its energy via a triplet state, [89] so that it may still be that, in solution at any rate, only singlet excited molecules may decompose to Type II products.

Photosensitization by hydrocarbon molecules has also been proposed for the decomposition of aldehydes and ketones in solution. The photolysis of solutions of diethyl ketone in i-propyl benzene [90] is stated to be of this kind. Ethane and carbon monoxide are the only gas products, although photolysis of solutions in cyclohexane and benzene, where sensitization cannot take place, yields ethylene and butane as well. It seems, however, that all ethyl radicals formed could become hydrogenated in i-propyl benzene, thus precluding the disproportionation and recombination reactions observed in less reactive solvents, and the difference in products observed may not necessarily be ascribed to sensitized reactions.

3.2. *Amines and amides.* We wish to discuss next the photochemistry of amino compounds. Farkas [91] irradiated solutions of ammonia in hexane and obtained hydrogen as the only gaseous product. The reaction was thought at that time to be a sensitized decomposition of hexane, although Bamford and Norrish [92] showed later that the hydrogen appeared not as a result of true sensitization, but from secondary reactions following the primary process:

$$NH_3 + h\nu \rightarrow NH_2 + H \tag{50}$$

A similar primary process was shown [93] to occur in the gas-phase photolysis of primary amines

$$RNH_2 + h\nu \rightarrow RNH + H \tag{51}$$

In solution the secondary processes are simplified, and Booth and Norrish [94] find that the only permanent gas evolved in the photolysis of amines in hydrocarbon solution is hydrogen. (The photolysis of ethyl methyl amine proves to be an exception: a little methane is found in the products, and some reaction must proceed by

$$C_2H_5NHCH_3 + h\nu \rightarrow C_2H_5NH + CH_3 \tag{52}$$

No corresponding step occurs for the fission of ethyl radicals, since ethane is not found in the reaction products.) Unsaturation develops in the solvent, and, provided that the absorbing substrate is present only in low concentration, the amount of unsaturation is equivalent to (and never exceeds) the amount of hydrogen evolved. The secondary processes in cyclohexane may, therefore, be written

$$H + C_6H_{14} \rightarrow H_2 + C_6H_{13} \tag{53}$$

$$RNH + C_6H_{14} \rightarrow RNH_2 + C_6H_{13} \tag{54}$$

$$2C_6H_{13} \rightarrow C_6H_{12} + C_6H_{14} \tag{55}$$

A decomposition of amines similar to the Type II process in carbonyl compounds is excluded, since the unsaturation never exceeds the hydrogen evolved. Subsequent studies of the photochemistry of amines in hexane solution [95] have indicated that these conclusions are substantially correct.

Amides combine some of the photochemical features of both carbonyl and amino groups. The photolysis of amide solutions [94] leads to the formation of both carbon monoxide and hydrogen, and the ratio $[CO]/[H_2]$ increases with an increase in the wavelength of the radiation. Some unsaturated hydrocarbon gas is evolved from propionamide, butyramide, valeramide and hexoamide: from the first two of these amides it has been established that the gas is ethylene. During the course of photolysis, primary amines develop in the solution, although aldehydes or $\alpha\beta$-diketones are not formed. It appears that Types I and II mechanisms operate in the primary photolytic step:

$$\text{Type I}$$
$$RCONH_2 + h\nu \longrightarrow RNH_2 + CO \tag{56}$$

$$\text{Type II}$$
$$C_3H_7CONH_2 + h\nu \longrightarrow CH_2{=}CH_2 + CH_3CONH_2 \tag{57}$$

For acetamide, of course, no Type II reaction is possible, and the absence of methane in the photolytic products of this amide indicates that the process written as (56) is of the *aldehyde* Type I, in which the radical fragments do not become free. Propionamide must yield ethylene by a Type III fission

$$CH_3CH_2CONH_2 + h\nu \rightarrow C_2H_4 + HCONH_2 \tag{58}$$

The decomposition, then, seems to be controlled by the carbonyl (or $-CONH_2$) group, rather than by the $-NH_2$ group taken alone.

Secondary reactions of the products from (56) and (57) are taken to be photochemical decompositions, e.g.

$$C_3H_7NH_2 + h\nu \rightarrow C_3H_7NH + H \tag{59}$$

$$CH_3CONH_2 + h\nu \rightarrow CH_3NH_2 + CO \tag{60}$$

$$CH_3NH_2 + h\nu \rightarrow CH_3NH + H \tag{61}$$

Thus the hydrogen evolved should be equivalent to that unsaturation developed in the solvent which is not accounted for by the Type II decomposition. This is true for butyramide and valeramide where ethylene and propylene can be completely removed from the solution.

The effect of variation of wavelength on the $[CO]/[H_2]$ ratio is looked upon as evidence for the photochemical reaction of the intermediate amines and amides in (59), (60) and (61) since, in general, the intermediates will have different absorption spectra from their precursors.

3.3. *Organic peroxides.* In our discussion of the photochemistry of hydrogen peroxide we indicated that the primary process was fission of the O—O bond to yield hydroxyl radicals. It appears that organic peroxides undergo substantially the same primary photochemical process. The photochemistry of organic peroxides is of particular interest in a consideration of the photochemical initiation of hydrocarbon combustion processes, since various peroxide species are postulated as reaction intermediates and may themselves be affected by light. An initial fission of the O—O bond is well established [96] for the thermal decomposition of organic peroxides, and an examination of the photochemistry of t-butyl hydroperoxide, dicumyl peroxide and cumene hydroperoxide by Norrish and co-workers [97,98] showed that this is also the primary photochemical process. The photolysis at $\lambda = 3{,}130$ Å of solutions of t-butyl hydroperoxide in carbon tetrachloride leads to the formation of t-butyl alcohol and oxygen as the main products together with smaller amounts of acetone, water, and products arising from the oxidation of methyl radicals. The quantum yield for the decomposition is 3·2 at 20 °C, and 5·3 at 50 °C, suggesting that a chain mechanism operates, and it is pointed out that the chain length may be considerably greater than the quantum yield if primary recombination or fluorescence makes the initiating step inefficient. Norrish and Martin [97] propose the following series of steps in the chain:

$$(CH_3)_3COOH + h\nu \rightarrow (CH_3)_3CO + OH \tag{62}$$

$$(CH_3)_3CO + (CH_3)_3COOH \rightarrow (CH_3)_3COH + (CH_3)_3COO \tag{63}$$

$$2(CH_3)_3COO \rightarrow 2(CH_3)_3CO + O_2 \tag{64}$$

and as termination reactions

$$(CH_3)_3COO + OH \rightarrow (CH_3)_3COH + O_2 \tag{65}$$

$$(CH_3)_3CO \rightarrow (CH_3)_2CO + CH_3 \tag{66}$$

Reaction (66) must be of only minor importance in the temperature region employed since 97 % of the t-butoxy radical goes finally to t-butyl alcohol. No methane is observed to be evolved. Solutions in n-hexane behave in a similar manner, although oxidation reactions involving the solvent give rise to alcohols which appear together with the products already mentioned. The quantum yield in hexane is 3·9, and independent of temperature. If the t-butyl hydroperoxide be dissolved in dioxan, then immediate hydrogenation of the radicals occurs, and the quantum yield drops to one. The interaction with the solvent is such that some etheric oxygen in the dioxan is converted into hydroxyl, and the fragmentation of the dioxan produces formaldehyde. In this solvent, photolysis by more energetic radiation ($\lambda = 2,450$–$2,800$ Å) increases the concentration of the dioxan decomposition products, although it does not alter their nature. The mode of decomposition of dicumyl peroxide [98] does seem rather more sensitive to changes of wavelength. In carbon tetrachloride solutions the main product is acetophenone at $\lambda = 3,130$ Å, and a little acetone is formed in addition at $\lambda = 2,537$ Å; in hexane solutions the main product at $\lambda = 3,130$ Å is $\alpha\alpha$-dimethyl benzyl alcohol, but a little acetophenone appears at the shorter wavelength. Since dicumyl peroxide is not itself subject to radical attack [99] the following scheme is proposed for the photolysis in hexane solutions ($R = C_6H_{13}$):

$$\phi(CH_3)_2CO \cdot OC(CH_3)_2\phi + h\nu \rightarrow 2\phi(CH_3)_2CO \tag{67}$$

$$\phi(CH_3)_2CO + RH \rightarrow \phi(CH_3)_2COH + R \tag{68}$$

$$2R \rightarrow \text{combine or disproportionate} \tag{69}$$

$$\phi(CH_3)_2CO \rightarrow \phi COCH_3 + CH_3 \tag{70}$$

$$CH_3 + RH \rightarrow CH_4 + R \tag{71}$$

$$2CH_3 \rightarrow C_2H_6 \tag{72}$$

In carbon tetrachloride solution the cumyl radical cannot abstract hydrogen and it must, therefore, be stabilized by ketone formation. At $\lambda = 3,130$ Å the process is

$$\phi(CH_3)_2CO \rightarrow \phi COCH_3 + CH_3 \tag{73}$$

while at $\lambda = 2,537$ Å the C—C bond to the phenyl group may also be broken

$$\phi(CH_3)_2CO \rightarrow \phi + CH_3COCH_3 \tag{74}$$

Cumene hydroperoxide undergoes photolysis in a similar manner. The product from the decomposition in n-hexane solution is mainly $\alpha\alpha$-dimethyl benzyl alcohol, while photolysis of carbon tetrachloride solutions yields acetone, acetophenone and a dark polymer. The quantum yield for the photolysis in hexane is $1 \cdot 57$ at $20\ ^{\circ}C$

($\lambda = 3{,}130$ Å); and for the photolysis in carbon tetrachloride the yield is 0·75. Again the results suggest a chain reaction propagated by radicals formed from hexane.

$$\phi(CH_3)_2COOH + h\nu \rightarrow \phi(CH_3)_2CO + OH \qquad (75)$$

$$\phi(CH_3)_2CO + RH \rightarrow \phi(CH_3)_2COH + R \qquad (68)$$

$$OH + RH \rightarrow H_2O + R \qquad (76)$$

$$R + \phi(CH_3)_2COOH \rightarrow ROH + \phi(CH_3)_2CO \qquad (77)$$

$$R + R \rightarrow \text{recombine and disproportionate} \qquad (69)$$

The solutions of cumene hydroperoxide show a strong fluorescence and so the kinetic chain length is probably considerably greater than the quantum yield implies.

It is noteworthy that while the major part of the reaction of the cumyloxy radical in inert solvents is to form ketones, only 8 % of the radicals form acetone in the photolysis of t-butyl hydroperoxide. The greater stability of the cumyloxy radical, compared with the t-butoxy radical, presumably arises from the influence of the benzene ring. Again, the cumene compounds, in contrast to t-butyl hydroperoxide, do not yield oxygen on photolysis.

Photosensitized decompositions of organic peroxides are also known. Thus solutions of tetralin hydroperoxide in chlorobenzene are but slowly decomposed by visible light, and the presence of a little fluorenone in the solution sensitizes the photolysis.[100] A similar photosensitization by fluorenone does not occur with benzoyl peroxide. Other molecules may, however, act as photosensitizers for the decomposition of this peroxide,[101] and the sensitization by both toluene and 2,5-diphenyloxazole has been studied. Toluene absorbs radiation at $\lambda = 2{,}650$ Å completely under the conditions of the experiments; in the wavelength region, 3,030–3,130 Å 2,5-diphenyloxazole absorbs completely, although neither toluene nor benzoyl peroxide absorbs at these wavelengths. The quantum yield for the decomposition of benzoyl peroxide solutions in toluene at $\lambda = 2{,}650$ Å, and for the decomposition of solutions in toluene containing 2,5-diphenyloxazole at $\lambda = 3{,}030$–3,130 Å, increases with increasing benzoyl peroxide concentration to a limit of 0·4. The same maximum quantum yield is obtained for the photolysis when the solvent is chloroform, which is transparent to both light sources. Decomposition in the 'opaque' solvents may, therefore, be attributed to energy transfer from the solvent to the benzoyl peroxide.

Emission of light from the thermal decomposition of organic peroxides has frequently been observed. Bowen and Lloyd[102] have investigated the chemiluminescence in the decomposition of benzoyl peroxide, cumene hydroperoxide, benzene hydroperoxide and acetaldehyde peroxide, and find that the emission may be divided into two parts. First, there is a blue luminescence from the bulk of the solution: this appears to arise from electronically excited products of the decomposition. Secondly, there is a long wavelength radiation, which seems to be partly, but not entirely, connected with the surface of the reaction vessel. In the case of benzoyl peroxide the emission arises only when daughter products of the reaction, probably hydroperoxidic in nature, have built up. The presence of oxygen appears to be necessary for the long wavelength chemiluminescence to develop, and Bowen and Lloyd suggest that the radiation is that of singlet oxygen. We have referred previously to the emission of this radiation in a different system (cf. ref. [46]). Bowen[103] suggests that since the reaction

$$
\begin{array}{ccc}
\mathrm{H} \;\; \mathrm{O-OH} & & \mathrm{O} \\
\diagdown \; \diagup & & \parallel \\
\mathrm{C} & \rightarrow & \mathrm{C} \; + \; \mathrm{H_2O} \\
\diagup \; \diagdown & & \diagup \; \diagdown \\
& (\mathrm{K}) &
\end{array}
\qquad (78)
$$

is about 70 kcal/mole exothermic, the ketone (K), is formed as a triplet. The process

$$
{}^3\mathrm{K} + {}^3\Sigma_g^- \, \mathrm{O_2} \rightarrow {}^1\mathrm{K} + {}^1\mathrm{O_2}
\qquad (79)
$$

would account for the formation of excited oxygen. The exact electronic state of the oxygen molecules is disputed,[104,48] both $^1\Sigma_g^+$ and $^1\Delta_g$ states having been proposed.

The peroxy radicals to which we have had occasion to refer in the present section are, of course, believed to play a part in many oxidation reactions; and a study of these reactions can lead to a fuller understanding of the behaviour of peroxy radicals in peroxide photochemistry. Thus the autoxidation of cumene[105] and of tetralin[106] using isotopically labelled oxygen and a mass spectrometric analytical technique, has provided a considerable amount of kinetic data concerning the reactions of the two appropriate peroxy radicals. Lloyd and Lange[107] have extended the earlier work of Hammond, Boozer, Hamilton and Sen[108] on the kinetics of reactions of peroxy radicals in tetralin autoxidation by a study of weakly inhibited systems. The autoxidation studies described have all used thermal initiation systems (azobisisobutyronitrile and bi-*tert*.-butyldiperoxy-oxalate), and are not properly considered further in a discussion of liquid-phase photochemistry. Some investigations, however, have utilized photochemical initiation in the presence of weak inhibitors to obtain kinetic information about peroxy radicals. Perhaps we may best introduce these studies by a mention of textile 'tendering'. 'Tendering' is the term applied to the loss of strength of a textile fibre which occurs as a result of chain degradation of the fibre-forming polymer. Such degradation can be brought about in a number of ways, e.g. photochemically, when a textile dyed with certain dyes is exposed to light, or by the action of oxidizing agents during bleaching. Much work has been carried out on the tendering of cellulosic and other textiles by dyes, and it is generally agreed that the process involves oxidation of the textile material photosensitized by the dye. There is by no means general agreement on the mechanism by which the excited dye initiates the tendering process. For example, transfer of energy to oxygen, direct oxidation of the textile substrate, and oxidation of hydroxyl ions to ·OH have been postulated. A full discussion of this topic here would take us too far afield, and we wish to refer only to the work of Bamford and Dewar[109,110] on the photosensitized oxidation of tetralin, a reaction which these workers considered[109,111] to be useful as a model for the photoxidation of cellulose textiles. They claimed that, for a series of vat dyes 'there is a rough correspondence between the quantum yields in the tetralin oxidation and tendering activity'. Further, they stated that: 'There is no such correlation for the sensitized polymerization of styrene or the quenching of the excited dyes by oxygen.'

The work on tetralin led to the evaluation of the absolute rate constants of the participating reactions for the first time. The kinetic scheme was essentially that of Bolland and Gee[112] and differs only in the mode of initiation.

$$
\begin{aligned}
D + h\nu &\rightarrow D^* & I_{abs} \\
D^* &\rightarrow D & k_f \\
D^* + O_2 &\rightarrow D + O_2 & k_1 \\
D^* + TH &\rightarrow DH + T\cdot & k_2 \\
T\cdot + O_2 &\rightarrow TO_2\cdot & k_3 \\
TO_2\cdot + TH &\rightarrow TO_2H + T\cdot & k_4 \\
2T\cdot &\rightarrow & k_5 \\
T\cdot + TO_2\cdot &\rightarrow & k_6 \\
2TO_2\cdot &\rightarrow & k_7
\end{aligned}
\tag{80}
$$

Here DH and TH represent dye and tetralin respectively; $T\cdot$ is a radical derived from tetralin by loss of a hydrogen atom and $TO_2\cdot$ the corresponding peroxy radical. The second and third reactions in this scheme allow for spontaneous (or solvent) deactivation and deactivation by oxygen, respectively. For long chains, with the approximation introduced by Bolland and Gee ($k_6 = \sqrt{[k_5 k_7]}$), the mechanism leads to the kinetic expression

$$
-\frac{d[O_2]}{dt} = \left\{\frac{k_2 I_{abs}[TH]}{k_f + k_1[O_2]_l + k_2[TH]}\right\}^{\frac{1}{2}} \frac{k_3 k_4 [O_2]_l [TH]}{k_3 \sqrt{(k_7)}\,[O_2]_l + k_4 \sqrt{(k_5)}\,[TH]}
\tag{81}
$$

where $[O_2]_l$ is the concentration of oxygen in the liquid. At high oxygen pressures the second term on the right of (81) becomes effectively constant; the physical implication is that all $T\cdot$ radicals react with oxygen before they are able to enter into termination reactions. With a constant rate of initiation, the rate of absorption of oxygen should, therefore, be independent of oxygen pressure. This is found to hold with initiation by benzoyl peroxide, but not with dye photosensitization; in the latter case the rate at high pressure falls off as the oxygen pressure increases, showing that the rate of initiation is a function of the oxygen pressure. These considerations justify the first term on the right of (81). Bamford and Dewar[110] evaluated a relative quenching coefficient q for a number of dyes; q is defined by the relation

$$q = \frac{k_1}{k_f + k_2[\text{TH}]} \qquad (82)$$

It was found that q is large (of the order 10^3) for Caledon Golden Yellow GK and Caledon Brilliant Orange 6 R and zero for anthraquinone and 1- and 2-chloroanthraquinones.

The rate of chain starting as a function of light intensity was measured by an inhibitor technique at high oxygen pressures, and hence it was possible to evaluate $k_4/\sqrt{k_7}$. By measuring the lifetime of the kinetic chains by the rotating sector technique under the same conditions k_7 and hence k_4 could be evaluated absolutely.

At low oxygen pressures the rate of solution of oxygen becomes rate-determining, and allowance for this must be made in the kinetics. When this is done, the ratio $k_3/\sqrt{k_5}$ may be determined. The rotating sector cannot be used under conditions where the solution of oxygen is rate-determining (indeed the observed rate of oxygen absorption may be much the same in continuous and intermittent light), but it was found possible to measure the photochemical after-effect, and thus to evaluate k_3 and k_5 absolutely.

TABLE I

$k_3 \quad 6 \cdot 76 \times 10^7$	$E_3 = 0 \cdot 0$ kcal mole^{-1}	$A_3 = 6 \cdot 8 \times 10^7$
$k_4 \quad 13 \cdot 3$	$E_4 = 4 \cdot 5$ kcal mole^{-1}	$A_4 = 2 \cdot 5 \times 10^4$
$k_5 \quad 7 \cdot 10 \times 10^6$	$E_5 = 2 \cdot 6$ kcal mole^{-1}	$A_5 = 5 \cdot 5 \times 10^4$
$k_7 \quad 2 \cdot 15 \times 10^7$	$E_7 = 0 \cdot 4$ kcal mole^{-1}	$A_7 = 4 \cdot 2 \times 10^7$

Table 1 shows the values obtained for the various velocity coefficients at 25 °C (in mole^{-1} l. s^{-1}), together with activation energies and frequency factors (in mole^{-1} l. s^{-1}).

3.4. Photosensitization of polymerization. The use of photosensitizers, such as benzoyl peroxide and azobisisobutyronitrile, as initiators for vinyl polymerization is familiar, and the value of photochemical techniques is well established. We should like to conclude this chapter with a discussion of some rather unusual photochemical initiators of polymerization. We have already noted, in the previous section, that photosensitization of polymerization occurs when monomers are irradiated in the presence of suitable dyes. Bamford and Dewar[109,111] studied the polymerization of styrene photosensitized by vat dyes at 14 °C, using wavelengths longer than 3,600 Å. Among the most active sensitizers were Caledon Yellow 5G and Caledon Dark Blue G, while Caledon Jade Green XN and Cibanone Orange 6R were found to be much less active. It was suggested that the excited dyes may photosensitize by a direct transfer of energy to the monomer. Oster[113] has obtained interesting results by using dyes as photosensitizers in the presence of oxygen together with a mild reducing agent. Both the nature of the reducing agent and the concentration of oxygen are critical. Under these conditions the quantum yield of initiation may be considerably greater than that obtained with the dye alone. The chemistry of the processes occurring is by no means clear. One function of the oxygen is presumably to oxidize the leuco form of the dye (which is formed in the absence of oxygen) to the semiquinone form, with production of hydroxyl radicals which initiate polymerization.

Finally, we wish to discuss the use of metal carbonyls as photochemical initiators of vinyl polymerization. It is found that the carbonyls of the group VII metals manganese and rhenium [$Mn_2(CO)_{10}$ and $Re_2(CO)_{10}$] act as efficient photosensitizers of vinyl polymerization in the presence[114,115] of small concentrations of carbon tetrachloride. The initiating species is believed to be CCl_3 derived from the carbon tetrachloride, and radioactive tracer studies[116] confirm this view. Further evidence for the participation of the halide in the initiation process is adduced from the occurrence of photogelation when a polymeric halide (polyvinyltrichloracetate) is substituted for the carbon tetrachloride. The rate of polymerization is highly sensitive to the carbon tetrachloride concentration when the latter is very small, although a limiting rate is reached at quite low halide concentrations: the exact dependence is related to the absorbed light intensity (not the incident intensity), and some experimental points for manganese carbonyl[114] at $\lambda = 4,358$ Å are shown in Fig. 3. At 'high' halide concentration the quantum yield for initiation is almost exactly unity, and the experimental observations suggest the reaction scheme for initiation:

$$\text{carbonyl} + h\nu \to F_1(+F_2) \tag{83}$$

$$F_1 + CCl_4 \to (I) \tag{84}$$

$$(I) \to CCl_3 \tag{85}$$

$$(I) + F_1 \to \text{inactive products} \tag{86}$$

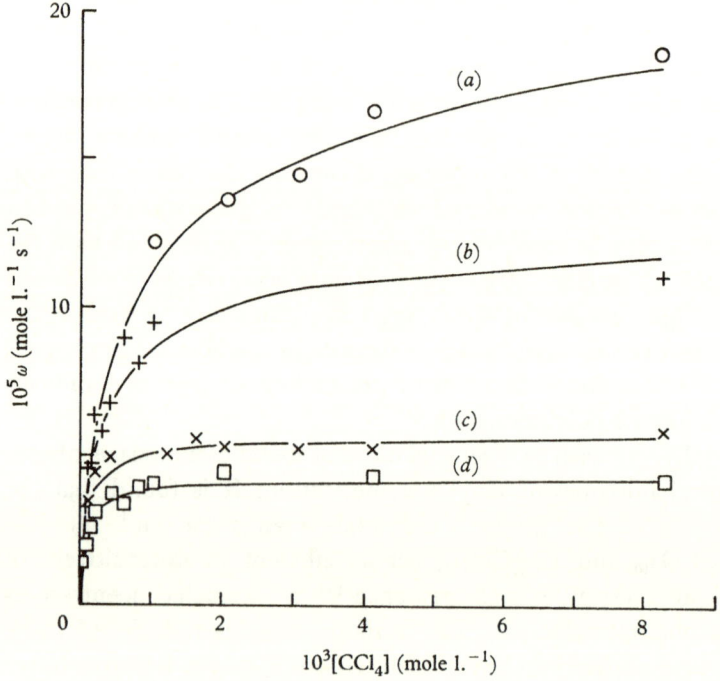

Fig. 3. Rate of polymerization as a function of carbon tetrachloride concentration
(a) O, $[Mn_2(CO)_{10}] = 2\cdot57 \times 10^{-4}$ mole l.$^{-1}$.
$\qquad I_{abs.} = 18\cdot1 \times 10^{-4}$ einstein l.$^{-1}$ s^{-1}.
(b) +, $[Mn_2(CO)_{10}] = 2\cdot57 \times 10^{-4}$ mole l.$^{-1}$.
$\qquad I_{abs.} = 6\cdot04 \times 10^{-4}$ einstein l.$^{-1}$ s^{-1}.
(c) ×, $[Mn_2(CO_{10})] = 0\cdot514 \times 10^{-4}$ mole l.$^{-1}$.
$\qquad I_{abs.} = 1\cdot27 \times 10^{-8}$ einstein l.$^{-1}$ s^{-1}.
(d) □, $[Mn_2(CO)_{10}] = 2\cdot57 \times 10^{-4}$ mole l.$^{-1}$.
$\qquad I_{abs.} = 0\cdot664 \times 10^{-8}$ einstein l.$^{-1}$ s^{-1}.

where F_1 and I are active intermediates. This scheme leads to predicted rates of polymerization which are displayed as the curves in Fig. 3.

When rhenium carbonyl [115] is the photosensitizer (at $\lambda = 3,650$ Å) the situation is more complex. After irradiation has ceased polymerization continues for several hours, and it appears that with rhenium carbonyl the second photolytic fragment F_2 must also react with carbon tetrachloride, but much more slowly than F_1. Thus during irradiation the concentration of F_2 builds up, and after turning off the light polymerization continues at a slowly decreasing rate. The experimental observations of the decay of polymerization rate in the dark are in accord with a rate constant for the reaction

$$F_2 + CCl_4 \rightarrow CCl_3 \qquad (87)$$

of about $2 \cdot 5 \times 10^{-4}$ l. mole^{-1} s^{-1}.

The general conclusions about the photochemical reactions have been confirmed [117] by experiments in which irradiation of the carbonyl dissolved in monomer is carried out in the absence of carbon tetrachloride. If the halide is added after the light is turned off polymer appears in relatively large amounts when rhenium carbonyl is used, and in much smaller amounts with manganese carbonyl, which suggests that in the former case a long lived intermediate is produced. Group VI metal carbonyls [$Mo(CO)_6$, $Cr(CO)_6$, $W(CO)_6$] are much less active [118] photochemically, the quantum yield being not more than $0 \cdot 1$. The difference possibly arises from the absence of the metal–metal bond which may be broken in the photolysis of the Group VII carbonyls.

REFERENCES

1 Rudberg, E. *Z. Physik.* 1924, **24**, 247.
2 Evans, M. G. and Polanyi, M. *Trans. Faraday Soc.* 1936, **32**, 1333.
3 Rabinowitch, E. and Wood, W. C. *Trans. Faraday Soc.* 1936, **32**, 1381.
4 Franck, J. and Rabinowitch, E. *Trans. Faraday Soc.* 1934, **30**, 120.
5 Noyes, R. M. *J. Am. Chem. Soc.* 1955, **77**, 2042.
6 Noyes, R. M. *Prog. Reaction Kinetics*, 1961, **1**, 129.
7 Goodeve, C. F. and Stein, N. O. *Trans. Faraday Soc.* 1931, **27**, 393.
8 Herzberg, G. *Trans. Faraday Soc.* 1931, **27**, 402.
9 Barrett, J. and Baxendale, J. H. *Trans. Faraday Soc.* 1960, **56**, 37.
10 Thomas, J. K. and Hart, E. J. *J. phys. Chem.* 1964, **68**, 2414.
11 Fricke, H. and Hart, E. J. *J. chem. Phys.* 1936, **4**, 418.
12 Allen, A. O. and Holroyd, R. A. *J. Am. Chem. Soc.* 1955, **77**, 5852.
13 Baxendale, J. H. *Radiation Res.* 1962, **17**, 312.
14 Markham, M. C. and Laidler, K. J. *J. phys. Chem.* 1953, **57**, 363.
15 Heyrovsky, M. and Norrish, R. G. W. *Nature, Lond.*, 1963, **200**, 880.
16 Norrish, R. G. W. and Rideal, E. K. *Proc. Roy. Soc.* A, 1923, **103**, 366.
17 Titlestadt, N. *Z. physik. Chem.* 1910, **72**, 257.
18 Norrish, R. G. W. and Paszyc, S. *Rocz. Chem. Pol.* 1963, **37**, 1305.
19 Heidt, L. J. *J. phys. Chem.* 1942, **46**, 624.
20 Volman, D. H. and Seed, R. *J. Am. Chem. Soc.* 1964, **86**, 5095.
21 Copeland, A. W., Black, O. D. and Garrett, A. B. *Chem. Rev.* 1942, **31**, 177.
22 Baxendale, J. H. and Wilson, J. A. *Trans. Faraday Soc.* 1957, **53**, 344.
23 Volman, D. H. and Chen, J. C. *J. Am. Chem. Soc.* 1959, **81**, 4141.
24 Hunt, J. P. and Taube, H. *J. Am. Chem. Soc.* 1952, **74**, 5999.
25 Fujimoto, M. and Ingram, D. J. E. *Trans. Faraday Soc.* 1958, **54**, 1304.
26 Daniels, M. *J. phys. Chem.* 1962, **66**, 1473.
27 Buxton, G. and Wilmarth, W. K. *J. phys. Chem.* 1963, **67**, 2835.
28 Hochanadel, C. J. *Radiation Res.* 1962, **17**, 286.
29 Currie, D. J. and Dainton, F. S. *Trans. Faraday Soc.* 1965, **61**, 1156.
30 Heidt, L. J. and Landi, V. R. *J. chem. Phys.* 1964, **41**, 176.
31 Vedeneev, V. I., Gerasimov, G. N. and Purmal, A. P. *Zhur. Fiz. Khim*, 1957, **31**, 1216.
32 Marshall, J. G. and Rutledge, P. V. *Nature, Lond.*, 1959, **184**, 2013.
33 Taube, H. *Trans. Faraday Soc.* 1957, **53**, 656.
34 Kistiakowsky, G. B. *Z. physik. Chem.* 1925, **117**, 337.
35 Castellano, E. and Schumacher, H. J. *J. chem. Phys.* 1962, **36**, 2238.
36 Heidt, L. J. and Forbes, G. S. *J. Am. Chem. Soc.* 1934, **56**, 1617.
37 Heidt, L. J. and Forbes, G. S. *J. Am. Chem. Soc.* 1934, **56**, 2365.
38 Heidt, L. J. *J. Am. Chem. Soc.* 1935, **57**, 1710.
39 Norrish, R. G. W. and Wayne, R. P. *Proc. Roy. Soc.* A, 1965, **288**, 200, 361.

40 McGrath, W. D. and Norrish, R. G. W. *Proc. Roy. Soc.* A, 1960, **254**, 317.
41 McGrath, W. D. and Norrish, R. G. W. *Proc. Roy. Soc.* A, 1957, **242**, 265.
42 Basco, N. and Norrish, R. G. W. *Proc. Roy. Soc.* A, 1961, **260**, 293.
43 DeMore, W. B. and Raper, O. F. *J. chem. Phys.* 1962, **37**, 2048.
44 DeMore, W. B. and Raper, O. F. *Can. J. Chem.* 1963, **41**, 808.
45 Raper, O. F. and DeMore, W. B. *J. chem. Phys.* 1964, **40**, 1053.
46 Bowen, E. J. and Lloyd, R. A. *Proc. chem. Soc.* 1963, p. 305.
47 Khan, A. U. and Kasha, M. *Nature, Lond.*, 1964, **204**, 241.
48 Arnold, J. S., Browne, R. J. and Ogryzlo, E. A. *Symposium on Chemiluminescence*, Durham, N.C., March 1965 (preprints, p. 35).
49 Lipscomb, F. J., Norrish, R. G. W. and Thrush, B. A. *Proc. Roy. Soc.* A, 1955, **233**, 172.
50 Bowen, E. J. and Cheung, W. M. *J. chem. Soc.*, 1932, p. 1200.
51 Bowen, E. J. *J. chem. Soc.* 1925, p. 1199.
52 Nagai, Y. and Goodeve, C. F. *Trans. Faraday Soc.* 1931, **27**, 508.
53 Bodenstein, M., Harteck, P. and Padelt, E. *Z. anorg. Chem.* 1925, **147**, 233.
54 Bowen, E. J. *Trans. Faraday Soc.* 1931, **27**, 513.
55 Dainton, F. S. *Disc. Faraday Soc.* 1964, **37**, 210.
56 Doering, W. E., Buttery, R. G., Laughlin, R. G. and Chaudhuri, N. *J. Am. Chem. Soc.* 1956, **78**, 3224.
57 Bradley, J. N. and Ledwith, A. *J. chem. Soc.* 1961, p. 1495.
58 Bamford, C. H., Casson, J. C. and Wayne, R. P. *Proc. Roy. Soc.* A, 1966, **289**, 3224.
59 Setser, D. W., Littrell, R. and Hassler, J. C. *J. Am. Chem. Soc.* 1965, **87**, 2065.
60 Frey, H. M. and Kistiakowsky, G. B. *J. Am. Chem. Soc.* 1957, **79**, 6373.
61 DeMore, W. B. and Benson, S. W. *Adv. Photochem.* 1964, **2**, 219.
62 Lyon, R. K. *J. Am. Chem. Soc.* 1964, **86**, 1907.
63 Kirkbride, F. W. and Norrish, R. G. W. *J. chem. Soc.* 1932, p. 1518.
64 Akeroyd, E. I. and Norrish, R. G. W. *J. chem. Soc.* 1936, p. 890.
65 Norrish, R. G. W., Crone, H. G. and Saltmarsh, O. D. *J. chem. Soc.* 1933, p. 1533.
66 Norrish, R. G. W. and Appleyard, M. E. S. *J. chem. Soc.* 1934, p. 874.
67 Bloch, B. M. and Norrish, R. G. W. *J. chem. Soc.* 1935, p. 1638.
68 Saltmarsh, O. D. and Norrish, R. G. W. *J. chem. Soc.* 1935, p. 455.
69 Bamford, C. H. and Norrish, R. G. W. *J. chem. Soc.* 1935, p. 1504.
70 Bamford, C. H. and Norrish, R. G. W. *J. chem. Soc.* 1938, p. 1521.
71 Bamford, C. H. and Norrish, R. G. W. *J. chem. Soc.* 1938, p. 1544.
72 Norrish, R. G. W. and Bamford, C. H. *Nature, Lond.*, 1937, **140**, 195.
73 Davies, W., Jnr. and Noyes, W. A., Jnr. *J. Am. Chem. Soc.* 1959, **81**, 5061.
74 Srinivasan, R. *J. Am. Chem. Soc.* 1959, **81**, 5061.
75 McMillan, G. R., Calvert, J. G. and Pitts, J. N., Jnr. *J. Am. Chem. Soc.* 1964, **86**, 3602.

76 Zahra, A. and Noyes, W. A., Jnr. *J. phys. Chem.* 1965, **69**, 943.
77 Borrell, P. and Norrish, R. G. W. *Proc. Roy. Soc.* A, 1961, **262**, 19.
78 Ausloos, P. *Can. J. Chem.* 1958, **36**, 383.
79 Sigal, P. *J. phys. Chem.* 1963, **67**, 2660.
80 Guillett, J. E. and Norrish, R. G. W. *Proc. Roy. Soc.* A, 1955, **233**, 153.
81 Leermakers, P. A. and Vesley, G. F. *J. Am. Chem. Soc.* 1963, **85**, 3776.
82 Leermakers, P. A. and Vesley, G. F. *J. Am. Chem. Soc.* 1964, **86**, 1768.
83 Srinivasan, R. *J. Am. Chem. Soc.* 1959, **81**, 1546.
84 Srinivasan, R. and Cremer, S. E. *J. Am. Chem. Soc.* 1965, **87**, 1647.
85 Srinivasan, R. *Adv. Photochem.* 1963, **1**, 83.
86 Borkowski, R. P. and Ausloos, P. *J. Am. Chem. Soc.* 1962, **84**, 4044.
87 Norrish, R. G. W. and Wayne, R. P. *Proc. Roy. Soc.* A, 1965, **284**, 1.
88 Borrell, P. *J. Am. Chem. Soc.* 1964, **86**, 3156.
89 Berman, J. D., Stanley, J. H., Sherman, W. V. and Cohen, S. G. *J. Am. Chem. Soc.* 1963, **85**, 4010.
90 Jarvie, J. M. and Laufer, A. H. *J. phys. Chem.* 1964, **68**, 2557.
91 Farkas, L. *Z. Phys. Chem.* B, 1933, **23**, 89.
92 Bamford, C. H. and Norrish, R. G. W. *J. chem. Soc.* 1938, p. 1531.
93 Bamford, C. H. *J. chem. Soc.* 1939, p. 17.
94 Booth, G. H. and Norrish, R. G. W. *J. chem. Soc.* 1952, p. 188.
95 Pouyet, B. *Bull. Soc. chim. fr.* 1964, p. 2582.
96 Bell, E. R., Raley, J. H., Rust, F. F., Senbold, F. H. and Vaughan, W. E. *Disc. Faraday Soc.* 1951, **10**, 242.
97 Martin, J. T. and Norrish, R. G. W. *Proc. Roy. Soc.* A, 1953, **220**, 322.
98 Norrish, R. G. W. and Searby, M. H. *Proc. Roy. Soc.* A, 1956, **237**, 464.
99 Bailey, H. C. and Godin, G. W. *Trans. Faraday Soc.* 1956, **52**, 68.
100 Ueberreiter, K. and Bruns, W. *Makromol. Chem.* 1963, **68**, 24.
101 Vasil'ev, I. N. and Kronganz, V. A. *Kin. i. Kat.* 1963, **4**, 204.
102 Bowen, E. J. and Lloyd, R. A. *Proc. Roy. Soc.* A, 1963, **275**, 465.
103 Bowen, E. J. *Nature, Lond.*, 1964, **201**, 180.
104 Khan, A. U. and Kasha, M. J. *J. chem. Phys.* 1963, **39**, 2105.
105 Bartlett, P. D. and Traylor, T. G. *J. Am. Chem. Soc.* 1963, **85**, 2407.
106 Traylor, T. G. *J. Am. Chem. Soc.* 1963, **85**, 2411.
107 Lloyd, W. G. and Lange, C. E. *J. Am. Chem. Soc.* 1964, **86**, 1491.
108 Hammond, G. S., Boozer, C. E., Hamilton, C. E. and Sen, J. N. *J. Am. Chem. Soc.* 1955, **77**, 3238.
109 Bamford, C. H. and Dewar, M. J. S. *Nature, Lond.*, 1949, **163**, 214.
110 Bamford, C. H. and Dewar, M. J. S. *Proc. Roy. Soc.* A, 1949, **198**, 252.
111 Bamford, C. H. and Dewar, M. J. S. *Proceedings of Symposium on Photochemistry in Relation to Textiles* (Society of Dyers and Colourists), 1949, p. 90.
112 Bolland, J. L. and Gee, G. *Trans. Faraday Soc.* 1946, **42**, 236, 244.

113 Oster, G. *Nature, Lond.*, 1954, **173**, 300.
114 Bamford, C. H., Crowe, P. A. and Wayne, R. P. *Proc. Roy. Soc.* A, 1965, **284**, 455.
115 Bamford, C. H., Crowe, P. A., Hobbs, J. and Wayne, R. P. *Proc. Roy. Soc.* A, 1966, **292**, 153.
116 Bamford, C. H., Eastmond, G. C. and Robinson, V. J. *Trans. Faraday Soc.* 1964, **60**, 751.
117 Bamford, C. H., Hobbs, J. and Wayne, R. P. *Chemical Communications*, 1965, p. 469.
118 Bamford, C. H., Brumby, S. and Wayne, R. P. *Nature, Lond.*, 1966, **209**, 292.

4

GASEOUS PHOTOCHLORINATION

F. S. DAINTON AND P. B. AYSCOUGH

1. Introduction

The term photochlorination in its widest sense means the introduction of chlorine atoms into molecules through the agency of
light, but in a more restricted sense is used to describe the reactions
of chlorine atoms formed by photochemical dissociation of chlorine
molecules. These reactions do not differ in any important respect
from those of chlorine atoms produced thermally or following interaction with ionizing radiation, though the photolytic method is
more convenient and more readily controllable.

Since chlorine atoms can react with most organic molecules and
with many inorganic species the field of study is potentially very
wide. In practice, however, intensive studies have been made on a
rather small number of simple gaseous systems involving reactions
with hydrogen, oxygen, ozone, hydrocarbons and organic halogen
compounds. As will be apparent later, the adjective 'simple' is
rather inappropriate because one characteristic feature of all photochlorinations is their great complexity except under the most
favourable experimental conditions. By way of compensation such
studies yield an immense amount of detailed information concerning the reactions of chlorine atoms and simple chlorinated
radicals. Many important features of these reactions have been
found to have significance far beyond the limits of photochlorination and the progress of modern reaction kinetics owes much to
experimental and theoretical procedures developed in this field.

As examples of such fruitful developments one may quote studies
of photosensitization, induction periods, explosion limits, and
mechanisms of catalysis and retardation, all of which were initiated
by work on photochlorination systems. Similarly, some developments in experimental procedures such as actinometry, gas
chromatography and mass spectrometry have been stimulated by
problems encountered in photochlorination, as were theoretical

studies of methods for handling complex rate equations for chain reactions involving numerous competitive and consecutive reactions.

Today the accumulation of accurate values of frequency factors and activation energies for reactions of chlorine atoms and chlorinated radicals with various groups of structurally related molecules provides some of the best data for examining theoretical interpretations of reaction rates. Among recent work, current studies of excited chloroalkyl radicals promise to yield significant information regarding the old problem of the relation between reaction rate and energy content.

The historical approach adopted in this review emphasizes the slow but continued development of understanding of these complex reaction mechanisms and draws attention to the digressions and controversies, some fruitful and some not, with which this subject has always been connected.

Absorption of light by chlorine molecules is continuous below 4,785 Å with λ_{max}. at about 3,300 Å corresponding to the transition $^1\Pi_u$ or $^3\Pi_u \leftarrow {}^2\Sigma_g$ and the molecule then dissociates into two normal ($^2P_{\frac{3}{2}}$) atoms. Dissociation can also occur from an O_u^+ level into one $^2P_{\frac{3}{2}}$ atom and one $^2P_{\frac{1}{2}}$ atom (excitation energy 2·5 kcal mole^{-1}) caused by induced predissociation during collisions. However, there are no observable kinetic consequences of this difference in energy content. Both dissociation processes leave the chlorine atoms with considerable excess kinetic energy since the Cl_2 dissociation energy is only 59·4 kcal mole^{-1}, corresponding to $\lambda \sim 4,800$ Å. It is interesting that this interpretation of the visible and ultraviolet absorption spectrum of chlorine by Franck was the first satisfactory account of the transitions involved in molecular electronic spectra.

Almost all photochemical studies of chlorine have employed mercury lamps as the irradiation source, using either the 3,660 or 4,358 Å emission lines. No differences in behaviour are observable as a result of the difference in excitation energy. Glass or quartz vessels are suitable and conventional methods of actinometry (e.g. using ferrioxalate solutions) are applicable. However, the advantages of the essentially simple experimental set-up are more than offset by the extraordinary sensitivity of all these reactions to traces of impurities. The necessity of 'ageing' the reaction vessel by pretreatment with reaction mixtures in order to obtain reproducible rates is commonplace in gas-phase kinetic studies, but the standard of purity required for the chlorine, necessitating repeated fractional distillation in grease-free systems, is unusually exacting. Although numerous semi-quantitative studies of the photolytic reactions of chlorine were made before 1930 it was only at this time that the sensitivity to impurities became fully apparent so that much early work is quantitatively invalid. This was also the time at which the true nature of photochlorination reactions began to emerge, mainly as a consequence of intensive studies of the reaction between chlorine and hydrogen which were carried out in Norrish's laboratory.

2. Classical studies on photochlorination

2.1. *The hydrogen–chlorine reaction.* By the early 1920's the main features of photochemical reactions had been established. The nature of the primary act was known, the relation between the number of molecules decomposed and the number of light quanta absorbed, expressed as the quantum efficiency, was used to characterize photochemical reactions and establish their mechanisms, and the idea of chain reactions in which 'chain carriers' were alternately destroyed and regenerated in repetitive reaction sequences was firmly established. Thus the reaction between hydrogen and chlorine could be written†

$$Cl_2 + h\nu \rightarrow 2Cl \tag{1}$$

$$Cl + H_2 \rightarrow HCl + H \tag{2}$$

$$H + Cl_2 \rightarrow HCl + Cl \tag{3}$$

followed by chain termination by reaction of H and Cl atoms. Under certain conditions the quantum yield was as high as 10^5 and the reaction was clearly recognized as an example of a Nernst chain reaction. The sensitivity to traces of water, oxygen, etc., could be

---

† In this chapter, and in succeeding chapters where there are many reaction schemes and rate equations, reaction steps have been numbered sequentially through the chapter, and rate equations have been designated by bold type letters where it is necessary to refer to them in the text.

attributed, in general terms, to interruptions in the sequence of reactions involved in chain propagation.

Neither the initiation nor the termination step was clearly understood at this time. The situation was confused by the preoccupation with intensive 'Baker' drying which had produced some extraordinary changes in physical and chemical properties. When the effect of intensive drying of the reagents in the hydrogen–chlorine reaction was examined, the change in reaction rate was pronounced. [1] The rate of formation of HCl was found to decrease markedly at partial pressures of water below about 10^{-4} torr and become immeasurable at 10^{-7} torr or less. This led to the suggestion that the reaction chains were initiated by chlorine atoms formed from chlorine dissolved in the water adsorbed on the surface of the reaction vessel rather than from dissociation in the gas phase. [2] The controversy surrounding this suggestion lasted several years, and was finally settled when it was found that the drying agents themselves were acting as retarders and that the water itself had no effect at all.

Although this particular controversy was unproductive it was at this point that Norrish first took an interest in photochlorination. His first major contributions in this field concerned the retarding effect of oxygen on the hydrogen–chlorine reaction and the nature of the induction period caused by adding ammonia to mixtures of hydrogen and chlorine. [3] It was shown that H_2O_2 was a product of the illumination of mixtures of hydrogen, chlorine and oxygen, [4] and the suggestion was made that this arose from combination of OH radicals produced by the reaction of H atoms and oxygen, thus accounting for the retarding effect of oxygen on the photochlorination. The reaction of Cl with O_2 later studied by flash photolysis [5] was not considered at this time. The detailed and precise studies of the inhibiting effect of ammonia were very significant in showing that the real inhibitor in this system was NCl_3 formed by reaction of ammonia and chlorine. The NCl_3 was shown to remove Cl atoms with great efficiency via reaction (4)

$$NCl_3 + Cl \rightarrow NCl_2 + Cl_2 \qquad (4)$$

which gives ultimately N_2 and Cl_2. Thus the induction period of this reaction was shown to be a period of very slow reaction in which almost all the Cl atoms were lost to NCl_3. Only when all the NCl_3 is removed does the reaction proceed at its normal fast rate. This study provides the key to the drastic retarding effect of a wide range of nitrogenous impurities on chlorine atom reactions. Very recently the radical NCl_2 and its decomposition product NCl have been identified by Norrish and Briggs[6] as intermediates in the chlorine photosensitized decomposition of nitrogen trichloride. The existence of reaction (4) is thus confirmed.

Another contribution of great significance in the field of photochemistry was the demonstration of sharply defined explosion limits in the reaction of hydrogen and oxygen, photosensitized by chlorine. [7] By varying the chlorine pressure at a given temperature Norrish showed that there are sharp limits of pressure, varying markedly with temperature, below which the reaction is slow and above which it becomes explosive. Observations of this kind were to become numerous and lead to the development of important theories of adiabatic explosions based on branching chain mechanisms.

However, these observations, valuable as they were, contributed little to the solution of the central problem which was to devise a mechanism for the photosynthesis of hydrogen chloride, in the presence and absence of oxygen. The controversy surrounding this reaction continued unabated, the main points at issue at this stage being the inhibiting effect of HCl, observed by Chapman but rejected by Bodenstein, and the exponent of the absorbed light intensity in the rate expression. Chapman[8] found a half-power dependence which led him to propose a reaction scheme similar to that of the hydrogen–bromine reaction with mutual termination of the reaction chains. Conversely, Bodenstein's observations[9] necessitated accepting wall termination and/or reactions such as $(5a)$ and $(5b)$. The resolution of this problem by Norrish and Ritchie[10] removed one of the major obstacles in the field of photochemistry and added an important new experimental technique based on continuous photometric analysis.

$$H + H_2 + Cl_2 \rightarrow H + 2HCl \qquad (5a)$$
$$H + H \rightarrow H_2 \qquad (5b)$$

The conflicting observations on which these interpretations were based arose from fundamentally different methods of following the course of the photolysis. Bodenstein followed the reaction by measuring the pressure fall when sufficient water was added to dissolve all the HCl formed, whereas Chapman interrupted the reaction periodically by freezing out the chlorine and HCl and measuring the residual hydrogen. The objections to both methods are readily apparent, and these were obviated by Norrish and Ritchie's method in which a transverse beam of light was used to monitor the concentration of molecular chlorine directly. By this means, these authors demonstrated in an unequivocal manner: (1) the marked retarding effect of HCl in both oxygen-free and oxygen-rich systems, [10] (2) the proportionality of the overall rate to $I_a^{\frac{1}{2}}$ in oxygen-free mixtures, indicating mainly mutual termination of reaction chains in the gas phase [10] and (3) the increase in the exponent of I_a with increasing oxygen pressure. [11] The reaction scheme then proposed for oxygen-free systems was

$$Cl_2 + h\nu \rightarrow 2Cl \qquad (1)$$

$$Cl + H_2 \rightarrow HCl + H \qquad (2)$$

$$H + Cl_2 \rightarrow HCl + Cl \qquad (3)$$

$$H + HCl \rightarrow H_2 + Cl \qquad (-2)$$

$$2Cl \rightarrow Cl_2 \qquad (6)$$

leading, by means of the usual stationary state assumptions, to the expression for the rate

$$d[HCl]/dt = \frac{2k_2 k_3 (k_1/k_6)^{\frac{1}{2}} [H_2][Cl_2]}{(k_3[Cl_2] + k_{-2}[HCl])} I_a^{\frac{1}{2}}$$

The experimental data were found to be in agreement with this equation with $k_{-2}/k_3 = 1.7$ and $2k_2(k_1/k_6)^{\frac{1}{2}} = 2.8 \times 10^{-3}$ over a wide range of pressures. Norrish and Ritchie recognized that reaction (6) was likely to involve a third body but found no dependence of quantum yield on the overall pressure. They therefore suggested that either the recombination took place on the surface or that the true termination reaction involved two Cl_3 radicals in equilibrium with Cl and Cl_2. This interpretation is close to current views considered in section 4.2. The influence of 'wall' and oxygen were attributed to the reactions:

$$Cl + wall \rightarrow \tfrac{1}{2}Cl_2 \tag{7}$$

$$Cl + O_2 \rightarrow ClO_2 \tag{8}$$

$$H + H_2 + O_2 \rightarrow H_2O + OH \tag{9}$$

$$H + HCl + O_2 \rightarrow H_2O + ClO \tag{10}$$

$$H + O_2 \rightarrow HO_2 \tag{11}$$

Reactions (9) and (10) in this scheme were needed to explain the reduced dependence on hydrogen pressure in mixtures containing oxygen and the considerable retarding effect of HCl.

An important feature of this analysis is the notion of reversal of a propagation step. Under the experimental conditions used by these workers, nearly two-thirds of the H atoms produced in (2) are converted back to molecular hydrogen by reaction (-2). Similar situations have been found in many other photochlorination systems, and much of the complexity of such systems arises from the multiplicity of reactions which may be available to the product of reaction (2) or its analogue. This is particularly important in the reactions of chlorine with hydrocarbons.

2.2. Classical studies of the photochlorination of olefins. Following the resolution of the major problems concerning the hydrogen–chlorine reaction, attention was directed towards the reactions of chlorine with hydrocarbons, saturated and unsaturated. The war-time work of Schumacher and his colleagues [12] on the photolysis of mixtures of chlorine and ethylene and its simple chloro-derivatives appears to have been the first systematic study of the variation of reaction rates with structure. It seemed that all the reactions proceeded by a radical chain mechanism similar to that proposed for the hydrogen–chlorine reaction but without the reverse reaction (-12) except in the case where the olefin denoted by A is ethylene, i.e.

$$Cl_2 + h\nu \rightarrow 2Cl \tag{1}$$

$$Cl + A \rightarrow ACl \tag{12}$$

$$ACl + Cl_2 \rightarrow ACl_2 + Cl \tag{13}$$

$$2ACl \rightarrow A_2Cl_2 \quad \text{or} \quad A + ACl_2 \tag{14}$$

For long chains this mechanism leads to the rate law

$$d[ACl_2]/dt = k_{13}.I_a^{\frac{1}{2}}[Cl_2]/k_{14}^{\frac{1}{2}}$$

which fits the experimental data provided [A] exceeds a certain value which varies with temperature and chlorine pressure. This is now known to correspond to the pressure above which other termination processes such as (15) and (16)

$$ACl + Cl \rightarrow ACl_2 \quad or \quad A + Cl_2 \tag{15}$$

$$2Cl + M \rightarrow Cl_2 + M \tag{16}$$

cease to be important. Consequently below this value the rate becomes progressively more dependent on [A] and less dependent on [Cl_2].

Although Schumacher's values for quantum yields are certainly too low, his observations that the quantum yield diminishes considerably and the overall activation energy increases with increasing chlorine substitution in the olefin are undoubtedly true. The causes of the effects will emerge from the discussion of the general mechanism for photochlorination which is considered later.

3. Measurement of absolute rate constants

None of the investigations described above led to the evaluation of absolute rate constants for any of the individual steps in the photochlorination mechanisms. To achieve this the reaction must be studied under non-steady-state conditions, i.e. *either* by studying the early post-irradiation phases of reactions, *or* by use of the rotating sector technique. The latter method had already been used by Berthoud and Bellerot[13] to study the reaction of bromine or iodine with oxalate ions, and had been treated theoretically by Briers, Chapman and Walters.[14]

Both methods involve the determination of the mean lifetime, τ, of the reaction chains which is equal to $(k_t \phi_1 I_a)^{-\frac{1}{2}}$, where k_t is the rate of the termination reaction and ϕ_1 the primary quantum yield. There have been numerous reviews of the rotating sector techniques,[15] and a recent discussion of the application of this method to photochlorination reactions.[16] The theory has also been modified to take into account such factors as two or more chain carriers, mixed termination, non-uniform production of chain centres, penumbra errors, etc.

3.1. Reactions of Cl atoms with CO, NOCl, and saturated hydrocarbons. The first photochlorination reaction to which the rotating-sector method was applied was the reaction between chlorine and carbon monoxide to form phosgene. Burns and Dainton,[17,18] working in Norrish's laboratory, showed that the mechanisms involved an equilibrium between Cl and COCl similar to that found for the photochlorination of ethylene and were able to evaluate the equilibrium constant

$$k_{-17}/k_{17} = 10^{2 \cdot 8} \exp\left(-6 \cdot 3 \text{ kcal}/RT\right) \text{ mole l.}^{-1}$$

and the rate constants for the propagation and termination steps, viz. $k_{18} = 10^{9 \cdot 4} \exp\left(-2 \cdot 9 \text{ kcal}/RT\right)$ and $k_{19} = 10^{11 \cdot 6} \exp\left(-0 \cdot 83 \text{ kcal}/RT\right)$ l. mole^{-1} s^{-1}.

$$Cl_2 + h\nu \to 2Cl \tag{1}$$

$$Cl + CO \to COCl \tag{17}$$

$$COCl \to CO + Cl \tag{-17}$$

$$COCl + Cl_2 \to COCl_2 + Cl \tag{18}$$

$$Cl + COCl \to COCl_2 \tag{19}$$

When NOCl was added the normal termination step was replaced by reactions (20) and (21) so that a study of the NOCl-retarded photochlorination of CO gave values of

$$k_{20} = 10^{10} \exp\left(-1 \cdot 06 \text{ kcal}/RT\right)$$

and $k_{21} = 10^{10 \cdot 7} \exp\left(-1 \cdot 14 \text{ kcal}/RT\right)$ l. mole^{-1}s^{-1}.

$$Cl + NOCl \to Cl_2 + NO \tag{20}$$

$$COCl + NOCl \to Cl_2 + NO + CO \tag{21}$$

Ashmore and Chanmugam,[19] also working in Norrish's laboratory, studied the slow thermal reaction between hydrogen and chlorine in the presence of NO and NOCl at about 300 °C. Under these conditions there is competition between NOCl (reaction 20) and H_2 (reaction 2) for Cl atoms, but the concentration of Cl atoms is almost identical with that established by the balanced reactions (-20) and (20). Since accurate values of k_{-20}/k_{20}

$$Cl + H_2 \rightarrow HCl + H \tag{2}$$

$$NO + Cl_2 \rightleftharpoons NOCl + Cl \tag{-20, 20}$$

are known k_2 could be determined directly. The value so obtained was in excellent agreement with earlier values obtained by Rodebush and Klingelhoefer[20] using an electrodeless discharge in chlorine–hydrogen mixtures, and by Steiner and Rideal[21] who measured the rate of the reverse reaction (-2) using the HCl-catalysed conversion of para- to ortho-hydrogen which involves (22)

$$H + HCl \rightarrow H_2 + Cl \tag{-2}$$

$$H + p.H_2 \rightleftharpoons o.H_2 + H \tag{22, -22}$$

as a rate-determining step, to measure the H atom concentration at about 1,000 °K. These three sets of values correspond to $k_2 = (8 \cdot 3 \pm 0 \cdot 6)\, 10^{10} \exp\left[-(5 \cdot 48 \pm 0 \cdot 14)/RT\right]$ over the range 300–1,000 °K, and provide a very reliable reference value from which the rate constants for reactions of Cl with other hydrocarbons and alkyl halides may be determined.[22]

For example, if mixtures of hydrogen, chlorine and hydrocarbon RH are illuminated, the consumption of H_2 and RH is regulated by the value of the ratio $k_{23}[Cl][RH]/k_2[Cl][H_2]$

$$Cl + H_2 \rightarrow HCl + H \tag{2}$$

$$Cl + RH \rightarrow HCl + R \tag{23}$$

Using this method, Trotman-Dickensen et al.[23] were able to draw up a list of rate constants and activation energies for a series of hydrocarbons from which the validity of empirical relations between bond dissociation energies and activation energies could be examined. The authors took into account the possible reversibility of the above reactions and further chlorination of the product RCl, but could not achieve great precision because of the problems associated with the separation of hydrocarbons by low-temperature distillation.

In the past decade the photochlorination of hydrocarbons and alkyl halides has been re-examined in a number of laboratories using more precise methods of product analysis such as gas chromatography and mass spectrometry. Intermittent illumination has been used where possible to determine individual rate constants, competitive methods have been used to derive others, and a number of additional cross-checks, using competition between, say, an olefin and an alkane yielding the same intermediate radical, have been applied. It is convenient to continue the discussion in terms of a general mechanism of photochlorination, the development of which has greatly facilitated recent studies.

4. General mechanisms for the photochlorination of hydrocarbons

Most of the relevant data for reactions between chlorine and alkenes (A), alkanes (RH) and hydrogen (RH = H_2), can be fitted into the framework of the following general mechanism:

$$Cl_2 + h\nu \rightarrow 2Cl \tag{1}$$

$$Cl + A \rightleftarrows ACl \tag{12, -12}$$

$$Cl + RH \rightleftarrows HCl + R \tag{12', -12'}$$

$$ACl + Cl_2 \rightleftarrows ACl_2 + Cl \tag{13, -13}$$

$$R + Cl_2 \rightleftarrows RCl + Cl \tag{13', -13'}$$

$$2ACl \rightarrow products \tag{14}$$

$$2R \rightarrow products \tag{14'}$$

$$ACl + Cl \rightarrow products \tag{15}$$

$$R + Cl \rightarrow products \tag{15'}$$

$$2Cl + M \rightarrow Cl_2 + M \tag{16}$$

$$Cl + wall \rightarrow \tfrac{1}{2}Cl_2 \tag{7}$$

The question of whether the reactions (14), (14'), (15) and (15') involve combination or disproportionation will not be considered at this stage. In all the photochlorinations under consideration the products from the termination reactions are either indistinguishable from the reactants or products of propagation reactions or the chains are so long that the products of the termination reaction are formed in undetectably small amounts.

This general scheme, when applied to olefins, under conditions in which wall termination (7) can be neglected, yields the rate expression (**A**)

$$d[ACl_2]/dt = I_a^{\frac{1}{2}} k_{13}[Cl_2]/(k_{14} + k_{15}\delta + k_{16}\delta^2[M])^{\frac{1}{2}} \qquad (\mathbf{A})$$

where $\delta = [Cl]/[ACl]$. For alkanes the same general rate expression applies, except that $\delta = [Cl]/[R]$.

This general mechanism is supported by studies involving comparison of the rates of two reactions with the same intermediate radical. For example, if $A = C_2Cl_4$ and $RH = C_2Cl_5H$,

$$ACl = R = C_2Cl_5$$

and, as expected, the rate constant ratios $k_{13}/k_{14}^{\frac{1}{2}}$ and $k_{13}'/k_{14}'^{\frac{1}{2}}$ are found to be identical within experimental error. [24,25]

On the other hand, alternative mechanisms involving Cl_3 as a chain carrier can now be ruled out. These suggestions are based on the argument that since $[Cl_2]$ is always comparable with and often greater than $[A]$ or $[RH]$ the chlorine atoms might be in equilibrium with Cl_3 (24, -24).

$$Cl + Cl_2 \rightleftharpoons Cl_3 \qquad (24, -24)$$
$$Cl_3 + A \rightarrow [ACl + Cl_2] \rightarrow ACl_2 + Cl \qquad (25)$$

Cl_3 would then be the true chain carrier, reacting with A to form ACl (reaction (25)). This hypothesis is unacceptable because whatever the nature of the termination step $2Cl_3 \rightarrow 3Cl_2$, $Cl_3 + Cl \rightarrow 2Cl_2$ or reaction (16), the rate would never be independent of $[A]$ which is contrary to observation.

4.1. *Limiting conditions.* To extract useful data from studies of such systems it is necessary to give careful consideration to the experimental conditions. The rate expression (**A**) contains the function $\delta = [Cl]/[ACl]$ or $[Cl]/[R]$ which can be evaluated only under certain limiting conditions.

We have already chosen to neglect reaction (7) by suitable choice of conditions and this is almost always experimentally possible. We now wish to determine the conditions under which reactions (14), (15) and (16) respectively are the main termination steps. It can be shown that in the photochlorination of olefins

$$\delta = [Cl]/[ACl] = (k_{13}[Cl_2] + k_{-12})/k_{12}[A] \qquad (\mathbf{B})$$

and the limiting conditions may be defined in the following manner. When [A] is very large δ is very small so that $k_{14} > k_{15}\delta > k_{16}\delta^2[M]$ and equation (A) reduces to equation (C).

$$d[ACl_2]/dt = (I_a/k_{14})^{\frac{1}{2}}k_{13}[Cl_2] \qquad \text{(C)}$$

At lower values of [A] such that $k_{15}\delta > k_{14}$ or $k_{16}\delta^2[M]$ the rate expression becomes equation (D).

$$d[ACl_2]/dt = (I_a/k_{15})^{\frac{1}{2}}k_{13}[Cl_2] \{k_{12}[A]/(k_{13}[Cl_2]+k_{-12})\}^{\frac{1}{2}} \qquad \text{(D)}$$

At very low values of [A] when $k_{16}\delta^2[M] > (k_{15}\delta + k_{14})$ the correct rate law will be given by equation (E).

$$d[ACl_2]/dt = (I_a/k_{16}[M])^{\frac{1}{2}}k_{12}[A]k_{13}[Cl_2]/(k_{13}[Cl_2]+k_{12}) \qquad \text{(E)}$$

On this basis the rate of photochlorination at a fixed chlorine concentration would be expected to increase proportionately to [A] at very low hydrocarbon concentrations and to become independent of [A] at high concentrations. This has been observed [24,25] for all the olefins in the series $C_2H_xCl_{4-x}$, where x varies from 0 to 3 inclusive and is illustrated in Fig. 1. Furthermore, it would be expected that the exothermic reaction (13) would have a frequency factor in the range 10^8 to 10^{10} l. mole^{-1} s^{-1}, typical of atom abstraction reactions, and a low energy of activation, whereas reaction (-12) would have a frequency factor of about 10^{13} s^{-1} and an energy of activation at least equal to the exothermicity of reaction (12), i.e. $E_{-12} \geqslant 17$ to 22 kcal mole^{-1}. Hence at temperatures less than 100 °C

$$k_{13}[Cl_2] \gg k_{-12}$$

and the term k_{-12} in the denominator of equations (D) and (E) can be neglected. Accordingly, at pressures of A where the rate is independent of [A] (equation ((C) the rate should be proportional to $[Cl_2]$ and this has been frequently shown to be the case. [24,25] However, as the pressure of A is reduced and the rate becomes increasingly dependent on [A] it should become less dependent on $[Cl_2]$. This prediction is experimentally less easy to verify but there is no doubt that where it has been tested the results are consistent with the prediction.

Clearly, by applying the rotating sector technique to photo-chlorinations at temperatures where k_{-12} is negligible it should be possible to determine A_{13}, E_{13}, A_{14} and E_{14} with some accuracy and the values are given in Table I. The agreement between different laboratories [24,25] has been excellent despite great differences in experimental technique.

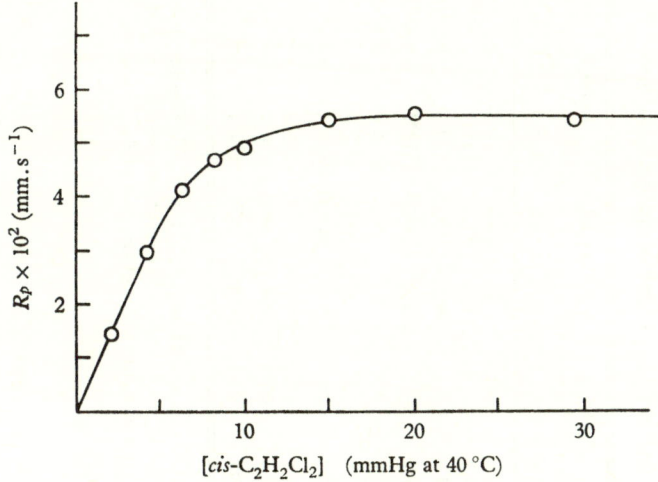

Fig. I. Variation of rate of photochlorination (R_p) with partial pressure of cis-dichloroethylene at 40 °C. Partial pressure of chlorine = 150 mmHg; $I_a = 2 \cdot 56 \times 10^{13}$ quanta cm^{-2} s^{-1}.

Attempts to obtain values of the Arrhenius parameters of reactions (12), (15) and (16), using the sector method, have been less successful because of the uncertainty, mentioned above, as to whether at the lowest convenient olefin concentrations, the condition for applicability of equation (E) has been attained. For example, with [Cl$_2$] ~ 200 mmHg and [A] ~ 2 mmHg for cis-1,2-C$_2$H$_2$Cl$_2$ for which E_3 ~ 2·5 kcal mole^{-1} it appears that termination by reaction (16) accounts for removal of about half the chlorine atoms but that first-order wall termination, i.e. reaction (7) and perhaps reaction (15) are also important. [25] When corrections are made for these termination processes and for the high concentration gradient of radicals in the cell, brought about by the necessarily high [Cl$_2$], the possible errors in k_{12} and k_{16} become very great. Moreover, as we show later in section 4.3, a significant fraction of the ACl radicals first formed revert to A + Cl and this

TABLE 1. *Arrhenius parameters for reactions in the photochlorination of olefins*

	C_2H_4 log A	E_a	C_2H_3Cl log A	E_a	$1,2\text{-}C_2H_2Cl_2$ log A	E_a	$1,1\text{-}C_2H_2Cl_2$ log A	E_a	C_2HCl_3 log A	E_a	C_2Cl_4 log A	E_a
(12) $A + Cl \rightarrow ACl$	10·2	0	10·2	0	10·0	0	—	—	9·8	0	9·4	0
	—	—	*10·3*	*1·5*	*10·3*	*1·5*	—	—	*10·3*	*1·5*	—	—
	—	—	**9·3**	**0**	**9·3**	**0**	**(9·3)**	**0)**	**9·3**	**0**	—	—
(−12) $ACl \rightarrow A + Cl$	13·9	23·6	13·8	23·8	13·7	20·3	13·7	20·3)	13·7	20·4	12·8	16·8
(13) $ACl + Cl_2 \rightarrow ACl_2 + Cl$	9·4	1·0	8·8	1·0	8·7	2·5	—	—	8·8	5·2	8·3	5·4
	—	—	*8·8*	*0·9*	*8·8*	*2·7*	*8·8*	*4·1*	*8·5*	*5·1*	—	—
(−13) $ACl_2 + Cl \rightarrow ACl + Cl_2$	(11·3)	(21·3)	(11·3)	(20·6)	(11·3)	(20·4)	(11·3)	(20·4)	(11·3)	(18·3)	(11·4)	(19·5)
(14) $2ACl \rightarrow products$	10·1	0	9·9	0·3	10·5	0·5	10·0	1·2	9·5	0·5	8·7	0
	—	—	**9·7**	**0**	**10·2**	**0**	**9·2**	**0**	**9·2**	**0**	—	—
(15) $ACl + Cl \rightarrow products$	(11·3)	(0)	(11·3)	(0)	(11·3)	(0)	(11·3)	(0)	(11·3)	(0)	(11·3)	(0)

Taken from ref. [16] with the exception of values in italics which are taken from refs. [25, 30, 31 and 36]. Figures in bold are calculated from experimental values in refs. [25, 30, 31, and 36] on the assumption that $E_a = 0$. Values in parentheses are theoretical estimates. Values of E_{12} and log A_{12} in ref. [16] are estimated from a mechanism which does not take into account the role of $ACl\ddagger$ radicals. Values of A are in l. mole^{-1} s^{-1} and E in kcal. mole^{-1}.

prevents the exact measurement of k_{12}. The values of k_{12} and k_{-12} given in Table 1 do not take this possibility into account although the errors in k_{12} due to this neglect are probably small. For cis-1,2-$C_2H_2Cl_2$ where a correction can be made because of the concurrent geometrical isomerization (see section 4.3) the values obtained were $\log A_{12} = 10 \cdot 3 \pm 0 \cdot 4$ (with A_{12} in l. mole^{-1} s^{-1}) and $E_{12} = 1 \cdot 2 \pm 0 \cdot 7$ kcal mole^{-1}, so the assumption that $E_{12} \sim 0$ for all chlorine addition reactions to these olefins is not unreasonable. The values of k_{-12} for non-excited ACl radicals are based on measured values of ΔH_{-12}, an assumed value of zero for E_{12} and the relation $E_{-12} = E_{12} + \Delta H_{-12}$. They are in good agreement with the results of pyrolysis experiments. [26]

The exothermicity of reaction (12) ensures that $E_{-12} > E_{13}$ and therefore as the temperature is increased k_{-12} increases more rapidly than k_{13} and equation (D) becomes increasingly more applicable than equation (C). Ultimately k_{-12} should exceed $k_{13}[Cl_2]$ and the rate law should tend towards equation (F). Moreover, since $\Delta H_{12} = E_{12} - E_{-12} = -(17 \text{ to } 22)$ and $2E_{12} - E_{15} < 11$ kcal mole^{-1} for all olefins examined, $(E_{-12} + E_{15}) > 2E_{13} + E_{12}$

$$d[ACl_2]/dt = k_{13}[Cl_2](I_a k_{12}[A]/k_{-12} k_{15})^{\frac{1}{2}} \qquad \text{(F)}$$

and the rate must pass through a maximum as the temperature is increased. This is illustrated in Fig. 2 for the case of $A = C_2Cl_4$. [24] Application of sector methods in this region would obviously permit evaluation of k_{15} and k_{12}/k_{-12} since k_{13} is already known. In fact the values of k_{15} quoted in Table 1 are derived from theoretical estimates.

4.2. Combination of chlorine atoms.

Although the most reliable data on the rate of reaction (6) were not obtained by studies of photochlorination, they are clearly of great importance in this context in view of the difficulty in obtaining such data by conventional photochemical methods. Flash photolysis, applied so successfully to the study of I atom recombination, cannot be used because of the absence of convenient absorption bands and because of the great efficiency of recombination on the walls. Linnett and Booth [27] and Bader and Ogryzlo [28] studied the recombination by means of an isothermal calorimetric method in which the atoms were generated by an electrodeless discharge and monitored by measuring the heat produced by combination on a catalytic wire. Hutton and Wright[29] used a similar method for generation but employed reaction (20) which takes place when NOCl is injected into the gas stream to monitor the Cl atom concentration.

$$Cl + NOCl \rightarrow Cl_2 + NO \qquad (20)$$

Their results were compatible with the reaction scheme

$$Cl + Cl_2 \rightleftharpoons Cl_3 \qquad (24, -24)$$

$$Cl_3 + Cl \rightarrow Cl_2 + Cl_2 \qquad (26)$$

$$Cl + wall \rightarrow \tfrac{1}{2}Cl_2 \qquad (7)$$

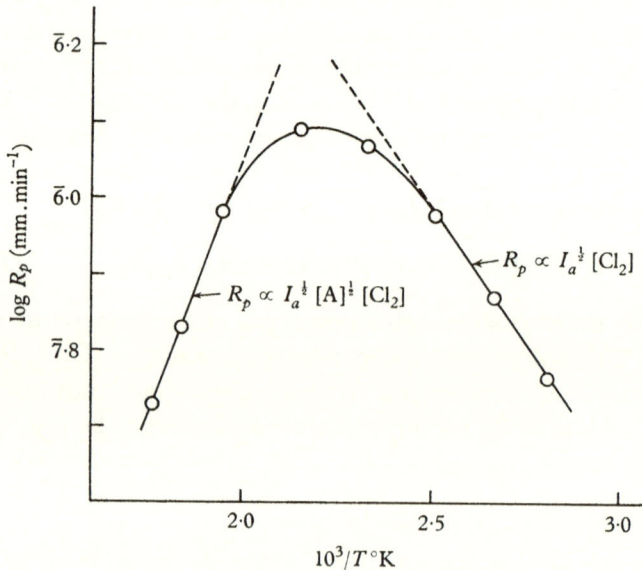

Fig. 2. Variation of rate of photochlorination (R_p) of tetrachloroethylene with temperature. (Redrawn from *Bull. Soc. Chim. Belg.*, 1956, **65**, 549.)

The reaction rate is given by equation (G)

$$-d[Cl]/dt = 2k_{24}k_{26}[Cl]^2[Cl_2]/(k_{-24} + k_{26}[Cl]) + k_7[Cl] \qquad (G)$$

and the complex rate constant $k_{24}k_{26}/k_{-24} = k_{16}$ which is valid at low [Cl] is 2×10^{10} l.^{2} mole^{-2} s^{-1}. This value is about half that obtained by Bader and Ogryzlo[28] and considerably less than that obtained by photochlorination of cis-1,2-$C_2H_2Cl_2$, but appears to be the most reliable estimate so far. The suggestion that a transient complex Cl_3 in very low concentration ($k_{24}/k_{-24} > 10^{-1}$ l. mole^{-1}) is an intermediate in the chain termination reaction is not in conflict with the earlier rejection of Cl_3 as a chain carrier.

4.3. Excited chloroalkyl radicals. The addition reaction (12) is exothermic to the extent of 20–22 kcal mole^{-1} and it is therefore pertinent to consider the possibility of observing a reaction

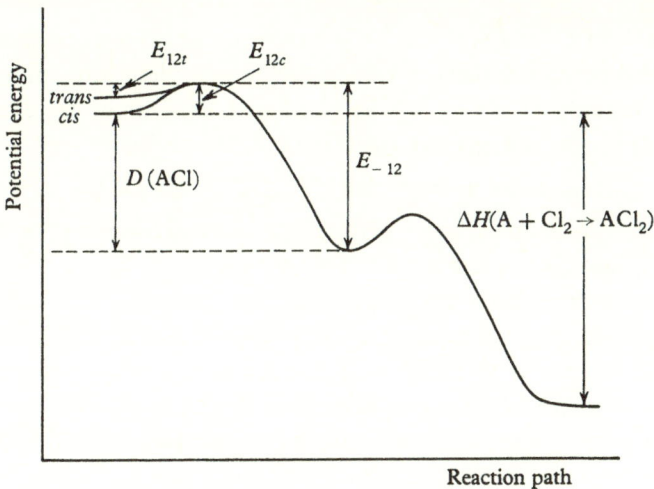

Fig. 3. Potential energy relationships for the propagation reaction in the photochlorination of A = $(CHCl)_2$. The first maximum relates the energy of ACl‡ to $trans$- or cis-A and to ACl, the energy of which corresponds to the minimum. The suffixes c and t refer to cis and $trans$ isomers. The second maximum relates the energy of the activated complex of reaction (13) to the initial and final state.

characteristic of the excited ACl radicals which must be the species first formed when a Cl atom adds to an A molecule. The energy relationships in the reaction sequences (12) and (13) are depicted in Fig. 3 from which it is evident that ACl‡ requires no further energy to decompose to A + Cl and this process will take place to a significant extent if $t_{\frac{1}{2}}^{\ddagger}$ is comparable to the time between collisions of ACl‡ with any other entity present which might de-energize it. It is also possible that ACl‡ might have higher values of k_{13} and k_{14} than ACl has, which raises the question of the significance of

6

the quoted values of these quantities. However, the most striking and irrefutable evidence of the reality of $ACl^{\ddagger}$ and of its decomposition is that observed in the photochlorination of cis- and trans-dichloroethylene. [30]

In this system the rates of isomerization predicted on the mechanism presented earlier but ignoring any possible rôle of $ACl^{\ddagger}$ should be given by equations (J) and (K) where x is the probability

$$R_i \text{ (cis} \rightarrow trans) = xk_{-12}[ACl] \tag{J}$$

$$R_i \text{ (trans} \rightarrow cis) = (1-x)k_{-12}[ACl] \tag{K}$$

that reaction (-12) will yield the trans-isomer. However, it was found that when mixtures of cis-1,2-dichloroethylene and chlorine are illuminated at 30–65 °C the yield of trans-isomer is several orders of magnitude greater than that calculated from

$$E_{-12} = 20\text{--}22 \text{ kcal mole}^{-1}, \quad A_{-12} \sim 10^{13} \text{ s}^{-1}.$$

In fact the observed variation with temperature of the rate gives the absurd values $E_{-12} = 4\cdot 2 \pm 2\cdot 1$ kcal mole^{-1} and $\log A_{-12} = 6\cdot 7 \pm 0\cdot 9$. It has therefore been proposed that the excited $ACl^{\ddagger}$ radical survives long enough to undergo either spontaneous dissociation into a Cl atom and an olefin molecule (either isomer) or collision with Cl_2 or olefin, resulting in either loss of a Cl atom with formation of olefin or deactivation to give a ground state radical. These steps are represented by the equations

$$A + Cl \rightarrow ACl^{\ddagger} \tag{$12^{\ddagger}$}$$

$$ACl^{\ddagger} \rightarrow A + Cl \tag{α}$$

$$ACl^{\ddagger} + Cl_2 \rightarrow A + Cl_3(\rightarrow Cl_2 + Cl) \tag{β}$$

$$ACl^{\ddagger} + Cl_2 \rightarrow ACl + Cl_2 \tag{γ}$$

$$ACl^{\ddagger} + A \rightarrow (ACl^{\ddagger} \text{ or } ACl) + A \tag{δ}$$

which replace (12) and (-12) in the general mechanism. The general form of the rate equation is still valid, i.e.

$$\delta = [Cl]/[ACl] = \frac{k_{13}[Cl_2] + k_{-12}}{k_{12}[A]}$$

but k_{-12} is now given by the expression $(k_\alpha + k_\beta[Cl_2])k_{13}/k_\gamma$ if it is assumed that collision of the $ACl^{\ddagger}$ with A results only in Cl atom transfer without geometrical isomerization. Consequently, the ratio of the rate of isomerization to that of photochlorination should be given by equation (L) where x is now the probability that reaction (α) will yield the *trans*-isomer and similarly y is the probability that reaction (β) will produce the *trans*-isomer.

$$\frac{R_i\ (cis \rightarrow trans)}{R_p} = \frac{(xk_\alpha + yk_\beta[Cl_2])[ACl^\ddagger]}{k_{13}[ACl][Cl_2]} = \frac{xk_\alpha + yk_\beta[Cl_2]}{k_\gamma[Cl_2]} \quad \text{(L)}$$

Consequently, a plot of this rate ratio against $[Cl_2]^{-1}$ should be linear in the pressure range in which x and y are constant. This is illustrated in Fig. 4.† It is found that the ratios $k_\alpha : k_\beta : k_\gamma$ and the

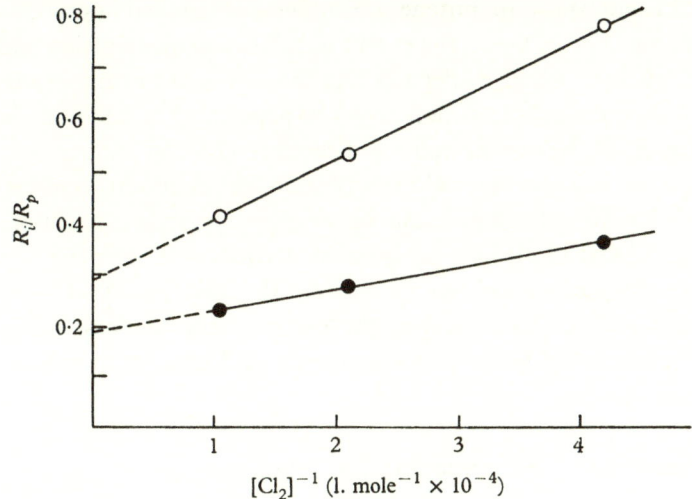

Fig. 4. Variation of ratio of rate of isomerization (R_i) to rate of photochlorination (R_p) of *cis*-dichloroethylene with chlorine concentration at 65 °C. ○ denotes [*cis*-(CHCl)$_2$]/[C$_2$H$_2$Cl$_4$]. ● denotes [*trans*-(CHCl)$_2$]/[C$_2$H$_2$Cl$_4$]. Olefin pressure = 40 mmHg; $I_a = 1\cdot19 \times 10^{12}$ quanta cm^{-2} s^{-1}.

values of x and y are almost independent of temperature, as expected because ACl‡ is the transition state for reactions (12) and (-12) and therefore already possesses all the necessary activation energy for decomposition. The lifetime of ACl‡ is about 10^{-8} s under the experimental conditions used.

Although a number of objections[16] have been made to this hypothesis, no reasonable alternative explanation of the well-founded observations on geometrical isomerization has been put

† Subsequent work[31b] has shown that at higher chlorine pressures the experimental values of R_i/R_p fall below the dotted lines in Fig. 4 for both *cis*- and *trans*-(CHCl)$_2$.

forward. Confirmation of the existence of vibrationally excited chloroalkyl radicals in several chlorine–olefin systems has recently been provided by observations of changes in both overall reaction rate and in the ratio R_i/R_p in the presence of propane [31a] and of inert gases such as He, CO_2, and SF_6. [31b] Further, Poutsma and Hinman [32] have shown that no geometrical isomerization occurs in inert solvents where deactivating collisions are much more numerous. Finally Oldershaw and Cvetanovic [33] have found it necessary to adopt the same hypothesis in the bromine-atom-induced isomerization of butene-2. Problems of detailed interpretation remain. For instance, the extent of collisional deactivation and isomerization by olefin molecules (reaction δ) is not firmly established. Also, the effect of high chlorine pressures (> 200 mm) in reducing R_i/R_p below the value indicated in (L) may indicate the participation of a precursor of the vibrationally excited ACl radical, possibly a π-complex which can dissociate without isomerization.

These observations are in general accord with evidence of vibrationally excited radicals formed by the addition of $-O-$, [34] $-CH_2-$, [35] etc., to olefins, and might lead to further understanding of the relationship between reaction rate and energy content of molecular species.

5. Competitive photochlorination reactions

According to the general scheme for photochlorination outlined above, provided the reaction chains are long, it is possible to set up competitive systems with various combinations of hydrocarbons and halogen compounds which yield values of relative rate constants in a simple manner. There will be considered as (a) addition–addition, (b) abstraction–abstraction, (c) addition–abstraction, (d) addition–dehydrochlorination competitions.

5.1. Addition–addition. When mixtures of chlorine and two olefins A and A' are illuminated, the propagation steps are as follows

$$Cl + A \rightarrow ACl \tag{12}$$

$$Cl + A' \rightarrow A'Cl \tag{12'}$$

$$ACl + Cl_2 \rightarrow ACl_2 + Cl \tag{13}$$

$$A'Cl + Cl_2 \rightarrow A'Cl_2 + Cl \tag{13'}$$

and for long chains we find, neglecting (-12) and $(-12')$,

$$\frac{d[ACl_2]\,dt}{d[A'Cl_2]\,dt} = \frac{-d[A]/dt}{-d[A']/dt} = \frac{k_{12}[A]}{k'_{12}[A']} \tag{M}$$

Thus measurement of the changes in [A] and [A'] or [ACl$_2$] and [A'Cl$_2$] should yield values of k_{12}/k'_{12}. However, in those systems in which the product of (12) or (12') is an excited radical ACl‡ which can dissociate before being thermalized, e.g. the chloroalkyl radicals, equations (12) and (12') must be replaced by those denoted (12‡), (α), (β), (γ), etc., in the previous section, for each olefin. The equation for the ratio of the rates then becomes (N).

$$\frac{d[ACl_2]/dt}{d[A'Cl_2]/dt} = \frac{k_2[A]}{k'_2[A']} \times \frac{1 + (k_{\alpha'} + k_{\beta'}[Cl_2]/k_{\gamma'}[Cl_2])}{1 + (k_{\alpha} + k_{\beta}[Cl_2]/k_{\gamma}[Cl_2])}, \tag{N}$$

Since the ratio $k_{\alpha}:k_{\beta}:k_{\gamma}$ is obtainable only when the rate of dissociation can be measured *either* by observing the rate of concurrent geometrical isomerization *or* by introducing, say, ^{36}Cl into the chlorine, ordinary competitive studies in this series of olefins cannot yield explicit values of k_{12}. Nevertheless, the competitive photochlorinations of *cis*-1,2-dichloroethylene and vinyl chloride and trichloroethylene have been studied by following the changes in ACl$_2$ and A'Cl$_2$ concentration with reasonable assumptions about the appropriate values of $k_{\alpha}:k_{\beta}:k_{\gamma}$ for the various ACl‡ radicals, and the Arrhenius parameters for all three olefins (see Table 1) have been shown to lie in the ranges $E_{12} = 1\cdot5 \pm 1\cdot0$ kcal mole^{-1} and log $A_{12} = 10\cdot3 \pm 0\cdot6$.

5.2. *Abstraction–abstraction.* In a mixture of chlorine and two alkanes RH and R'H the propagation steps are (23), (23'), (27) and (27')

$$Cl + RH \rightarrow R + HCl \tag{23}$$

$$Cl + R'H \rightarrow R' + HCl \tag{23'}$$

$$R + Cl_2 \rightarrow RCl + Cl \tag{27}$$

$$R' + Cl_2 \rightarrow R'Cl + Cl \tag{27'}$$

and the relationships (**O**) may be readily established, again with the

$$\frac{d[RCl]/dt}{d[R'Cl]/dt} = \frac{-d[RH]/dt}{-d[R'H]/dt} = \frac{k_{23}[RH]}{k'_{23}[R'H]} \qquad (O)$$

provisos of long reaction chains. Neglect of reverse reactions (-23), ($-23'$), (-27), ($-27'$) is also implied and is generally justified by the metathetical and exothermic nature of these reactions. Consequently these conditions apply generally to alkanes at moderate temperatures though (-23) can be significant at higher temperatures when RH = methane and $\Delta H_{23} \simeq 0$. Pritchard, Pyke and Trotman-Dickenson [23] used the rate of disappearance of RH and RH' to estimate $k_{23}/k_{23'}$ in their studies on alkanes and alkyl halides but a more accurate method is to measure the initial rate of formation of RCl and R'Cl. This was done by Knox, [37] by Anson, Frederick and Tedder [38] and by Knox and Nelson [39] over the temperature range 200–600 °K for a large number of hydrocarbons. The two methods agree within experimental error when proper account is taken of further chlorination by reactions (28) and (29).

$$Cl + RCl \rightarrow R''Cl + HCl \qquad (28)$$

$$R''Cl + Cl_2 \rightarrow R''Cl_2 + Cl \qquad (29)$$

For example, when RCl = CH_3Cl, R''Cl = CH_2Cl, and R''Cl_2 = $C_2H_2Cl_2$, these reactions have rates which are comparable with (23) and (23') above, so that rate measurements based on integration over a period of time are not reliable and it is necessary to apply corrections for the consumption of RCl and to derive k_{23} and $k_{23'}$ by successive approximations. For reasonable accuracy it is desirable that $k_{23}[RH]$ and $k_{23'}[R'H]$ do not differ by more than one order of magnitude.

Competition between saturated hydrocarbons and alkyl halides can clearly be treated in the same manner and it is possible to devise cross-checks to establish the validity of the mechanisms for most reaction mixtures. Most of the hydrocarbons C_1 to C_5 and the simpler alkyl chlorides have been studied in this manner and Arrhenius parameters for the abstraction reactions (23) established. (See Table 2.)

In contrast to the addition of Cl to olefins, the abstraction of H from alkanes varies considerably in rate according as the C—H bond broken is primary, secondary, or tertiary. Differences in A factors are small, but E_a varies from 3·8 kcal for CH_4 to 0·1 kcal for the tertiary C—H bond in isobutane.

TABLE 2. *Arrhenius parameters for reactions of Cl with hydrogen and hydrocarbons (taken from ref. [22])*

Standard reaction: $Cl + H_2 = HCl + H$. $E = 5,480 \pm 140$ kcal mole^{-1}.
$A = (4·1 \pm 0·3) \times 10^{10}$ l. mole^{-1} s^{-1}.
Secondary standard: $Cl + CH_4 = HCl + CH_3$. $E = 3,830 \pm 250$ kcal mole^{-1}.
$A = (0·6 \pm 0·3) \times 10^{10}$ l. mole^{-1} s^{-1}.

Compound	A per H atom. (10^{-10} l. mole^{-1} s^{-1})	E_a (kcal mole^{-1})
H_2	4·1	5·48
HD	3·3 ±0·2	5·97±0·02
HT	3·0 ±0·2	6·04±0·02
D_2	3·0 ±1·0	6·60±0·15
Primary C–H bonds		
CH_4	0·6	3·83±0·18
C_2H_6	1·5 ±0·3	1·02±0·13
C_3H_8	1·7 ±0·3	0·98±0·13
nC_4H_{10}	1·4 ±0·3	0·77±0·14
$isoC_4H_{10}$	1·3 ±0·3	0·80±0·14
$neoC_5H_{12}$	1·4 ±0·3	0·90±0·14
Secondary C–H bonds		
C_3H_8	3·6 ±0·7	0·66±0·13
nC_4H_{10}	2·4 ±0·5	0·30±0·14
$cycloC_3H_6$	0·9 ±0·2	4·12±0·14
$cycloC_4H_8$	3·2 ±0·7	0·80±0·15
$cycloC_5H_{10}$	2·3 ±0·6	0·60±0·20
Tertiary C–H bond		
$isoC_4H_{10}$	2·1 ±0·5	0·10±0·16
Chlorinated hydrocarbons		
CH_3Cl	1·1 ±0·3	3·28±0·20
CH_2Cl_2	1·3 ±0·2	2·96±0·10
$CHCl_3$	0·69±0·06	3·32±0·09
$CDCl_3$	0·5 ±0·2	4·00±0·20
C_2H_5Cl	0·7 ±0·25	1·50±0·25
C_2HCl_5	0·5 ±0·15	3·40±0·20

5.3. *Addition–abstraction.* It is possible to write down rate equations for systems under appropriately restricted conditions in which chlorine atoms can either add to an olefin or abstract from an alkane or alkyl halide, and to apply these to the determination of relative rate constants. Such studies provide useful cross-checks between different homologous series. A system of special interest is that in which the intermediate radical is common to the addition and the abstraction reaction. An example is the mixture of C_2Cl_4 and C_2HCl_5 for which $ACl = R = C_2Cl_5$. This system was studied by Goldfinger and colleagues[40] and provided most useful confirmation of the validity of the general photochlorination mechanisms. In general when $ACl \neq R$ the most interesting question is whether termination is the same as in the pure addition or pure abstraction system or whether the new chain-breaking step (30) replaces them.

$$ACl + R \rightarrow products \qquad (30)$$

Thus three distinct situations can be defined in which the rate of addition is either unchanged by the concurrent abstraction, e.g. $C_2Cl_4 + CH_4$, CH_3Cl or $C_2H_2Cl_2$ or vice versa, or when both reactions are retarded, e.g. $C_2Cl_4 + CHCl_3$. The latter observation shows that termination is by combination of C_2Cl_5 and CCl_3. These systems are discussed in more detail elsewhere.[16]

5.4. *Addition–dehydrochlorination.* The photochlorination of trichloroethylene has been very extensively studied.[24,25] At high temperatures the product of the addition, pentachloroethane, loses HCl[41] so that another olefin, C_2Cl_4, is formed. Despite the complexity of the resulting competitive reactions, significant data have been obtained for the dehydrochlorination reactions, and extensions of this work are clearly possible. Such extremes will be greatly facilitated by the rapid-scanning mass spectrometric analysis procedure recently developed and applied to this system.[42]

6. The theoretical interpretation of reaction rates

The attempts to apply transition-state theory to the interpretation of rates of reaction of chlorine atoms and molecules have been recently reviewed.[16,22] Despite the considerable attention paid to this problem theoretical treatments have not yet reached the stage where detailed correlations between structure and reaction rates are possible. Absolute rates of the correct order of magnitude are predicted quite readily and trends in rates within a limited group of related reactions may also be predicted with the aid of semiempirical relationships but further improvement is unlikely without a much more detailed knowledge of the structure of the transition-state complex.

It will be recalled that the evaluation of A factors on the transition-state theory requires values of $\Delta S^{\ddagger}_{tr.}$, $\Delta S^{\ddagger}_{rot.}$ and $\Delta S^{\ddagger}_{vib.}$, the translational, rotational and vibrational entropies of activation. Though the former can be precisely calculated from the molecular weights of reactants and complex, the estimation of the latter two quantities involves assumptions about the structure of the complex which can have a critical effect on the outcome of the calculation. The vibrational term is particularly sensitive to assumptions about bond lengths and angles, and the relatively good agreement between $A_{calc.}$ and $A_{obs.}$ for a number of reactions of Cl with hydrocarbons is rather fortuitous, since uncertainties in $\Delta S^{\ddagger}_{rot.}$ and $\Delta S^{\ddagger}_{vib.}$ partially cancel. The range of $A_{calc.}$ (0·8 to 4×10^{10}) is almost identical with the range of $A_{obs.}$ (0·7 to 4·1 $\times$ 10^{10}) for saturated hydrocarbons but there are many individual discrepancies. [43]

Knox and Trotman-Dickensen [44] used scale models of complexes R–H–X, etc., to obtain accurate values of $\Delta S^{\ddagger}_{rot.}$, whereas Johnson and Goldfinger [45] assumed that reactants and complexes were always linear in a series of reactions involving (a) hydrogen-abstraction by Cl atoms, or (b) Cl atom transfer. Despite the over-simplification involved, there is generally satisfactory agreement for both A and E_a in the series

$$(CH_4 \text{ to } CHCl_3) + Cl \quad \text{and} \quad (CH_3 \text{ to } CCl_3) + Cl_2.$$

Calculations for the most heavily chlorinated radicals are least accurate. In the same paper a large number of reactions with zero activation energy (mainly radical and atom combinations) are considered in terms of a rotating complex of the two reactants. They calculate the rate at which the complex crosses to a state where the attractive forces are greater than the centrifugal forces so that a close encounter, leading to combination, can occur. Although values of A of the correct order are obtained, some quite significant trends in $A_{obs.}$ are not predicted, e.g. the decrease of 10^{-2} in the rate of combination of CCl_3 compared with CH_3. It is suggested that the assumption of a central attractive potential is inadequate and that non-central forces between Cl atoms must be taken into account.

Correlations between $A_{obs.}$ and bond dissociation energies have been more successful though these remain largely empirical in nature. Expressions such as that proposed originally by Evans and Polanyi [46]

$$E_a = B\Delta H_0^0 + C$$

where ΔH_0^0 is the standard enthalpy of reaction and B and C are adjustable parameters and $1 > B > 0$ have been found to give good general agreement with several series of related reactions of halogen atoms. [22] The use of such relations to predict activation energies by interpolation or, conversely, to estimate bond dissociation energies from observed values of E_a, seems justified by experience for Cl atom reactions if not for methyl radicals. [47, 48]

Even after sixty years of intensive study, photochlorination reactions remain a challenge to theoretical and experimental chemists everywhere. As in the reactions themselves, the induction period was followed by a period of rapid propagation. Since the initial studies in Cambridge numerous new reactive centres have been produced—and termination seems unlikely!

REFERENCES

1   Cohen, A. and Jung, G. Z. phys. Chem. 1924, 110, 705.
2   Norrish, R. G. W. J. chem. Soc. 1925, p. 2316.
3   Griffiths, J. G. A. and Norrish, R. G. W. Trans. Faraday Soc. 1931, 27, 451.
4   Norrish, R. G. W. Trans. Faraday Soc. 1931, 27, 461.
5   Lipscomb, F. J., Norrish, R. G. W. and Porter, G. Nature, Lond., 1954, 174, 785.
6   Briggs, A. G. and Norrish, R. G. W. Proc. Roy. Soc. A, 1964, 278, 27.
7   Norrish, R. G. W. Proc. Roy. Soc. A, 1932, 135, 334.
8   Chapman, M. C. C. J. chem. Soc. 1923, p. 3002; Chapman, D. L. and Gibbs, F. B. Nature, Lond., 1931, 127, 854.
9   Bodenstein, M. and Dux, W. Z. phys. Chem. 1913, 85, 297; Bodenstein, M. and Unger, W. Z. phys. Chem. 1931, B 11, 253.
10 Ritchie, M. and Norrish, R. G. W. Proc. Roy. Soc. A, 1933, 140, 99; Proc. Roy. Soc. A, 1933, 140, 112.
11 Norrish, R. G. W. and Ritchie, M. Proc. Roy. Soc. A, 1933, 140, 713.

12 Müller, K. and Schumacher, H. J. *Z. phys. Chem.* 1937, B **35**, 285;
    1937, B **35**, 455; Schmitz, R. and Schumacher, H. J. *Z. phys. Chem.*
    1941–2, B **51**, 281; 1942, B **52**, 72.
13 Berthoud, A. and Bellerot, H. *Helv. Chem. Acta*, 1924, **7**, 307.
14 Chapman, D. L., Briers, F. and Walter, B. *J. chem. Soc.* 1926, p. 562.
15 E.g. *Chain Reactions*, Dainton, F. S. Methuen, 1956.
16 Chiltz, G., Goldfinger, P., Huybrechts, G., Martens, G. and
    Verbeke, G. *Chem. Rev.* 1963, p. 355.
17 Burns, W. G. and Dainton, F. S. *Trans. Faraday Soc.* 1950, **46**, 411.
18 Burns, W. G. and Dainton, F. S. *Trans. Faraday Soc.* 1952, **48**, 39.
19 Ashmore, P. G. and Chanmugam, J. *Trans. Faraday Soc.* 1954, **49**,
    254; 1954, **49**, 270.
20 Rodebush, W. H. and Klingelhoefer, W. C. *J. Am. Chem. Soc.* 1933,
    55, 130.
21 Steiner, H. and Rideal, E. K. *Proc. Roy. Soc.* A, 1939, **173**, 503.
22 Fettis, G. C. and Knox, J. H. *Progress in Reactions Kinetics*, 1964,
    2, 1.
23 Pritchard, H. O., Pyke, J. B. and Trotman-Dickenson, A. F. *J. Am.
    Chem. Soc.* 1954, **76**, 120; 1955, **77**, 2629.
24 Goldfinger, P. *Bull. Soc. Chim. Belg.* 1956, **65**, 549; 1959, **68**, 5.
25 Dainton, F. S., Lomax, D. A. and Weston, M. *Trans. Faraday Soc.*
    1957, **53**, 460; 1962, **58**, 308; Ayscough, P. B., Cocker, A. J.,
    Dainton, F. S. and Hirst, S. *Trans. Faraday Soc.* 1962, **58**, 295.
26 Barton, D. H. R. and Howlett, K. E. *J. chem. Soc.* 1949, p. 155;
    1949, p. 165; 1951, p. 2033; Barton, D. H. R. and Onyon, P. F.
    J. Am. Chem. Soc. 1950, **72**, 988.
27 Linnett, J. W. and Booth, M. H. *Nature, Lond.*, 1963, **199**, 1181.
28 Bader, L. W. and Ogryzlo, E. A. *Nature, Lond.*, 1964, **201**, 491;
    J. chem. Phys. **41**, 2926.
29 Hutton, E. *Nature, Lond.*, 1964, **203**, 835; Hutton, E. and Wright, M.
    Trans. Faraday Soc. 1965, **61**, 78.
30 Ayscough, P. B., Cocker, A. J., Dainton, F. S., Hirst, S. and
    Weston, M. *Proc. chem. Soc.* 1961, 244; Ayscough, P. B., Cocker,
    A. J. and Dainton, F. S. *Trans. Faraday Soc.* 1962, **58**, 284.
31 (*a*) Knox, J. H. and Riddick, J. *Trans. Faraday Soc.* 1966, **62**, 1190.
    (*b*) Ayscough, P. B., Dainton, F. S. and Fleischfresser, B. *Trans.
    Faraday Soc.* 1966, **62**, 1838.
32 Poutsma, M. L. and Hinman, R. L. *J. Am. Chem. Soc.* 1964, **86**,
    3807.
33 Oldershaw, G. A. and Cvetanovic, R. J. *J. chem. Phys.* 1964, **41**,
    3639.
34 Cvetanovic, R. J. *Adv. Photochemistry*, 1963, **1**, 115.
35 Frey, H. M. *Progress in Reaction Kinetics*, 1964, **2**, 131.
36 Ayscough, P. B., Cocker, A. J., Dainton, F. S. and Hirst, S. *Trans.
    Faraday Soc.* 1962, **58**, 318.
37 Knox, J. H. *Trans. Faraday Soc.* 1962, **58**, 275; *Bull. Soc. chim. Belg.*
    1962, **71**, 764.
38 Anson, P. C., Fredericks, P. S. and Tedder, J. M. *J. chem. Soc.* 1959,
    p. 918; 1960, p. 144.

92   F. S. DAINTON AND P. B. AYSCOUGH

39 Knox, J. H. and Nelson, R. L. *Trans. Faraday Soc.* 1959, **55**, 937.
40 Dusoleil, S., Goldfinger, P., Martens, G., Mahieu-Van der Auwera,
    A. M. and Van der Auwera, D. *Trans. Faraday Soc.* 1961, **57**,
    2197, 2210.
41 Goldfinger, P. *Bull. Soc. chim. Belg.* 1956, **65**, 561.
42 Goldfinger, P., Huybrechts, G. and Verbeke, G. *Symposium on Mass
    Spectrometry*, Pergamon Press, 1962; Huybrechts, G., Meyers, L.
    and Verbeke, G. *Trans. Faraday Soc.* 1962, **58**, 1128.
43 Fettis, G. C., Knox, J. H. and Trotman-Dickenson, A. F. *J. chem.
    Soc.* 1960, p. 1064, 4177; *Can. J. Chem.* 1960, **38**, 1643.
44 Knox, J. H. and Trotman-Dickenson, A. F. *J. phys. Chem.* 1956, **60**,
    1367.
45 Goldfinger, P. and Johnston, H. S. *J. chem. Phys.* 1962, **37**, 700.
46 Evans, M. G. and Polanyi, M. *Trans. Faraday Soc.* 1938, **34**, 11.
47 Johnston, H. S. and Rapp, D. *J. Am. Chem. Soc.* 1961, **82**, 1.
48 Dainton, F. S., Ivin, K. J. and Creak, G. A. *Trans. Faraday Soc.*
    1962, **58**, 326.

5

FLASH PHOTOLYSIS

GEORGE PORTER

1. Introduction

Flash photolysis is an experimental technique for the rapid initiation of change and for the direct observation of the transient intermediates involved. It utilizes an intense flash of visible or ultraviolet radiation whose duration is short compared with that of the changes which are to be observed. By this means, a material of short life may be prepared in quantities great enough for study by absorption spectroscopy and other physical methods.

The original flash photolysis apparatus was constructed in Cambridge in 1947 and applied to the spectroscopic detection and kinetic study of gaseous free radicals. The use of very high intensity flash sources in photochemistry was described in 1949[1] and the first account of the method of flash photolysis and spectroscopy appeared at the beginning of 1950.[2] Since that time it has been applied in many laboratories to a variety of spectroscopic, chemical and biochemical problems.

A detailed account of the method and of the several hundred publications which describe its applications is neither necessary nor appropriate in this volume since several reviews have appeared quite recently.[3,4] It may, however, be useful at this time to take a retrospective view of the progress which has been made, to examine the limitations and capabilities as they are now, and to look at future potentialities.

2. Flash photolysis and other fast reaction techniques

In order to observe a particular substance or condition, some physical measurement must be carried out in a time which is comparable with or less than the lifetime of the substance under the conditions of observation. Most of the conventional measurements of chemistry require that the lifetime of the substance studied is at least several seconds and the science of chemistry has therefore been concerned mainly with the description of 'stable' materials. There is nothing fundamentally different about 'stable' molecules, nearly all of them are unstable in the thermodynamic sense; and the only property which distinguishes most 'unstable' species is that their rate of reaction is so rapid that somewhat more sophisticated methods must be utilized to study them.

The development of special techniques for this purpose has been very rapid over the last fifteen years. Before this, the detection and study of fast reactions was mainly dependent on stationary-state flow techniques such as the flame, for combustion reactions, and the technique of Hartridge and Roughton for fluid systems. [5] Modern developments of the latter, particularly the stopped-flow technique which is essentially a pulsed-mixing method, have improved the time resolution to 1 or 2 ms.

Apart from mixing, the principal methods of bringing about a chemical change are by radiation or by heating. Very shortly after the development of flash photolysis, pulsed heating methods were introduced; shock-tube techniques for gases and the temperature-jump method for liquids. The latter, developed by Eigen and his school, utilizes a short pulse of electrical current through a conducting liquid and Eigen has introduced a variety of perturbation methods using, for example, pressure and electric field changes both as single pulses and as periodic perturbations. [5,6] In these latter methods, known as relaxation techniques, the system studied is usually near to equilibrium and the rapid relaxation to equilibrium at the new condition is followed after the perturbation. Together these methods have revolutionized the science of reaction kinetics and few reactions are now beyond the time resolution of existing techniques.

The characteristic which principally distinguishes the flash photolysis method is that the perturbation produced by the light pulse transfers the system far from the equilibrium condition. The equilibrium concentration of most unstable species, such as free radicals and electronically excited states, is quite negligible under all attainable conditions so that small perturbations from equilibrium are unable to produce detectable amounts of such species. Very large temperature perturbations, such as those produced in shock tubes, result in a transfer from one Boltzmann equilibrium to another and large concentrations of labile species far removed from equilibrium conditions are not obtained. The light pulse may, of course, also be used to produce a small perturbation from equilibrium since, after the initial relaxation of the excited states, the equilibrium is often displaced only slightly from its original condition, either by a temperature rise or displacement of a protolytic equilibrium for example. Other relaxation techniques are, however, usually simpler of application for these near-equilibrium processes and the particular value of flash photolysis, and the application for which it is unique, is the direct observation of highly unstable free radicals and excited states. It has been responsible for most of the postwar developments in free radical spectroscopy, the absorption spectroscopy of excited states and the direct kinetic study of these species.

3. The experimental method

For the observation and study of a species under conditions where it has a short lifetime, two experimental requirements must be met. First, the species must be prepared in a concentration high enough to allow observation in a time comparable with or shorter than its lifetime. Secondly, some physical parameter must be measured over a similar period of time. In the technique of flash photolysis the first requirement is met by using a brief, intense flash of visible or ultraviolet light and the second is usually met by recording the optical absorption spectrum at short time intervals afterwards. Many arrangements are possible, but most of them employ a cylindrical reaction vessel of glass or quartz, between 10 and 100 cm in length, alongside which lie one or more electronic flash tubes, the whole being enclosed in a reflector. [3]

Observation of the transient absorption spectra is carried out in one of two ways:

3.1. Flash spectroscopy. The spectrum is recorded photographically, at any required time after the photolysis flash has been discharged, by means of a second flash lamp, the light from which passes through the reaction vessel to the slit of the spectrograph. A range of wavelengths is therefore recorded at one single time delay and kinetic studies are carried out by repeating the experiment at a series of delay times.

3.2. *Kinetic spectrophotometry.* The spectrograph is replaced by a monochromator or filter which selects a narrow wavelength interval and the optical density of the reaction mixture at this wavelength is recorded by means of a continuous source, the light from which is monitored photoelectrically after passing through the reaction vessel and monochromator. A single experiment then records the optical density at one wavelength but at all times after the flash.

The first method is most suitable for the recording of spectra and preliminary survey of a reaction, whilst the second method is more convenient and accurate for kinetic studies.

At first it might appear that an even more useful approach would be that which combined the two above recording methods by recording the whole spectrum repeatedly after a single flash and indeed during the development of the method in 1947 this was the approach first used. A rapid-scanning grating spectrometer was constructed which recorded the whole spectrum oscillographically with a repetition rate of 50/s. Not only was this repetition rate far too slow for most transients but the resolution of the apparatus was too low for vapour phase spectroscopy of free radicals. It is clear that such an increase in the amount of information recorded can only be achieved by a corresponding loss in sensitivity and it also became clear that such a method would have few advantages. In practice, one requires the whole spectrum of a certain transient at high resolution only once for analysis, and a few more times to establish that it is constant over the conditions of the investigation. Thenceforth, for kinetic work, the interest lies in the variation in the intensity with time and this is usually measured far more accurately and conveniently by photoelectric measurements at one or two selected wavelengths. It is therefore not surprising that rapid scanning spectrometers have still found little application to flash photolysis except in special cases such as infra-red recording [7] where high resolution is less important and flash photographic recording techniques are not possible.

From the point of view of information, the two-flash photographic technique is more powerful than the photoelectric method since intense monochromatic sources for the latter are not available, and to record at only one wavelength is, therefore, to discard information which is intrinsically available. The recorded information, and the related factors of precision, signal/noise, etc., can of course be increased by multiplication of either the apparatus or the number of experiments. In the photographic method this is often necessary when using high resolution instruments in order to obtain sufficient plate density and the photographic plate provides the integrating device. In photoelectric recording, electronic methods of integration of repetitive data are now well developed and commercially available (for example, the computer of average transients or CAT) and such methods have been applied to the study of small transients following flash photolysis.[8]

Given spectroscopic recording equipment of sufficient sensitivity and time resolution, the capabilities of a flash photolysis apparatus are determined by the characteristics of the flash, in particular its energy and duration. Other properties, which are less important, are spectral distribution and reproducibility; the former permits of little control when using high energy electronic flash tubes since the output is principally a recombination and retardation continuum which is not greatly affected by the gas filling or operating characteristics; only at relatively low current densities does the line spectrum become significant. The output from flash to flash is very reproducible and usually well within the reproducibility of the reaction mixture and intensity measurements.

The two important parameters of energy and duration are related since, when other factors have been optimized, an increase in energy is always accompanied by an increase in flash duration. There is no difficulty in obtaining flashes of nanosecond duration but the energies will then be measured in microjoules; conversely flashes which dissipate hundreds of thousands of joules may be used, with durations of the order of milliseconds. The design is therefore always a compromise; first one must decide on the flash energy required (one may assume about 10% efficiency of conversion to light in the 200–500 mμ region) to produce the desired effect and then one designs a condenser bank and flash lamp with the lowest possible inductance or the shortest possible current pulse when critically damped. What can be done is best indicated by reference to the characteristics of the wide variety of flash discharge lamps described in the literature.[9]

In practice, the variation in characteristics of flash equipment used for photolysis is not very great. The lower energy limit is one or two joules with duration of one or two microseconds and is set by the lower limit of observable photochemical change. Such lamps are quite adequate for strongly absorbing systems at low concentration with quantum yields near unity, of which chlorophyll is a good example. The higher limit is usually about 10,000 J with a flash duration of about 100 μs. Higher energies have been used but little is gained; the life of the lamps becomes short so that alternative methods such as open sparks have to be used and then the efficiency of light production falls.

At this point it may be worth while to correct an impression, commonly held, that flash photolysis is a somewhat elaborate and expensive technique requiring considerable know-how. For most applications, for which a few hundred joules flash is quite adequate, the apparatus occupies a few feet of bench space and costs very little, given an oscilloscope and spectrograph. As to know-how there have been numerous visitors to our laboratory who, without previous experience of the technique, have been able to obtain meaningful results within a few hours. But it is also a common experience that the recording of a change lasting 1 ms often entails several days of preparation of the reaction mixture and even longer for interpretation.

4. Applications of flash photolysis

4.1. *General.* The primary act of light absorption is to raise the absorbing species to a higher electronic state. Such a state has different physical and chemical properties from the ground state and, therefore, if the system were originally in thermodynamical equilibrium, this equilibrium will be displaced towards the new equilibrium of the excited state, the extent of displacement depending on the relative rates of establishment of the new equilibrium and of deactivation of the excited state by radiative or other processes. If a change is to be observed, these processes must also be fast compared with the re-establishment of the original equilibrium in the ground state. It is not necessary for the original system to be in thermodynamic equilibrium and a common type of system which is studied by flash photolysis is one in metastable equilibrium which is induced to return to a position of lower free energy by the light flash which serves to overcome the activation energy necessary for this change to occur. All atoms and molecules absorb visible or ultraviolet light so that, in principle, all systems can be displaced from their position of stable or metastable equilibrium in this way.

Whether one is concerned with absorption by a transient species, or merely an optical density change in the spectrum of a stable species, optical pumping must be fast compared with the rate of decay. This is the only requirement for observation of the effect since flashes of duration longer than that of the transient merely lead to a stationary state concentration which can be observed within the flash if the photographic method is used. If the photoelectric method is to be used, or kinetic studies are required, it also becomes necessary for the flash to be short compared with the transient lifetime, since scattered light from the photolysis flash prevents measurement until the output of this flash has decayed to at most a few per cent of its maximum value. If the rate of production of transient by light absorption is $k_0 I_{abs}$. mole l.$^{-1}$ s^{-1}, then for a first-order decay of transient of specific rate k_1 the stationary state concentration will be $k_0 I_{abs}./k_1$ and, for a second-order decay of transient with specific rate k_2, the stationary state concentration will be $(k_0 I_{abs}./k_2)^{\frac{1}{2}}$. Estimates of $k_0 I_{abs}$. can be made from a knowledge of the spectra and the efficiency of the flash lamps [3] and are typically of the order of magnitude of 1 einstein s^{-1} in a reaction cell of volume $\frac{1}{10}$ l. If k_0, the quantum yield, is unity these figures lead to a stationary concentration for a transient, with first-order decay, of $10/k_1$ mole l.$^{-1}$, and, with second-order decay, of $3/k_2^{\frac{1}{2}}$. It is clear that detection within the flash may be possible even when the decay is too rapid to be followed.

For kinetic work, the limiting condition is approximately that the flash half-height duration should be less than the transient half-life. For first-order decays this sets an upper limit to rate constants which can be measured of $k_1 = 10^6$ s^{-1}, and this limit has been very nearly reached in practice. The limiting rate constant for second-order reactions is a function of the extinction coefficient at the measured wavelength. It has been shown that, provided the extinction coefficient exceeds 100 l. mole^{-1} cm^{-1}, the fastest rate constants, with values up to 10^{11} l. mole^{-1} s^{-1} are accessible. [9] There are very few molecules which do not have extinction coefficients in an accessible region greater than this and rate constants exceeding 10^{10} l. mole^{-1} s^{-1} have been frequently measured by flash photolysis.

Some of the principal fields of application of flash photolysis techniques will now be outlined.

4.2. Free radical spectroscopy. In 1949 the spectra of unstable polyatomic free radicals were almost unknown and none could be assigned with certainty. Today over 100 such spectra have been observed, including nearly all the ones of principal chemical and structural interest, and these observations have been made almost exclusively in absorption by flash photolysis. Some interesting work, which has been very useful for assignment purposes has been carried out by the newer technique of radical trapping or matrix isolation [10] and a few of the simpler radicals have also been studied by means of their gas-phase emission spectra.

In the field of radicals with three or four atoms the most interesting results have been those of Herzberg [11] who, working in the first place in the vacuum ultraviolet, was able to observe the methylene and methyl radicals, to say something of their structure and to resolve, to a great extent, the problem of the triplet and singlet states of methylene. These, and other high resolution studies on radicals having few atoms, are described in more detail in the next chapter. It is important to note that high resolution instruments must be used for such studies, not merely for the purpose of interpretation but also for detection where fine structure is present, since sensitivity increases with resolving power up to the point where the strongest lines are fully resolved. On the other hand, sensitivity is very low whatever the resolving power if the spectra are completely diffuse, and this is probably the explanation of the fact that one of the most commonly postulated radicals, HO_2, is still undetected despite many attempts and despite the fact that its spectral position can be predicted with some confidence. This is very probably due to the fact that, even at long wavelengths, the spectrum is predissociated owing to the weak bond and is therefore diffuse and uncharacteristic. It is interesting that the first polyatomic absorption spectrum to be observed, [12] that of the analogous radical HS_2, is predissociated throughout.

The other particularly fruitful field under this heading has been that of the aromatic free radicals. It seems that, given the right conditions, almost every aromatic molecule can be made to yield a free radical spectrum by flash photolysis, and already we have detected over one hundred of these. The spectra are, of course, far more complex than those referred to above but many of them are nevertheless quite sharp and show much fine structure in the vapour phase. In some cases, such as the recently discovered phenyl radical, [13] a partial analysis is possible and leads to useful structural information, in others, such as benzyl, [14] even a partial vibrational analysis cannot be made with any certainty. These radicals are, however, equally interesting from other points of view such as the electronic transitions involved and the primary photochemical processes which occur. Owing to the wide variety of aromatic compounds, it is possible to assign the spectra by means of a systematic study of related series, a procedure which is not possible with simple aliphatic radicals. Finally, and in the long run perhaps most important of all, these radicals are readily observed and studied kinetically in solution as well as in the gas phase, where their behaviour is of great chemical and biochemical relevance. [15]

4.3. *Excited state absorption spectra.* If a molecule has a reasonably strong absorption spectrum then the lifetime of the upper level is correspondingly short, usually 10^{-6} s or less. This is the case with most transitions which are spin allowed and, therefore, absorption from upper singlet states has not yet been observed, though in some cases this would appear to be feasible even with present techniques. The lowest triplet levels of organic molecules have, on the other hand, very long radiative lifetimes and the correspondingly weak absorption spectra to these states from the ground state do not preclude their population, since this can occur by intersystem crossing from upper singlet levels.

The first studies of triplet states were made by means of their phosphorescence emission spectra. [16] Such spectra of long-lived metastable states are, however, generally observed only in solids or very viscous media, since the states are very rapidly depopulated by radiationless processes in fluids and gases. Just how rapidly this depopulation occurred was not known but it appeared feasible that the lifetime of triplet states in solution, although it must be a small fraction of the phosphorescence lifetime (of the order of seconds in aromatic hydrocarbons), might be within the time resolution of flash photolysis. This was found to be the case by Porter and Windsor, [17] who detected the triplet absorption spectra of many aromatic molecules in fluid solvents, such as hexane, at room temperature, and found lifetimes of the order of 1 ms.

The energies of upper triplet levels observed in this way were in good agreement with the calculations of Pariser. [18] One notable omission was benzene and, indeed, all single-ring aromatic and heterocyclic compounds, for which no triplet–triplet spectra could be observed. Early in 1966, however, this state of affairs was rectified, and the triplet absorption spectrum of benzene has been observed quite near to the predicted position. [19] The failure of many previous attempts to detect it seems to be due to an accumulation of difficulties; it is diffuse and not easily characterized, it appears in the same region as the parent singlet absorption, and its lifetime is too short for detection in ordinary fluid solvents so that flash photolysis has to be carried out at liquid nitrogen temperature in a rigid glass. Although these low-temperature flash photolysis techniques have been perfected recently, the light scattering and other difficulties associated with a 20 cm path of low-temperature glass introduce considerable difficulties in the detection of a weak transient at short wavelengths. The observation resolves a number of problems which have recently been accumulating, associated with the fact that photochemical processes of benzene and its homologues are biphotonic at 2,537 Å, and it was necessary that triplet benzene absorbed quite strongly at this wavelength if these results were to be rationalized.

4.4. *Gas kinetics.* The gap between the observation of the spectra of transient species in the gas phase and the use of these spectra to study their kinetics could hardly be wider. The difficulties of this field are very great, many years of work may be necessary for the complete study of a single species, but the possibilities, particularly for studies at a higher level of detail and sophistication than conventional methods, are almost inexhaustible.

The first kinetic study by flash photolysis was the recombination of iodine atoms. [20] This system has been investigated over a period of nearly 15 years by many workers and, although still holding some secrets, it is perhaps the most thoroughly investigated of fast chemical reactions. The rate constants and the temperature coefficients (negative) have been measured using a wide variety of chaperon molecules and the papers giving theoretical interpretations of the reactions are nearly as numerous as the experimental papers. There have been a number of difficulties, particularly those associated with the adiabatic temperature rise (see later) and a number of surprises such as the finding that I_2 was 1,000 times more efficient than helium and, more recently, that nitric oxide was 20 times more efficient still. [21] Indeed the whole of the findings about the very wide range of efficiencies and temperature coefficients are somewhat unexpected and can, it seems, only be explained in terms of the formation of charge transfer complexes between iodine atom and chaperon, [22] though whether this interpretation is to be carried so far as to include also the inert gases is still in dispute. [23]

Another gas reaction which has been studied in detail by photoelectric recording is the triplet–triplet annihilation reaction in anthracene and naphthalene which has been shown to occur with nearly unit collision efficiency. [24] Oxygen quenching of triplets in the gas phase occurs with a rate constant about 10 times less.

Most other gas-phase kinetic studies have been carried out by the photographic, point-by-point method. The reasons for this are two-fold. First, since the species concerned, simple free radicals mostly, have discrete fine structure, relative intensity measurements present a very difficult problem by any method and plate photometry, usually a rather inaccurate procedure, is, in this case, probably less susceptible to error, especially if the two-path method is used. Secondly, many of these studies are concerned, not merely with the overall concentration of a single species, but with the relative changes in several species and, what is particularly interesting, with relative changes within the same spectrum caused by non-Boltzmann distributions.

It is in this latter application that flash photolysis studies have shown their most original and pregnant contribution to gas kinetics. The discovery was made, quite accidentally, during our studies on the photolysis of chlorine dioxide,[25] that a new spectrum appeared which was eventually, but not immediately, identified as the Schumann–Runge system of oxygen with such an abnormal intensity distribution that few of the bands had been previously observed. The oxygen spectrum had a normal room-temperature rotational distribution but anomalously high vibrational levels were populated. Further study[26] showed that the source of the excitation was the chemical reaction

$$ClO_2 + O = ClO + O_2^*$$

and many other examples of this type of phenomenon have since been discovered by Norrish and his colleagues, who have also applied flash photolysis to the production of non-Boltzmann distributions in a number of other ways.[4] In principle, and already to some extent in practice, it is therefore possible to record, during the course of a reaction, the population of every individual rotational and vibrational level and to derive the rate constants of reaction for each of them—a long step from the averaged rates which are all that can be derived from conventional kinetic studies. The method has been developed at a time which is most appropriate, coinciding, as it does, with rapid developments in the theoretical study of this field.

All the above reactions are carried out, as nearly as possible, under isothermal conditions by using the thermal capacity of an excess of inert gas. In our very early work, before the effects of the adiabatic temperature rise were fully appreciated, many reactions were studied without diluent, with some remarkable effects. The first encouraging photochemical observation in a flash photolysis apparatus was the preparation of carbon, in fine cobweb form throughout the reaction vessel, from the flash photolysis of a substance whose normal photochemical behaviour was rather well known—acetone! We also observed apparently almost complete destruction of some molecular species such as ketene but there was a reversal of the reaction at longer times, contrary to all chemical expectation. These were all thermal effects, the latter resulting from the concentration gradient, which mirrors the temperature gradient, leaving a relatively high concentration near the walls and a low concentration at the centre of the vessel where observations are made. In kinetic studies, such as those of iodine recombination, these effects are so important that a 1,000-fold dilution with inert gas is essential for accurate work.

The flash-heating effect was utilized quite early for the study of combustion and some quite interesting results were obtained, [27] particularly regarding the mechanism of carbon formation. [28] The results were only semiquantitative, however, and, in view of the wealth of data which can be obtained about many different species and their energy distributions throughout the course of reaction, further studies of this kind will probably prove of value in the combustion field. The system is extremely complex, with changing temperature, chemical and concentration gradients, but this is characteristic of most real combustion systems and should now prove tractable with modern methods of computation. The flash-heating effect has been applied more recently to condensed systems with much originality by Nelson. [29]

4.5. *Solution kinetics.* Kinetic studies in solution, by flash photolysis methods, present more possibilities and fewer difficulties than those in the gas phase. Not only are the troublesome temperature effects virtually eliminated, but the rapid establishment of Boltzmann equilibria and the broadening of spectral bands greatly facilitate the interpretation and the application of accurate photometric measurements. It has already been mentioned that nearly all aromatic molecules produce observable transients on flash photolysis and the field of application is very wide indeed.

The cage effect often reduces the quantum yield of radical formation in solution but this is compensated for by the much higher concentration of higher molecular weight compounds which can be obtained in solution. In addition to radical production by dissociation, they may be formed by reaction of the parent molecule with solvent or other substrate and a large variety of radicals of the semiquinone and ketyl type, produced by reaction of triplet carbonyl compounds, has been studied in this way. [30] If the molecule itself does not absorb light in a convenient region, the excitation may be brought about by energy transfer from another molecule which has been electronically excited.[31] Excitation of vinyl compounds, and their *cis-trans* isomerization, provides an important example of this method.

The study of triplet states in solution has already been referred to, and the kinetic study of triplet decay has been a major activity over the last few years. The radiationless mechanisms whereby the triplet decays in fluid media provide a complex and challenging problem which seems now to have been solved in its major aspects though problems of detail remain.[32] One of the morals of these studies is that solvents are never absolutely pure and that when we are concerned with rate constants near to 10^{10} l. mole^{-1} s^{-1}, corresponding to diffusion control in typical fluids, impurity concentrations of one part in 10^8 can be all important. One such impurity, which nearly always plays a predominant part if it is present, is oxygen.

The long life of the triplet state makes it a very attractive subject for detailed study of electronically excited states. It is possible to carry out a determination of many of the principal chemical and physico-chemical properties which characterize a molecule, and direct measurements of such quantities as acidity constants have been obtained which afford very interesting comparisons with similar determinations in the ground and first excited singlet states.[33] Rate constants of protonation in the triplet state, and acidity constants and protonation constants of free radicals, are also readily obtained in suitable cases.

Protonation and deprotonation rate constants in the ground state may also be determined by flash photolysis methods, and so may electron transfer rates. The principle here is similar to that used in other perturbation and relaxation techniques; a system at equilibrium is perturbed by the flash but instead of observing relaxation to a new equilibrium the relaxation to the original condition is observed. The method depends on the fact that protonic equilibria, for example, are usually very different in electronically excited states and establishment of this excited state equilibrium is often very rapid. During the lifetime of the excited state, therefore, the new protonic equilibrium is established and, after electronic relaxation which occurs from the singlet state in about 10^{-8} s, a non-equilibrium protonic situation remains in the ground state. The relaxation of this situation to the ground-state equilibrium can be followed in appropriate cases by suitable choice of pH.[34] In such cases the method is in every way as convenient as the temperature-jump and other relaxation methods, but again the particular power of the flash technique is its application to excited-state protonation and to other systems far removed from equilibrium.

4.6. *Photobiology.* Photobiology is applied photochemistry, but the application is of such importance and current interest that it deserves special mention. It might appear that, in view of the complexity of kinetic study of most triatomic and other simple molecules in the gas phase, the flash photolysis of proteins and nucleic acids is a problem of hopeless complexity. In fact, the interpretation of such systems is often more certain and convincing than that of the simpler molecules, owing to the rapid relaxation of excess vibrational excitation in complex molecules in solution and the ease of photometric study of diffuse spectra.

It happens that the principal chromophores in proteins are those of the aromatic amino acids and that their behaviour is little changed from that of the acids themselves. Thus, one of the principal transients in the flash photolysis of egg albumin is the phenoxyl radical from tyrosine, the absorption spectrum being little changed from that of phenoxyl obtained by flash photolysis of phenol itself.[35] The flash photolysis of riboflavin, lumiflavin and similar molecules is entirely similar to that of simpler quinones, and the triplet state, as well as the semiquinone intermediates, have been well characterized.[36] As an example of an even more complex system, flash photolysis measurements, coupled with stopped-flow investigations, have been able to elucidate the four separate rate constants of the haemoglobin–oxygen reaction.[37]

As would be expected, the application of flash photolysis to photosynthesis was a fairly early development. Studies are principally of two kinds: the study of chlorophyll and related porphyrins in dilute solution and the study of the chloroplasts themselves. The former have revealed the importance of the environment, the effect of solvent and of dimerization on the efficiency of triplet formation,[38] the latter have revealed a whole series of transients, most of them very weak, which, if they can be satisfactorily interpreted, may provide the most direct and potent method for the final elucidation of this complex process. [39] But there is a wide gap between the dilute solution of chlorophyll and the highly structured chloroplast and, in an attempt to bridge this gap, we have recently developed a method of microscopic flash photolysis, incorporating just the same principles but utilizing a cell volume only a few microns in diameter, which seems to us to provide the most promising approach to the study of transient photobiological phenomena in the biostructures themselves. [40]

5. Recent and future developments

Two or three years ago, several laboratories simultaneously and successfully replaced the light flash of the flash photolysis technique by a pulse of electrons and so introduced the powerful technique of pulse radiolysis. [41] In some ways this method has advantages over the light pulse technique; the electron pulse is more readily shaped and can be a clean-edged pulse lasting only 1 μs or, in recent apparatus, even a few nanoseconds. Furthermore, the efficiency of absorption is almost unity so that, although the energy per pulse is far less than in the flash apparatus, being typically only a few joules, the absorbed energy is not very much less. These statements apply to liquids; it is doubtful whether anything like the same efficiency will be attained in the gas phase. One of the major achievements of the pulse radiolysis technique has been in the detection of solvated electrons, but many of the radicals already characterized by flash photolysis have also been observed. The main application of the pulse radiolysis technique is to the study of radiation chemical processes and, as such, it is probably the most powerful technique introduced since the war. For the study of the individual processes it is intrinsically more complicated than photochemical pulsation, though it is surprising how very similar are the transient effects produced by light and by electron pulses in many systems.

A few years ago most people would have said that there seemed little hope of a major improvement in light sources for flash photolysis. As far as conventional sources are concerned this has been borne out and flash durations of pulsed-gas discharges are about the same as they were 10 years ago. But the introduction of the laser, and in particular of the giant-pulsed laser, has brought new thoughts and new hope to the problem to such an extent that it would now be surprising if satisfactory, high power, nanosecond sources are not available for flash photolysis work in the near future. At present a single pulse of up to 100 J can be produced in about 10 ns in the red or infra-red region and, by second harmonic generation, this can be used to produce a similar pulse, of a few joules, in the near ultraviolet. Already this is a useful source for flash photolysis; for example the fundamental is suitable for chlorophyll studies and the first harmonic for benzophenone, surely two of the most 'popular' molecules at present. But it can be only a matter of time before the wavelength range is extended and the advantages of the pulsed laser, with its directional output so useful for quantum yield studies, its monochromaticity and polarization which have obvious applications for special purposes but, above all, its nanosecond pulse of extremely high intensity, make it the routine source for flash photolysis studies.

The detection method used in conjunction with flash photolysis has been almost exclusively electronic absorption spectroscopy. It is easy to suggest that other systems should be developed, but less easy to develop them. In the opinion of the author, and precluding unforseen technical developments, this method will remain the most powerful one, not only because it is obviously complementary to a method which uses optical excitation of the molecule but because of the superior sensitivity of the photographic plate and the photomultiplier cell to other physical detectors of fast response time. Nevertheless, other detectors will undoubtedly prove useful and already successful work has been carried out with a rapid scanning infra-red flash photolysis apparatus [7] and with time of flight mass spectrometric detection. [42] At present no really useful microsecond magnetic resonance apparatus seems to be available though, in view of the obvious incentives and usefulness of such a detector, and the fact that its development seems within the power of existing technology, its appearance is to be expected in the near future. There are other physical measurements which would seem to be immediately applicable to flash photolysis measurements but which have not yet been adapted to this purpose, of which one obvious example is optical rotatory dispersion or circular dichroism.

The study of light in all its aspects has been the basis of most of the great fundamental developments of physics in the present century. Its application to chemistry is relatively new and the technique of flash photolysis is only one part of this application. Its potentialities are, however, so varied, and its capacity for development so great, that it is likely to play a significant part in the future development of chemistry.

REFERENCES

1 Norrish, R. G. W. and Porter, G. *Nature, Lond.*, 1949, **164**, 658.
2 Porter, G. *Proc. Roy. Soc.* A, 1950, **200**, 284.
3 Porter, G. *Technique of Organic Chemistry*. Interscience, 1963, vol. VIII, part II, 1055.
4 Norrish, R. G. W. Faraday Lecture, *Chemistry in Britain*, 1965, 1, 289.
5 *Technique of Organic Chemistry*. Interscience, 1963, vol. VIII, part II.
6 Eigen, M. *Disc. Faraday Soc.* 1954, **17**, 194.
7 Herr, K. C. and Pimental, G. C. *Proceedings 7th Int. Symp. on Free Radicals*, Padua, 1965.
8 Rüppel, von H., Bültemann, V. and Witt, H. T. *Ber. Bunsenges.* 1964, **68**, 340.
9 Porter, G. *Z. Elektrochem.* 1960, **64**, 59.
10 Norman, I. and Porter, G. *Proc. Roy. Soc.* A, 1955, **230**, 399.
11 Herzberg, G. *Proc. Roy. Soc.* A, 1961, **262**, 291.
12 Porter, G. *Disc. Faraday Soc.* 1950, **9**, 60.
13 Porter, G. and Ward, B. *Proc. chem. Soc.* 1964, p. 288; *Proc. Roy. Soc.* A, 1965, **287**, 457.
14 Porter, G. and Ward, B. *J. chim. phys.* 1964, p. 1517; Porter, G. and Wright, F. J. *Trans. Faraday Soc.* 1955, **51**, 1469.
15 Land, E. J. and Porter, G. *Trans. Faraday Soc.* 1963, **59**, 2016, 2027.
16 Lewis, G. N. and Kasha, M. *J. Am. Chem. Soc.* 1944, **66**, 2100.
17 Porter, G. and Windsor, M. *J. chem. Phys.* 1953, **21**, 2088; *Disc. Faraday Soc.* 1954, **17**, 178; *Proc. Roy. Soc.* A, 1958, **245**, 238.
18 Pariser, R. *J. chem. Phys.* 1956, **24**, 250.
19 Godfrey, T. S. and Porter, G. *Trans. Faraday Soc.* 1966, **62**, 7.
20 Christie, M. I., Harrison, A. J., Norrish, R. G. W. and Porter, G. *Proc. Roy. Soc.* A, 1952, **216**, 152; 1955, **231**, 446; Russell, K. E. and Simons, J. *Proc. Roy. Soc.* A, 1953, **217**, 271; Marshall, R. and Davidson, N. *J. chem. Phys.* 1953, **21**, 659.
21 Porter, G., Szabo, Z. G. and Townsend, M. G. *Proc. Roy. Soc.* A, 1962, **270**, 493.
22 Porter, G. and Smith, J. A. *Proc. Roy. Soc.* A, 1961, **261**, 28.

23 Porter, G. *Disc. Faraday Soc.* 1962, **33**, 198.

24 Porter, G. and West, P. *Proc. Roy. Soc.* A, 1964, **279**, 302.

25 Lipscomb, F. J., Norrish, R. G. W. and Porter, G. *Nature, Lond.*, 1954, **174**, 785.

26 Lipscomb, F. J., Norrish, R. G. W. and Thrush, B. A. *Proc. Roy. Soc.* A, 1956, **233**, 455.

27 Norrish, R. G. W. and Porter, G. *Proc. Roy. Soc.* A, 1952, **210**, 439; Norrish, R. G. W., Porter, G. and Thrush, B. A. *Proc. Roy. Soc.* A, 1953, **216**, 165; 1955, **227**, 43.

28 Knox, K., Norrish, R. G. W. and Porter, G. *J. chem. Soc.* 1952, p. 1477; Porter, G. 4th Combustion Symposium, 1953, p. 248.

29 Nelson, L. S. and Lundbergh, J. L. *J. phys. Chem.* 1959, **63**, 433.

30 Bridge, N. K. and Porter, G. *Proc. Roy. Soc.* A, 1958, **244**, 259, 276; Porter, G. and Wilkinson, F. *Trans. Faraday Soc.* 1961, **57**, 1686; Beckett, A., Osborne, A. D. and Porter, G. *Trans. Faraday Soc.* 1963, **59**, 2038, 2051; 1964, **60**, 873.

31 Porter, G. and Wilkinson, F. *Proc. Roy. Soc.* A, 1961, **264**, 1.

32 Porter, G. and Wright, M. R. *Disc. Faraday Soc.* 1959, **27**, 18; Jackson, G. and Livingston, R. *J. chem. Phys.* 1961, **35**, 2182; Linschitz, H., Steel, C. and Bell, J. A. *J. phys. Chem.* 1962, **66**, 2574; Hilpern, J. W., Porter, G. and Stief, L. J. *Proc. Roy. Soc.* A, 1964, **277**, 437.

33 Jackson, G. and Porter, G. *Proc. Roy. Soc.* A, 1961, **260**, 13.

34 Godfrey, T. S., Porter, G. and Suppan, P. *Disc. Faraday Soc.* 1965, **39**, 194.

35 Grossweiner, L. I. and Mulac, W. A. *Radiation Res.* 1959, **10**, 515.

36 Knowles, A. and Roe, E. M. F. *4th Int. Photobiology Congress, Oxford,* 1964.

37 Gibson, Q. H. *Progress in Reaction Kinetics,* Pergamon Press, Oxford, 1964, **2**, 321.

38 Livingston, R. and Ryan, V. *J. Am. Chem. Soc.* 1953, **75**, 2176; Livingston, R., Porter, G. and Windsor, M. *Nature, Lond.*, 1954, **173**, 485; Linschitz, H. and Sarkanan, K. *J. Am. Chem. Soc.* 1958, **80**, 4826.

39 Müller, A., Rumberg, B. and Witt, H. T. *Proc. Roy. Soc.* B, 1963, **157**, 313.

40 Porter, G. and Strauss, G. *Proc. Roy. Soc.* A, 1966, **295**, 1.

41 Matheson, M. S. and Dorfman, L. M. *Progress in Reaction Kinetics,* Pergamon Press, Oxford, 1965, **3**, 237.

42 Meyer, R. T. 149th A.C.S. Meeting, Detroit, 1965.

6

FLASH PHOTOLYTIC STUDIES OF FREE RADICALS IN THE GAS PHASE

B. A. THRUSH

1. Introduction

Prior to 1945, the properties of free radicals in the gas phase were known mainly from two largely unrelated fields of study. On one hand, much information about the chemical properties of free radicals had been obtained from the detailed kinetics of photochemical processes [1, 2] and of chain reactions. [3] On the other, studies of free radical spectra, mainly in emission from discharges, had provided detailed information about the structures and stabilities of many diatomic radicals, [4] but many of the species, particularly polyatomics, which are important in photochemistry and gas kinetics could not be studied by emission spectroscopy.

An important exception to this situation was provided by the work of Oldenberg and Riecke [5] on the OH radical and of White [6] on the CN radical. In these studies, the free radical was produced by a pulsed discharge (through water vapour or cyanogen respectively) and the kinetics of its disappearance were determined from the decay of its absorption spectrum observed with a pulsed light source. Lack of knowledge as to the other species present in the discharge products was one disadvantage of this technique.

The lack of overlap between the spectroscopic and kinetic methods of studying free radicals arose partly because the steady-state concentrations of free radicals present in photochemical systems and in chain reactions are far too low for their detection by absorption spectroscopy. The other difficulty was the failure to produce systems giving emission spectra which could be attributed to such chemically important species as CH_2 and CH_3. The study of such species by absorption spectroscopy has several potential advantages. In some cases where predissociation of the upper state prevents the spectrum appearing in emission, the absorption spectrum still shows resolvable rotational structure. The absorption normally involves many fewer levels than the emission spectrum; it is therefore easier to analyse and it yields information about the ground state of the species concerned. It provides information about the concentration and internal energy distribution of the species concerned, while the emission spectrum depends generally on the excitation conditions.

The development of the technique of flash photolysis [7-9] has gone a long way towards bridging this gap. The use of an intense flash of ultraviolet light to produce free radicals ensures that the species formed are of interest to photochemists. These species are frequently too unstable to be observed in emission from electric discharges, even from such specialized sources as the Schüler discharge. [10] In addition to the theoretical importance of the spectra of many of these species, the information about their structure and stability can be of great use to the kineticist.

By comparison with conventional photochemical studies which have to rely largely on product analysis, flash photolysis offers considerable advantages in providing a means of identifying many of the transient species involved. From this point of view, it provides an excellent method of ascertaining the mechanism of a reaction. A quantitative determination of the individual rate constants is less readily achieved, since it is often difficult to evaluate the absolute concentrations of free radicals. With the exception of a few species for which the radiative life and hence the f-value of the transition are known, the only method available for determining absolute concentration of transient species is by material balance in the system. Normally, the amount of decomposition of the parent molecule can be readily determined, but it is not easy to show that all the transient species have been identified. This is particularly true when free atoms are involved, since most of them can only be observed by vacuum spectroscopy. The recent development of quantitative methods for studying their reactions in discharge flow systems [11, 12, 13] provides a method of obtaining kinetic data which complements the information obtained from studies by flash photolysis.

This chapter is not intended to be a comprehensive review of the applications of flash photolysis to studies of free radicals in the gas phase. The emphasis has been placed upon the problems that have attracted considerable attention, and the methods adopted for circumventing difficulties that arose in particular applications. The author would like to express his deep gratitude to Norrish for his continued help and encouragement over many years, and to draw the reader's attention to the high proportion of the research described here which was initiated by Norrish during the last fifteen years of a very distinguished career.

2. Kinetic studies of free radicals

2.1. *The halogens.* One of the first problems studied by flash photolysis was the third-order recombination of iodine atoms in the gas phase. This is potentially a straightforward system, since molecular iodine absorbs visible light strongly and thereby dissociates to give iodine atoms. The recombination of iodine atoms can readily be followed quantitatively by observing the return of the absorption by molecular iodine. For this purpose a continuous light source and photomultiplier cell with oscillograph recording are most suitable.

The first studies made in this way [14, 15, 16] were all in reasonable agreement with the earlier measurements of Rabinowitch and Wood [17] based on the steady-state iodine atom concentration which is present in continuously irradiated iodine-inert gas systems. There were, however, significant differences between these early studies using flash photolysis. These were resolved by the discovery that molecular iodine is a particularly efficient third body for iodine atom recombination [18] and by the identification and analysis of the effect of temperature rise during flash photolysis. This latter effect is present to some extent in all photochemical studies of gases, since some of the energy of the absorbed photons is degraded by internal conversion or quenching or appears as translational energy of the molecular fragments; heat is also released during the subsequent atom or radical reactions. The removal of this heat at the walls immediately sets up a radial temperature gradient; since the pressure is uniform throughout the vessel, the concentration of molecules must be greater near the walls than close to the axis of the vessel along which the monitoring beam passes. The temperature rise thus reduces the concentration of iodine molecules in the monitoring beam, below the average concentration in the vessel. The iodine atom concentration which is found from the decrease in iodine molecular absorption appears to be larger than its true value. Another possible difficulty is that the metastable excited $^2P_\frac{1}{2}$ iodine atoms which are one of the products of the photo-dissociation of I_2 in its continuum only relax slowly to the $^2P_\frac{3}{2}$ ground state. If this were a slow process, it could affect the rate of recombination of iodine atoms; however, recent observations show that it is unimportant. [19]

The general features which emerged from the early studies were that the relative efficiencies of different third bodies could not be explained on a three-body collision basis even for the inert gases [14] and that the observed negative temperature coefficients were too large to be explained in this way. [16] It was noted that there was an approximate relationship between the logarithm of the third-order rate constant and the boiling-point of the third body concerned; [16] even I_2 which is 470 times as efficient as argon [18] fits reasonably into this pattern, eliminating the need to postulate the molecule I_3 as an intermediate. These third-order rate constants can also be correlated satisfactorily with other properties of the third body such as its ionization potential.

Later, Porter and Smith [20] made careful measurements of the temperature coefficients of the third-order rate constant for a variety of third bodies; they showed that these could be expressed by a relation of the type

$$A \exp\left(+E/RT\right)$$

where A was remarkably constant for most of the third bodies studied. These results were explained by a 'chaperone' theory in which E corresponds to the binding energy between an iodine atom and the third body or chaperone, which presumably arises from exchange interactions. An exception to this behaviour is provided by nitric oxide, [21] where the very high rate constant for recombination is attributed to the presence of NOI in equilibrium with NO and I. The absorption spectrum of NOI was detected.

Fewer and less extensive studies have been made of bromine atom recombination. The general picture which emerges is closely similar to that for iodine atom recombination, although there is conflicting evidence about the importance of Br_2 as a third body.[22, 23] The dissociation of bromine in shock tubes[24] has been studied more thoroughly than that of iodine; the data so obtained and those from flash photolysis can be combined fairly satisfactorily either in a simple Arrhenius expression or in a T^n form.

Further studies are needed, particularly to fill the gap between the temperatures used in the two types of measurements. An interesting approach to this rather difficult problem was the use of flash photolysis to study bromine atom recombination in shock-heated mixtures of bromine with inert gases.[23]

The kinetics of the recombination of chlorine atoms are not readily studied by flash photolysis, owing to the weakness of absorption by molecular chlorine, which limits both the degree of dissociation and the accuracy of monitoring it. Briggs and Norrish[25] have shown that the flash photolysis of chlorine, particularly at high pressures, yields a short-lived absorption spectrum in the ultraviolet due to the partially metastable $^3\Pi_{0u}$ state of chlorine. This state which is the upper state of the weak visible absorption of chlorine is populated during chlorine atom recombination either directly or by transfer of energy to a chlorine molecule which acts as third body. A similar phenomenon occurs with molecular bromine.

One of the first free radicals to be studied by flash photolysis was ClO,[8] a species which appears readily in systems containing chlorine and oxygen. The formation and decay of ClO in mixtures of these two gases is interesting since the process is reversible, chlorine and oxygen being the only products. Porter and Wright[26] showed that the formation of ClO from chlorine atoms must involve an intermediary of the form Cl—O—O since the spectroscopic determination of the bond energy in ClO[26, 27] showed that the reaction

$$Cl + O_2 \rightarrow ClO + O$$

is too endothermic to occur to any extent. The formation of ClO in the overall process

$$Cl + Cl + O_2 \rightarrow ClO + ClO$$

competes effectively with the homogeneous recombination of chlorine atoms by other third bodies.

The recombination of ClO radicals was found to be a second-order process, the rate of which was independent of temperature. This was explained in terms of the intermediate formation of Cl_2O_2 which could decompose to yield Cl_2 and O_2.

$$ClO + ClO \rightleftarrows Cl_2O_2 \rightarrow Cl_2 + O_2$$

The overall rate constant can only be found by determining the extinction coefficient of ClO; this was done in the flash photolysis of ClO_2, which unlike Cl_2 is almost completely destroyed by the photolytic flash. [28] Assuming that ClO_2 is photolysed to yield stoichiometric amounts of ClO

$$ClO_2 + h\nu \rightarrow ClO + O$$

$$O + ClO_2 \rightarrow ClO + O_2$$

A rate constant of approx. 6×10^{10} cm^3 mole^{-1} s^{-1} was obtained for the bimolecular decay of ClO. A somewhat lower value of $2 \cdot 4 \times 10^{10}$ cm^3 mole^{-1} s^{-1} was obtained in the chlorine-sensitized and straight photolysis of chlorine monoxide, [29] and it was suggested that the value obtained in the photolysis of ClO_2 could have been affected by the occurrence of the reaction

$$O + ClO \rightarrow Cl + O_2$$

It is probable that systems involving ClO are more complex than assumed in these studies. In particular, the reaction forming ClO from chlorine atoms and molecular oxygen cannot go to completion, and its equilibrium position favours chlorine atoms rather than ClO unless the molecular oxygen pressure approaches one atmosphere. The stability of Cl—O—O provides an unknown factor which could affect the interpretation.

The formation of BrO and IO observed by Durie and Ramsay [27] in the flash photolysis of other halogen–oxygen systems is of interest. The bond energies of BrO and IO are sufficiently lower than that of ClO to make the formation of these species from two ground-state halogen atoms and an oxygen molecule too endothermic to occur. It was originally suggested that the extra energy needed was obtained by the participation of metastable $^2P_{\frac{1}{2}}$ halogen atoms. More recent experiments [30] have cast doubt on this view and indicate that the formation of BrO is largely associated with light absorption by molecular oxygen around 2,000 Å.

The chlorine-sensitized decomposition of nitrogen trichloride is a related process in which the NCl and NCl_2 radicals have been detected,[31] as predicted by earlier kinetic studies,[32] where it was deduced that NCl_2 is produced by the action of chlorine atoms on NCl_3.

An interesting observation is that the NCl_2 radical which has a strong diffuse absorption spectrum is itself decomposed by the photolytic flash.

2.2. *Other kinetic studies.* Kinetic spectroscopy has been used to study the flash photolysis of many other comparatively simple molecules in the gas phase. These include nitrogen dioxide,[28] ozone,[33], water,[34] nitrosyl halides,[35] carbon disulphide and oxysulphide,[36,37] sulphur oxides,[38] hydrogen azide,[39] phosphine,[40] cyanogen and its derivatives,[41,42] halogenated methanes,[43,44] and many aromatic and unsaturated hydrocarbons and their derivatives.[45-48]

In most cases the observed free radical spectra are those which would be expected as primary fragments on the basis of the known products of the steady photolysis; for instance, CS from CS_2, SO from SO_2, NH from NH_3, CN from cyanogen derivatives. The species formed in many secondary reactions, such as S_2 from CS_2, NH_2 and N_3 from HN_3, and NCO from cyanogen plus oxygen are also usually consistent with mechanisms based on the products of steady photolysis and their quantum yields. Great care must, however, be exercised in the identification of a free radical solely on the basis of the occurrence of an unanalysed transient spectrum in the photolysis of a particular molecule. Partial and complete deuteration of the parent molecule provides a convenient method of determining the number of hydrogen atoms present in the transient species, and isotopic substitution is also important for compounds containing carbon, nitrogen and/or oxygen, even when the rotational structure can be analysed, since the similarity in the masses of these atoms can lead to difficulties in identification of the radical concerned. If a spectrum cannot readily be analysed, the normal method of identifying its carrier as a radical R is to obtain this spectrum in the photolysis of a series of compounds RX containing this group, but not from a related series R'X. A good example of this approach is the identification of the benzyl radical ($C_6H_5CH_2$) spectrum from benzyl but not benzal compounds[47] although this spectrum is also obtained strongly from cycloheptatriene.[48] Substituted benzenes show somewhat different behaviour; benzene and mono halo-benzenes yield the phenyl radical,[49] but aniline, phenol and nitrobenzene[49] give the cyclopentadienyl radical.[46] Other examples of the detection of somewhat unexpected products are C_3, NCN and HNCN, in addition to CH_2, CH and CH_3, in the photolysis of diazomethane under conditions where there is sufficient inert gas to make the system isothermal.[50] Considering the difficulties involved, few errors have been made in identifying radical spectra and even these have been rectified quickly.

Most of the studies mentioned in the first paragraph of this section have been essentially qualitative in the sense that the absolute concentrations of the free radicals involved have not been determined. An interesting exception is the study of the reactions of the CN radical where the two groups of workers concerned have used different methods for determining the absolute concentration of CN. Paul and Dalby[51] recorded the absorption of the (0, 0) band of the CN violet system photo-electrically, and the calibration factor was determined from the known radiative life (f-value) of this transition, allowing for the Doppler, natural and collisional broadening of the rotational lines.[52] Only collisional broadening is difficult to assess other than by calibration, but it is not important except at high total pressures. This method has considerable potentialities for diatomic species whose radiative lives can be determined by independent methods, but is not yet suitable for more complicated molecules where there are considerable discrepancies between the measured radiative lives and those estimated from integrated absorption coefficients.

Basco and co-workers[41] estimated the concentration of CN by assuming that its rapid reaction with nitric oxide[42] is stoichiometric. They concluded that the decay of CN is second order in CN, but did not establish whether a third body is required. By contrast, Paul and Dalby found that the CN disappearance was first order and involved reaction with the parent molecule (C_2N_2 or ClCN); as their concentrations of CN were about 100 times lower than in the work of Basco *et al.* [42] these observations are not contradictory. Both workers obtain rate constants for the reaction of CN with molecular oxygen which lie within 10 % of 5×10^{12} cm³ mole⁻¹ s⁻¹; this is excellent agreement, although it should be noted that this value would not be affected by any error in the absolute concentration of CN since the reaction is first order in this species.

The observation [41] of CN with up to six quanta of vibrational energy is of interest. This appears to occur by optical pumping, CN being raised to the $B^2\Sigma$ state by absorption in its strong violet system, and then fluorescing or being collisionally quenched into excited vibrational levels of the ground state. A similar phenomenon is observed with nitric oxide where the first two vibrational levels are populated. [53]

The production of nitric oxide with up to ten quanta of vibrational energy in the photolysis of nitrosyl bromide and chloride [35] does not occur by this mechanism, and it appears to be a direct product of the primary photochemical act, in which up to half the excess energy of the photon can appear as vibration in the NO molecule produced. This appears to be an isolated example of such excitation, since other cases of the formation of molecules with large amounts of vibrational energy in flash photolysis experiments are due to the occurrence of secondary exothermic atom-transfer reactions. Thus the formation of vibrationally excited O_2 molecules in the photolysis of chlorine dioxide, [28] nitrogen dioxide [28] and ozone [33] occurs in the second step.

$$OXO + h\nu \rightarrow O + XO$$
$$O + OXO \rightarrow O_2^* + XO$$

No vibrationally excited ClO or NO could be detected in these reactions.

Flash photolysis provides a useful method for studying vibrational relaxation, particularly from higher vibrational levels; this topic is discussed in the next chapter.

The photolysis of aldehydes and ketones has interested photochemists for many years. A great deal of information on the reactions of the free radicals involved has been obtained from the steady photolysis; for instance, the absolute rate constant for the recombination of methyl radicals has been determined by sector methods. In flash photolysis studies, the absorption spectra of the CH_3 and CHO radicals have been observed in the photolysis of those aldehydes and ketones which would be expected to yield them. [54,55] These absorptions are relatively short lived and are not particularly amenable to kinetic studies.

There is an alternative method of applying flash photolysis to the study of these problems—by comparing the products of flash photolysis with those of conventional photolysis. For molecules such as acetaldehyde where the photolysis can proceed by a chain mechanism, the increased light intensity in flash photolysis, which may be a million times greater than in conventional systems, increases the rate of radical–radical reactions as compared with radical–molecule reactions. Thus, ethane which is formed by the combination of two methyl radicals is increased at the expense of methane which is produced by methyl radical attack on acetaldehyde. This has been observed experimentally; [56] it was also found that the products of the flash photolysis of acetone and of diacetyl are similar to those of the steady photolysis, as might be expected since these photolyses do not involve a chain mechanism.

Very few studies have been made of the flash photolysis of aldehydes and ketones. [56-58] The main reason is the considerable dependence of their decomposition products upon the wavelength of the exciting light. The broad continuous nature of the emission from a flash source and the difficulty of isolating a narrow band without very considerable loss of intensity have therefore limited comparison with conventional photolysis. The production of monochromatic flashes with a magnesium spark proved very difficult. [57]

Flash photolysis is not very suitable for the study of polymerization reactions, since the high radical concentrations produced give a very large ratio of termination to chain propagation.

3. Spectroscopic studies of free radicals

It was mentioned in the introduction that high resolution spectroscopic studies of free radicals produced by flash photolysis have greatly extended our knowledge of the structure and properties of polyatomic free radicals. The application of flash photolysis to diatomic radical spectra has been largely an extension of known principles to species which have greater interest for the photochemist or kineticist. During the 1930's and 1940's, the difficulties associated with the analysis of the spectra of triatomic and larger molecules appeared to be ones of complexity; the spectra of such bent molecules as NO_2 and SO_2 contain an immense number of rotational lines which are very hard to resolve, whereas the spectrum of the linear species CO_2^+ had been almost completely analysed. The most obvious line of approach appeared to be the triatomic hydrides, where the smaller moments of inertia would significantly reduce the number of lines observed in the spectrum. The emission spectra of NH_2 (the ammonia α bands) and HCO (the hydrocarbon flame bands) were known, but neither proved amenable to analysis. The analyses of the absorption spectra of these species, [55,59,60] which have many fewer lines, and of the absorption spectra of many other triatomic species, such as CH_2, [54,61] HNO, [62,63] HCF, HCCl, [64] HSiCl, HSiBr, [65] C_3, [66] NCN, [50] BO_2, [67] NCO, [68] NCS [69] and N_3, [39,70] have strikingly confirmed two predictions about these species.

One of these is Walsh's use [71,72] of a comparatively simple molecular orbital treatment to predict the configuration of the ground and excited states of many of the smaller polyatomic species. His predictions have been tested most thoroughly for the triatomic molecules and have proved remarkably accurate. Walsh showed that the most important factor in determining the equilibrium configuration of triatomic species was the behaviour of the molecular orbitals derived from those p-orbitals on the central atom which lie perpendicular to the bond axis when the molecule is linear. When such an orbital does not involve in-phase (bonding) overlap with a p-orbital on one of the end atoms, bending of the molecule in the plane of the orbital sharply reduces its energy, since this orbital acquires the character of a lone-pair s-orbital on the central atom with which it shares a_1 symmetry. There are, of course, two such π-orbitals which are degenerate in the linear configuration. Only one of these decreases in energy on bending the molecule, and the energy of the other is virtually unaffected. These orbitals begin to be occupied in the ground state of triatomic molecules of types AH_2, HAB and ABC or AB_2 when they have more than four, ten and sixteen valency electrons respectively, that is for molecules with more valency electrons than BeH_2, HCN or CO_2.

This must be considered with the other prediction which comes from Renner and Teller's work on the interaction between the electronic orbital angular momentum (Λ) and the orbital angular momentum (l) associated with the bending vibration of a linear polyatomic molecule. Renner[73] showed that this interaction destroys the degeneracy of the Λ-doublets in the potential curve representing the dependence of the energy on bond angle, splitting it into two components (V^+ and V^-). The total angular momentum $K\hbar$ about the molecular axis, where $K = |l+\Lambda|$, plays an important role in determining the energy of different sublevels. This splitting depends on the magnitudes of the parameter $\epsilon = (V^+ - V^-)/(V^+ + V^-)$ which is a measure of separation of the two potential curves (V^+ and V^-). The behaviour of the vibronic levels of such linear triatomic molecules as BO_2[67] and NCO[68] which have fifteen valency electrons is adequately explained by small values of this parameter, $|\epsilon| < 1$, and the two potential curves for bending are essentially parabolae which lie close together and have a common vertex. It is interesting to note that this effect in linear molecules was first detected in absorption spectra obtained by flash photolysis, although it must have been present in the emission spectrum of CO_2^+, which had previously been analysed and contained some unexplained features. [74]

When the interaction between the two orbital angular momenta is large, one or both of the potential curves can cease to be stable in a linear configuration. This corresponds to large values of the parameter ϵ, and one has a situation in which one linear and one bent state or two bent states of a molecule correlate with the same degenerate electronic state in the linear configuration. This situation obtains in triatomic molecules with incompletely filled non-bonding π-orbitals which Walsh showed to be important in determining the equilibrium bond angle.

With one electron in this orbital, the molecule would be in a Π electronic state. Bending the molecule removes the degeneracy of the Π state, and in this case the energy of the molecule decreases when it is bent in the plane of the occupied orbital and increases if it is bent in a plane perpendicular to this. This behaviour is shown by HCO [55,59] (and presumably by NO_2 and BH_2) where the visible absorption spectrum corresponds to a transition from the bent ground state and a linear first excited state both of which correlate with the same Π state in the linear configuration.

Molecules with two electrons in these orbitals are of considerable chemical interest. These include CH_2, HNO, HCF, HCCl, HSiCl, HSiBr, CF_2 and SO_2. For a linear species the configuration $(\pi)^2$ yields $^3\Sigma^-$, $^1\Delta$ and $^1\Sigma^+$ states in order of increasing energy. Bending the molecule will favour energetically the state in which both π-electrons go into the orbital in the plane of the molecule. This state is derived from one component of the $^1\Delta$ state, as required by the Pauli Exclusion Principle for a fully occupied orbital. For all the species listed above except CH_2, the gain in energy on bending exceeds that lost due to pairing of the electron spins, and these molecules have bent singlet ground states. For CH_2 the situation is less clear [54,61] since absorption spectra from a bent singlet state ($^1A_1$), with a bond angle of $104°$, and from a linear triplet state ($^3\Sigma_g^-$) have been observed in the flash photolysis of diazomethane. The absorption spectrum of the singlet state decays more rapidly than that of the triplet, and hence the latter is presumably the ground state; their separation is unknown but cannot be large. The fact that methylene is present in both the singlet and triplet states in the photolysis of diazomethane agrees well with studies of its chemical behaviour in such systems. [76]

The species HNO, [62,63] HCF [64] and HCCl [64] have bond angles close to $108°$ in their ground states and show visible or near infra-red absorption spectra to an excited state in which the bond angle has increased to a value close to that of the ground state of HCO, which is $119°$. This transition clearly corresponds to the promotion of one electron from the lone-pair orbital to the π orbital which is perpendicular to the molecular plane, and shows clearly the dominant role of the lone-pair orbital in determining equilibrium bond angles.

This is further illustrated by NH_2, which has three electrons in these orbitals. In its ground state two are in the lone-pair orbital and the bond angle is $103°$, which is very close to that for the $^1A_1$

state of CH_2. The first excited state has one electron in the lone-pair orbital and two in the non-bonding π-orbital perpendicular to the molecular plane. These two states, which both correlate with a linear $^2\Pi$ state, were the first example of the Renner–Teller effect, [60] but it has not yet been definitely established whether the upper component is linear like HCO or bent like HNO, HCF and HCCl.

The other triatomic molecules for which the spectra obtained in flash photolysis have been analysed have less than sixteen valency electrons. C_3 has twelve, NCN has fourteen and NCO, NCS and N_3 have fifteen. All these are linear as predicted. [66,68–70]

Of the tetratomic species investigated, the methyl radical is probably planar or nearly planar. [54] The fifteen-electron free radical HNCN has a linear or very nearly linear heavy atom skeleton with the H—N bond at an angle of $116°$ to it. [77]

Rotational analyses have not yet been made of the spectra of larger free radicals, because of the dense rotational structure or diffuseness. Their geometries are therefore not known, although in some cases partial vibrational analyses have been made. Radicals definitely identified include allyl and vinoxy, [78] phenyl, [49] benzyl and related aromatic radicals, [47] cyclopentadienyl and cyclo-heptatrienyl. [46,48]

Although predissociations have been observed in many free radical spectra, in most cases they do not provide limits for bond energies which are as narrow as those obtained from kinetic studies or mass spectrometry. An exception is HNO, where the predissociation limit [63] which corresponds to D (H—NO) = 48·6 kcal mole^{-1} has been shown by chemiluminescence studies [79] to correspond to the true dissociation limit. HNO can be important in the pyrolysis of hydrocarbons in the presence of nitric oxide. Relationships between bond lengths and bond dissociation energies do not yet seem to have been applied to data obtained in flash photolysis. The bond lengths in the ground state of CH_2 and NH_2 are notably short, being between 1·02 and 1·03 Å. [54,60]

The ionization potentials of three free radicals have been determined from the convergence of Rydberg series observed by flash photolysis. The values obtained [54,48] are 10·396 eV for CH_2, 9·840 eV for CH_3 and 6·24 eV for C_7H_7 (tropyl). The agreement with mass spectrometric studies is excellent for the methyl radical, [80,82] and fair for the tropyl radical where the mass spectrometric value is 0·35 eV higher. [81] There is a large difference between the spectroscopic and the other mass spectral values for methylene, but it is likely that the systems studied in this work contained very little methylene. [82] Now that the ionization potentials of so many molecules have been determined from Rydberg series, by photoionization or electron bombardment, it is to be hoped that methods other than mass spectrometry will be applied to more free radicals.

4. The study of combustion by flash photolysis

The more labile intermediates in combustion reactions only reach high concentrations under conditions where the reaction is proceeding very rapidly; that is, during an explosion or within the narrow reaction zone of a flame front. Flash photolysis can provide a direct method of studying such processes, since such photochemical initiation should produce a homogeneous explosion. This can be achieved by using either a fuel or oxidant which absorbs ultraviolet light or by adding a photosensitizer. Aromatic hydrocarbons and substances such as carbon disulphide, hydrogen sulphide, hydrazine, ammonia and phosphine fall into the first category, while sensitizers are necessary for aliphatic hydrocarbons unless they contain several double or triple bonds. Nitrogen dioxide is a useful photosensitizer since it absorbs quite strongly in the near ultraviolet, where it decomposes to yield oxygen atoms. Alkyl nitrites have also been used successfully.

The most interesting feature of the study of the hydrogen/oxygen and hydrogen/nitrogen dioxide systems [83] was the acceleration of the decay of the hydroxyl radical by molecular hydrogen, which was attributed to the reaction

$$OH + H_2 \rightarrow H_2O + H$$

The general features of the explosive combustion of hydrocarbons were best illustrated by the acetylene/oxygen reaction sensitized by nitrogen dioxide. [84] In these experiments, virtually all the nitrogen dioxide was removed during the photolytic flash, and there followed an induction period, normally lasting less than a millisecond, at the end of which the concentrations of free radicals such as OH and CN had reached a detectable level. At this stage, very rapid changes in free radical concentration occurred, some radicals disappearing completely and others reaching relatively high concentrations and then decaying over a period of several milliseconds. The striking features of this stage were the rapidity with which the radical concentrations rose and the sharp changes of radical products with small changes in mixture composition. If the oxygen in nitrogen dioxide was regarded as equivalent to molecular oxygen, the stoichiometry could be expressed by the equation

$$C_2H_2 + O_2 = 2CO + H_2$$

For acetylene to oxygen ratios of less than unity, OH was the only free radical to reach a high concentration. When the acetylene pressure exceeded the total oxygen pressure, C_2, CH and CN were the dominant species; with increasing acetylene pressure, C_3 became detectable and carbon deposition was observed. Under these conditions a very strong continuous absorption with maximum intensity at 3,700 Å was observed. [85] The carrier of this spectrum, which appeared to be a precursor of carbon deposition, showed similar kinetic behaviour to C_3; the species responsible has not yet been identified.

The hydrocarbons, ethylene, ethane and methane, show somewhat similar behaviour. [85] The stoichiometry is less sharp, and the change-over from carbonaceous radicals to OH occurs at somewhat higher total oxygen to fuel ratios. Much larger quantities of nitrogen dioxide were needed as sensitizers in this work, and uncertainty as to how completely it was reduced could have affected the results. The observed stoichiometries must be partly those of the chain combustion reactions, and it is probably that the free radicals 'overshoot' their equilibrium concentrations. The other important factor must be the thermodynamic stability of the radical concerned in the combustion products at high temperatures. The extreme thermal stability of carbon monoxide means that carbonaceous radicals (C_2, C_3, CH, etc.) will be more stable for carbon to oxygen ratios greater than unity, while ratios less than unity will favour oxygenated species such as hydroxyl.

Attempts have been made to improve the time resolution in flash photolysis studies of combustion by using photomultiplier cells and oscillographic recording to follow the free radical spectra in emission and absorption. [86] These experiments were designed to determine the rate of rise of the free radical concentrations, but they disclosed some unexpected features, which were later shown to arise from inhomogeneity of the explosion. [87] The inherent slight lack of uniformity in the irradiation of the reaction vessel causes the induction period to be shorter in the centre of the vessel than at the ends. This sometimes produces shock (detonation) waves, which travel to the ends of the reaction vessel where they produce sharp pulses of light emission.

This apparent limitation of the flash photolysis technique has, however, provided a very convenient method of studying how substances such as lead tetra-ethyl inhibit 'knock' in internal combustion engines. The addition of lead tetra-ethyl to hydrocarbon–oxygen mixtures photosensitized by amyl nitrite lengthened the induction period and suppressed the light pulses associated with detonation. [88] Under these conditions, the spectrum of gaseous PbO was the only absorption to appear strongly during the induction period; this substance was rapidly and completely reduced to atomic lead at the end of the induction period. This provided clear evidence that molecular PbO in the gas phase strongly inhibited detonation and that the active species were in a molecular state rather than present as a colloidal smoke. In a subsequent study, [89] it was found that tellurium dimethyl and iodine also act by lengthening the induction period, but that tin tetra-ethyl, iron pentacarbonyl and ferrocene act in a different manner. These latter species form colloidal smokes of the metal oxide or metal which scatter light strongly during the induction period. The induction period is not lengthened, and it was therefore suggested that the anti-knock action is not caused by termination of reaction chains on these particles but by cooling of the system due to the intense light emission by these particles during the combustion reaction. It must be remembered, however, that surface recombination processes frequently have significant activation energies, and chain termination on the particles could become more important as the temperature rises sharply at the end of the induction period.

The combustion of many other substances can be investigated

by flash photolysis. Most of the fuels studied to date have comparatively strong absorption spectra, and a photosensitizer is not needed to initiate the reaction. Many hydrides have been investigated, notably hydrogen sulphide, [90] phosphine, [40] ammonia and hydrazine. [91] The behaviour of these reactions was in many ways similar to the hydrocarbon combustion, such radicals as SH, SO, PH, NH and NH_2 appearing as intermediates. In the oxidation of hydrogen sulphide, the sulphur dioxide afterglow emission associated with the chemiluminescent combination of O and SO provided a convenient method of studying the radical decays. [90] Norrish has recently shown [91,92] that the oxidation of these hydrides can be explained by a degenerate branching chain mechanism exactly analogous to that for hydrocarbon combustion. This uses the isoelectronic principle to compare the behaviour of the compounds and radicals from groups IV, V and VI elements. Thus HNO and SO are regarded as the degenerate branching intermediates analogous to formaldehyde.

REFERENCES

1 Steacie, E. W. R. *Atomic and Free Radical Reactions.* 2nd ed. Reinhold, 1954.
2 Kondratiev, V. N. *Chemical Kinetics of Gas Reactions.* Pergamon, 1964.
3 Semenov, N. N. *Some Problems of Chemical Kinetics and Reactivity.* Pergamon, 1958.
4 Herzberg, G. *Spectra of Diatomic Molecules.* 2nd ed. van Nostrand, 1950.
5 Oldenberg, O. and Riecke, F. F. *J. chem. Phys.* 1939, **7**, 485.
6 White, J. U. *J. chem. Phys.* 1940, **8**, 79, 459.
7 Norrish, R. G. W. and Porter, G. *Nature, Lond.*, 1949, **164**, 658.
8 Porter, G. *Proc. Roy. Soc.* A, 1950, **200**, 284.
9 Norrish, R. G. W. and Thrush, B. A. *Q. Rev.* 1956, **10**, 149.
10 Schüler, H. *Spectrochim. Acta*, 1950, **4**, 85.
11 Kaufman, F. *Prog. Reaction Kinetics*, 1961, **1**, 3.
12 Jennings, K. R. *Q. Rev.* 1961, **15**, 237.
13 Thrush, B. A. *Prog. Reaction Kinetics*, 1965, **3**, 63.
14 Christie, M. I., Norrish, R. G. W. and Porter, G. *Proc. Roy. Soc.* A, 1953, **216**, 152.
15 Marshall, R. and Davidson, N. *J. chem. Phys.* 1953, **21**, 659.
16 Russell, K. E. and Simons, J. *Proc. Roy. Soc.* A, 1953, **217**, 271.
17 Rabinowitch, E. and Wood, W. C. *J. chem. Phys.* 1936, **4**, 497.

18 Christie, M. I., Harrison, A. J., Norrish, R. G. W. and Porter, G. *Proc. Roy. Soc.* A, 1955, **231**, 446.
19 Donovan, R. J. and Husain, D. *Nature, Lond.*, 1965, **206**, 171.
20 Porter, G. and Smith, J. A. *Proc. Roy. Soc.* A, 1961, **261**, 28.
21 Porter, G., Szabo, Z. G. and Townsend, M. G. *Proc. Roy. Soc.* A, 1962, **270**, 493.
22 Givens, W. G. and Willard, J. E. *J. Am. Chem. Soc.* 1959, **81**, 4773.
23 Burns, G. and Hornig, D. F. *Can. J. Chem.* 1960, **38**, 1702.
24 Palmer, H. B. and Hornig, D. F. *J. chem. Phys.* 1957, **26**, 98.
25 Briggs, A. G. and Norrish, R. G. W. *Proc. Roy. Soc.* A, 1963, **276**, 51.
26 Porter, G. and Wright, F. J. *Disc. Faraday Soc.* 1953, **14**, 23.
27 Durie, R. A. and Ramsay, D. A. *Can. J. Phys.* 1958, **36**, 35.
28 Lipscomb, F. J., Norrish, R. G. W. and Thrush, B. A. *Proc. Roy. Soc.* A, 1956, **233**, 455.
29 Edgecombe, F. H. C., Norrish, R. G. W. and Thrush, B. A. *Proc. Roy. Soc.* A, 1957, **243**, 24; *Spec. Publ. Chem. Soc.* 1958, **9**, 121.
30 Burns, G. and Norrish, R. G. W. *Proc. Roy. Soc.* A, 1963, **271**, 289.
31 Briggs, A. G. and Norrish, R. G. W. *Proc. Roy. Soc.* A, 1964, **278**, 27.
32 Griffiths, J. G. A. and Norrish, R. G. W. *Trans. Faraday Soc.* 1931, **27**, 451.
33 McGrath, W. D. and Norrish, R. G. W. *Proc. Roy. Soc.* A, 1960, **254**, 317.
34 Porter, G. and Black, G. *Proc. Roy. Soc.* A, 1962, **266**, 185.
35 Basco, N. and Norrish, R. G. W. *Proc. Roy. Soc.* A, 1962, **268**, 291.
36 Wright, F. J. *J. phys. Chem.* 1960, **64**, 1648.
37 Callear, A. B. and Norrish, R. G. W. *Nature, Lond.*, 1960, **188**, 53.
38 Norrish, R. G. W. and Oldershaw, G. A. *Proc. Roy. Soc.* A, 1958, **249**, 498.
39 Thrush, B. A. *Proc. Roy. Soc.* A, 1956, **235**, 143.
40 Norrish, R. G. W. and Oldershaw, G. A. *Proc. Roy. Soc.* A, 1961, **262**, 1, 10.
41 Basco, N., Nicholas, J. E., Norrish, R. G. W. and Vickers, W. H. J. *Proc. Roy. Soc.* A, 1963, **272**, 147.
42 Basco, N. and Norrish, R. G. W. *Proc. Roy. Soc.* A, 1965, **283**, 291, 302.
43 Simons, J. P. and Yarwood, A. J. *Trans. Faraday Soc.* 1961, **57**, 2167.
44 Mann, D. E. and Thrush, B. A. *J. chem. Phys.* 1960, **33**, 1731.
45 Callomon, J. H. and Ramsay, D. A. *Can. J. Phys.* 1957, **39**, 129.
46 Thrush, B. A. *Nature, Lond.*, 1956, **178**, 155.
47 Porter, G. and Wright, F. J. *Trans. Faraday Soc.* 1955, **51**, 1469.
48 Thrush, B. A. and Zwolenik, J. J. *Bull. Soc. Chem. Belg.* 1962, **71**, 642; *Disc. Faraday Soc.* 1963, **35**, 196.
49 Porter, G. and Ward, B. *Proc. Chem. Soc.* 1964, p. 288.
50 Herzberg, G. and Travis, D. N. *Can. J. Phys.* 1964, **42**, 1658.
51 Paul, D. E. and Dalby, F. W. *J. chem. Phys.* 1962, **37**, 592.
52 Mitchell, A. C. G. and Zemansky, M. W. *Resonance Radiation and Excited Atoms.* Cambridge University Press, 1934.

53 Basco, N., Callear, A. B. and Norrish, R. G. W. *Proc. Roy. Soc. A*, 1961, **260**, 459.
54 Herzberg, G. *Proc. Roy. Soc. A*, 1961, **262**, 291.
55 Herzberg, G. and Ramsay, D. A. *Proc. Roy. Soc. A*, 1955, **233**, 34.
56 Khan, M. A., Norrish, R. G. W. and Porter, G. *Proc. Roy. Soc. A*, 1953, **219**, 312.
57 Mains, G. J., Roebber, J. L. and Rollefson, G. K. *J. chem. Phys.* 1955, **59**, 733.
58 Shilman, A. and Marcus, R. A. *J. chem. Phys.* 1963, **39**, 996.
59 Johns, J. W. C., Priddle, S. H. and Ramsay, D. A. *Disc. Faraday Soc.* 1963, **35**, 90.
60 Dressler, K. and Ramsay, D. A. *Phil. Trans. Roy. Soc. A*, 1959, **251**, 553.
61 Herzberg, G. and Johns, J. W. C. *Mém. Soc. r. Sci. Liège*, 1963, **7**, 117.
62 Dalby, F. W. *Can. J. Phys.* 1958, **36**, 1136.
63 Bancroft, J. D., Hollas, M. and Ramsay, D. A. *Can. J. Phys.* 1962, **40**, 322.
64 Merer, A. J. and Travis, D. N. *Can. J. Phys.* 1966, **44**, 525, and to be published.
65 Herzberg, G. and Verma, R. D. *Can. J. Phys.* 1964, **42**, 395.
66 Gausset, L., Herzberg, G., Lagerqvist, A. and Rosen, B. *Disc. Faraday Soc.* 1963, **35**, 113; *Astrophys. J.* 1965, **142**, 45.
67 Johns, J. W. C. *Can. J. Phys.* 1961, **39**, 1738.
68 Dixon, R. N. *Phil. Trans. Roy. Soc. A*, 1960, **252**, 165; *Can. J. Phys.* 1960, **38**, 10.
69 Dixon, R. N. and Ramsay, D. A. To be published.
70 Douglas, A. E. and Jones, W. J. *Can. J. Phys.* 1965, **43**, 2216.
71 Walsh, A. D. *J. chem. Soc.* 1953, pp. 2260–2331.
72 Ramsay, D. A. *Determination of Organic Structures by Physical Methods*, vol. II, p. 246. Academic Press, 1962.
73 Renner, R. *Z. Physik*, 1934, **92**, 172.
74 Bueso-Sanllehi, F. *Phys. Rev.* 1941, **60**, 556.
75 Mrozowski, S. *Phys. Rev.* 1947, **72**, 682, 691.
76 Frey, H. M. *Progress Reaction Kinetics*, 1964, **2**, 131.
77 Herzberg, G. and Warsop, P. A. *Can. J. Phys.* 1963, **41**, 286.
78 Currie, C. L. and Ramsay, D. A. *Seventh Symposium on Free Radicals*, Padua, 1965.
79 Clyne, M. A. A. and Thrush, B. A. *Disc. Faraday Soc.* 1962, **33**, 139.
80 Lossing, F. P., Ingold, K. U. and Henderson, I. H. S. *J. chem. Phys.* 1954, **22**, 621.
81 Harrison, A. G., Honnen, L. R., Dauben, H. J. and Lossing, F. P. *J. Am. Chem. Soc.* 1960, **82**, 5593.
82 Langer, A., Hipple, J. A. and Stevenson, D. P. *J. chem. Phys.* 1954, **22**, 1836.
83 Norrish, R. G. W. and Porter, G. *Proc. Roy. Soc. A*, 1952, **210**, 239.
84 Norrish, R. G. W. Porter, G. and Thrush, B. A. *Proc. Roy. Soc. A*, 1953, **216**, 165.
85 Norrish, R. G. W., Porter, G. and Thrush, B. A. *Proc. Roy. Soc. A*, 1955, **227**, 423.

86 Norrish, R. G. W., Porter, G. and Thrush, B. A. *Fifth Symposium on Combustion*, p. 651. Reinhold, 1955.

87 Thrush, B. A. *Proc. Roy. Soc.* A, 1955, **233**, 147.

88 Erhard, K. H. L. and Norrish, R. G. W. *Proc. Roy. Soc.* A, 1956, **234**, 178; 1960, **259**, 297.

89 Callear, A. B. and Norrish, R. G. W. *Proc. Roy. Soc.* A, 1960, **259**, 304.

90 Norrish, R. G. W. and Zeelenberg, A. P. *Proc. Roy. Soc.* A, 1957, **240**, 293.

91 Husain, D. and Norrish, R. G. W. *Proc. Roy. Soc.* A, 1963, **273**, 145.

92 Norrish, R. G. W. *Chem. in Britain*, 1965, **1**, 289.

7

ENERGY TRANSFER IN MOLECULAR COLLISIONS

A. B. CALLEAR

1. Introduction

By the 1930's it was recognized that the interconversion of translational and vibrational energy is slow, and especially if the vibrational frequency is high, a very large number of gas kinetic collisions are required for relaxation. [1] Ultrasonic dispersion provided a means of comparing the specific heat of a gas at low sound frequency to the specific heat at high frequency when vibration can no longer follow the periodic temperature changes accompanying sound propagation. The classical Landau–Teller [2] theory correctly interpreted the dependence of relaxation times on the reduced mass of the collision partners, on the vibrational frequency, and on the temperature. In 1931, Zener [3] published an elegant wave-mechanical solution for vibrational energy transfer in gases which forms the basis of modern theory. However, only in 1952 was the theory reduced to a form suitable for numerical comparison with experiment, [4] under the stimulus of shock-tube measurements of vibration-translation relaxation of simple molecules. Vibrational relaxation rates of a large number of molecules have now been measured, and the results are in general accord with theory.

Also by the 1930's it was realized that electronically excited species, e.g. Na ($3\,^2$P), Hg (6^3P), He (2^3S), may survive a very large number of gas-phase collisions by monatomic gases. [5,6] Stationary state and kinetic methods were employed to measure life-times, by producing transients with a short duration electric discharge, and by photoelectric record of the reversal of resonance lines. Few systematic correlations have been deduced from the behaviour of these highly excited species, which appear to show considerable chemical affinity for polyatomic gases. The kinetics are determined by the particular form of the potential curves of the initial and final states.

Energy transfer processes are fundamental to an understanding of reaction kinetics, photochemistry, and gas lasers. A unimolecular reaction is initiated by vibrational excitation leading to a transition complex, which breaks up to distribute energy in some specific manner amongst the available degrees of freedom. Finally the fragments may undergo electronic, vibrational and rotational relaxation by collision. Photochemical initiation similarly produces a highly energized complex which breaks up into excited fragments. The discussion which follows is not intended to be a general review including reactive processes, and is almost entirely confined to the true energy transfer process

$$A + B^* \rightarrow A^* + B$$

in which there is neither chemical reaction nor exchange of charge. Some of the lesser known types of energy transfer are given special emphasis in relation to recent spectroscopic experiments conducted at Cambridge and to other work which was started originally through the efforts of Professor Norrish. General applications are discussed in the final section.

In some energy transfer processes, the 'type' of energy is preserved, for example the reaction producing population inversion in the helium/neon laser (E–E transfer):

$$\mathrm{He}\,(2\,^3\mathrm{S}_1) + \mathrm{Ne}\,(2\,^1\mathrm{S}_0) \rightarrow \mathrm{He}\,(1\,^1\mathrm{S}_0) + \mathrm{Ne}\,(2s)$$

In other processes, the energy is converted from one form to another, for example the spin-orbit relaxation of atomic selenium by N_2 (E–V transfer):

$$\mathrm{Se}\,(4\,^3\mathrm{P}_0) + \mathrm{N}_2\,(v = 0) \rightarrow \mathrm{Se}\,(4\,^3\mathrm{P}_2) + \mathrm{N}_2\,(v = 1)$$

Considering the four types of molecular energy, it is convenient to classify ten different kinds of energy transfer, though except for the rare case of exact resonance some energy is always converted to translation.

Of these ten types, we have a detailed understanding only of V–T relaxation, because of the circumstance that vibrational quanta are comparable with kT in a workable temperature range. [7, 8] Also discussed below under specific headings are R–T, R–V, V–V, E–T, E–V and E–E energy transfer. Of the remainder, T–T energy transfer can be investigated with ultrasonic dispersion at very high frequency and low pressure, and together with transport properties, provides a means of investigating intermolecular forces. [9] Resonant transfer of rotational quanta between identical molecules may be one of the causes of pressure broadening in microwave spectroscopy. [10] There is presently no information about R–R transfer between unlike molecules. Except for one possible example in section 8, pure R–E transfer is unknown.

2. Theory

The following general theoretical arguments will be of value in considering the experimental data.

2.1. Landau and Teller [2] pointed out that the probability of vibration–translation energy transfer depends on the ratio of the period of vibration to the duration of the collision. If the repulsive potential is steep, the colliding molecule rebounds before the oscillator has completed a single cycle, and energy transfer has a high probability. The other extreme is a shallow interaction potential which changes only slowly with distance. In this case the collision occurs over a long time interval, during which there is a gradual conversion of kinetic energy into potential energy. The process ultimately reverses and the particles fly apart. In the latter case the probability of excitation or de-excitation of vibration is small, because the oscillator has time to adjust itself as the colliding molecule approaches. The interaction is on the centre of gravity of the vibrator. Landau and Teller employ classical arguments to show that the probability of relaxation is proportional to

$$\exp\left[(-l/v)/(1/2\pi\nu)\right]$$

where ν is the vibrational frequency in s^{-1} units, v is the relative velocity, and l is a length characteristic of the interaction potential. Its relation to the more rigorous wave mechanical model is seen from the following discussion.

2.2. We now give a simple outline of Zener's distorted wave treatment, following the refinements of Jackson and Mott. [11] The model consists of a beam of particles impinging on to a set of oscillators which may be considered to be equivalent to a vibrating plate. The object will be to calculate the probability of excitation from the zero point to the first vibrational level. The centre of gravity of BC is at the origin, B is situated X from the origin, and A is at a point x (Fig. 1). The gas-phase problem is simulated by employing the reduced mass (μ) of the system in the translational wave functions, and by averaging the relative velocity through one translational degree of freedom.

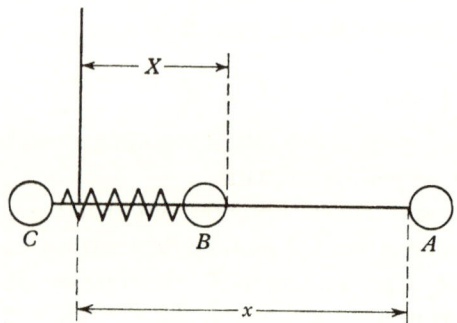

Fig. 1. Collision of atom A with diatomic molecule BC.

In the absence of a field a beam of particles is correctly described by the function

$$\psi = A(\cos\ (kx - \omega t) + i \sin\ (kx - \omega t)) = A\ e^{i(kx - \omega t)}$$

which is a wave motion moving from left to right. It is complex, and the number of particles per unit volume $|\psi\psi^*|$, usually written $|\psi^2|$, is equal to $|A^2|$. The independence on x and t arises because the real and complex parts are π out of phase. A wave moving from right to left is represented by

$$\psi = A\ e^{i(-kx - \omega t)}$$

The frequency of the beam is not affected when the speed is changed by application of a field, and it is therefore convenient to omit the time factor. The wavelength is $2\pi/k$, and because of the De Broglie relationship $mv\lambda = h$,

$$k = mv/\hbar$$

The above wave functions are solutions of the wave equation

$$\left(\frac{\partial^2}{\partial x^2}+k^2\right)\psi = 0, \quad \text{or} \quad \left(\frac{\partial^2}{\partial x^2}+\frac{2m(\frac{1}{2}mv^2)}{\hbar^2}\right)\psi = 0$$

If the beam enters a retarding field due to a potential $V(x)$, the wavelength at all points is given by the well-known equation

$$\left(\frac{\partial^2}{\partial x^2}+\frac{2m}{\hbar^2}(W-V(x))\right)\psi = 0$$

where W is the kinetic energy at zero potential. As the beam is slowed down by a repulsive field, the wavelength and the amplitude must increase, because the number of particles crossing unit area, $k\hbar|A^2|/m$ (i.e. $v|A^2|$) is independent of x.

For the problem of vibrational excitation illustrated in Fig. 1, at large positive values of x we require a solution of the form

$$f_0(x) = e^{-ik_0x}+A_0e^{ik_0x}$$

$$f_1(x) = A_1e^{ik_1x}$$

where $f_0(x)$ is the sum of the incident and elastically scattered translational waves, and $f_1(x)$ is the inelastically scattered translational wave.

2.3. After eliminating the centre of gravity of the complete system, the wave equation becomes

$$\left\{\frac{\hbar^2}{2\mu_{BC}}\frac{\partial^2}{\partial X^2}+\frac{\hbar^2}{2\mu}\frac{\partial^2}{\partial x^2}+E-V(X)-V(X, x)\right\}\Psi = 0$$

where $V(X)$ is a Hooke's Law or Morse potential, etc., for the isolated vibrator, $V(X, x)$ is the interaction potential, E is the total energy, μ_{BC} is the reduced mass of the oscillator and μ is the reduced mass of the system $A-BC$. It is usually assumed that the solution can be expanded in the form

$$\Psi = \sum_n \psi_n(X)f_n(x)$$

where $\psi_n(X)$ are the eigenfunctions of the isolated vibrator, and $f_n(x)$ is the associated translational wave function. This separation holds only if the amplitude of vibration is small compared to the range of the repulsive forces. The distortion of the oscillator during the collision is discussed in section 2.8, on page 145.

If $\psi_n(X)$ are independent of x, then

$$\sum_n \left(\frac{\partial^2}{\partial x^2} + k_n^2 - \frac{2\mu}{\hbar^2} V(x, X)\right) \psi_n(X) f_n(x) = 0 \tag{i}$$

where k_n is related to the velocity at $x \to \infty$. Multiply by $\psi_i(X)$, and integrate with respect to X from $-\infty$ to $+\infty$. The vibrational wave functions are orthogonal, and therefore

$$\left(\frac{\partial^2}{\partial x^2} + k_i^2\right) f_i(x) = \frac{2\mu}{\hbar^2} \sum_n f_n(x) \int V(x, X) \psi_n(X) \psi_i(X) \, dX$$

It is shown below that the terms on the right-hand side can be neglected unless either $i = n$ or $i = n \pm 1$. Since we are considering excitation from $v = 0$ to $v = 1$, the following two equations are required:

$$\left(\frac{\partial^2}{\partial x^2} + k_0^2\right) f_0(x) = \frac{2\mu}{\hbar^2} f_0(x) \int V(x, X) \psi_0(X) \psi_0(X) \, dX$$

$$+ \frac{2\mu}{\hbar^2} f_1(x) \int V(x, X) \psi_0(X) \psi_1(X) \, dX \tag{ii}$$

$$\left(\frac{\partial^2}{\partial x^2} + k_1^2\right) f_1(x) = \frac{2\mu}{\hbar^2} f_0(x) \int V(x, X) \psi_0(X) \psi_1(X) \, dX$$

$$+ \frac{2\mu}{\hbar^2} f_1(x) \int V(x, X) \psi_1(X) \psi_1(X) \, dX \tag{iii}$$

Provided the probability of vibrational excitation is small, the second term on the right-hand side of equation (ii) is small compared to the first. Therefore solution of equation (ii) yields $f_0(x)$, which may be substituted in (iii) to solve for $f_1(x)$. Thus the remaining problem is simply a mathematical one.

2.4. An intermolecular repulsive force between an oscillator and a colliding molecule approximates to the form

$$V(x, X) = C \exp\left(-(x - X)/l\right) = U(x) \exp\left(X/l\right) \tag{iv}$$

Therefore

$$\int V(x, X) \psi_i(X) \psi_n(X) \, dX = U(x) \int \exp\left(X/l\right) \psi_i(X) \psi_n(X) \, dX$$

$$= U(x) V_{in}$$

From the harmonic oscillator functions given below, it is shown that since l is generally large compared to the amplitude of vibration, the diagonal elements $V_{nn} \simeq 1$. Equations (ii) and (iii) then take the form

$$\left(\frac{\partial^2}{\partial x^2} + k_0^2 - \frac{2\mu}{\hbar^2} U(x)\right) f_0(x) = 0 \tag{v}$$

$$\left(\frac{\partial^2}{\partial x^2} + k_1^2 - \frac{2\mu}{\hbar^2} U(x)\right) f_1(x) = \frac{2\mu}{\hbar^2} U(x) V_{01} f_0(x) \tag{vi}$$

Equation (vi) can be solved by letting $F_n(x)$ denote the solution of the equation

$$\left(\frac{\partial^2}{\partial x^2} + k_n^2 - \frac{2\mu}{\hbar^2} U(x)\right) F_n(x) = 0 \tag{vii}$$

normalized to have the asymptotic form

$$F_n(x) = \cos\left(k_n x + \eta\right)$$

as $x \to \infty$. If the probability of energy transfer is small, the incident and elastically scattered waves have the same amplitude:

$$f_0(x) = 2F_0(x)$$

Writing $f_1(x) = yF_1(x)$, where y is a function of x, and substituting in (vi) it is seen that

$$\frac{d}{dx}\left[F_1(x)^2 \frac{dy}{dx}\right] = \frac{4\mu}{\hbar^2} V_{01} U(x) F_0(x) F_1(x)$$

or

$$F_1(x)^2 \frac{dy}{dx} = \frac{4\mu}{\hbar^2} V_{01} \int_{-\infty}^{x} U(x) F_0(x) F_1(x) \, dx$$

the integral being taken from $-\infty$ where $F_n(x)$ must vanish. At large x

$$\frac{dy}{dx} = \frac{4\mu}{\hbar^2} V_{01} T \sec^2\left(k_1 x + \eta\right)$$

where

$$T = \int_{-\infty}^{+\infty} U(x) F_0(x) F_1(x) \, dx$$

and becomes independent of x. Therefore at large x

$$y = \frac{4\mu V_{01}}{\hbar^2 k_1} T[\tan\left(k_1 x + \eta\right) + \text{const.}]$$

Since as $x \to \infty$, $yF_1(x)$ must have the form $A_1 e^{i(k_1 x + \eta)}$ the constant of integration is equal to $-i$.

$$f_1(x) = -\frac{4i\mu V_{01}}{\hbar^2 k_1} T\, e^{i(k_1 x + \eta)}$$

and

$$|A_1| = \frac{4\mu V_{01} T}{\hbar^2 k_1}$$

The probability of vibrational excitation per collision is therefore

$$|A_1^2|\, v_1/|A_0^2|\, v_0 = \frac{16\mu^2 V_{01}^2 T^2}{\hbar^4 k_0 k_1} \tag{viii}$$

Below we derive the matrix element V_{01} for the harmonic oscillator, but evaluation of T involves a lengthy integration of equation (vii) to obtain $F_n(x)$, and we accept the result of Jackson and Mott. [11]

2.5. We can regard

$$T = \int_{-\infty}^{+\infty} C\, e^{-x/l} F_0(x)\, F_1(x)\, dx$$

as the 'overlap' of the translational wave functions on the interaction potential. In the absence of a field, two beams of different wavelength must cancel when the product is integrated over a large distance, so that $F_1(x)$ and $F_0(x)$ only contribute to T, where the potential is finite, and the main contribution is near the classical turning point.

Jackson and Mott solved equation (vii) with the exponential interaction potential and thereby evaluated T. Following a development similar to that of Schwartz, Slawsky and Herzfeld, [4]

$$\frac{T^2}{k_0 k_1} = \pi^2 [\Delta E]^2\, l^4 \exp\left(-2\pi l(k_0 - k_1)\right)$$

$$= \pi^2 [\Delta E]^2 l^4 \exp\left[-\frac{4\pi l |\Delta E|}{\hbar(v_0 + v_1)}\right] \tag{ix}$$

where $\Delta E = h\nu$ for the case of vibrational excitation. The variations of T with l, v, and ΔE in the exponential term, are similar to the Landau–Teller classical formulation. The form of T should be similar for all types of energy transfer in which the dependence of the potential on the intermolecular distance x is $e^{-x/l}$ in both eigenstates (tunnelling between parallel potential curves) and the quantities corresponding to V_{nn} are equal. Since the variation of the overall probability with ΔE is dominated by the exponential term, we anticipate that different types of energy transfer may show similar trends, and the evidence for this is discussed in the sections which follow.

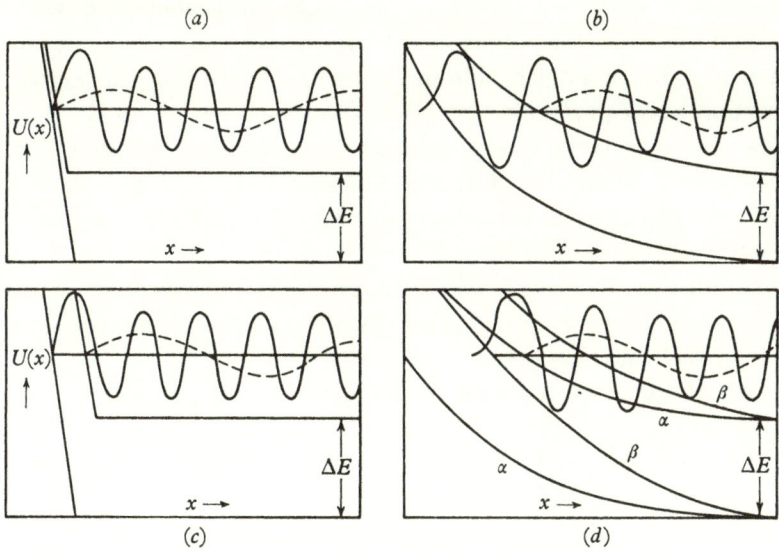

Fig. 2. Full lines represent initial, and dotted lines the final translational wave-functions (schematic).

The potential diagrams of Fig. 2 illustrate the overlap of the translational wave functions. Case (a) illustrates for hard spheres the substantial overlap where the interaction potential is finite. For vibrational excitation, $\Delta E = h\nu$. In example (b) with l large, the transition probability is very small because the final wave function overlaps the initial wave where the latter is executing regular S.H.M. in a region where the potential is finite; the positive and negative contributions to the integrand will largely cancel. (a) and (b) are cases of 'parallel potential curves', and the square of the overlap is given by equation (ix). Cases (c) and (d) illustrate the effect of the potential curves becoming non-parallel. In case (c), the interaction potential is greater in the upper state for a given value of x, i.e. the potential curves diverge. The overlap of the wave functions is thereby reduced. The mean internuclear separation of a vibrating molecule increases slightly with increase of vibrational quantum number because of anharmonicity, and Mies[12,13] has shown that this can lower the transition probability 10- to 100-fold. Case (d) illustrates the increase of transition probability which can occur if the potential curves converge or cross. According to Nikitin,[14] when two nitric oxide molecules ($^2\Pi$) collide, the inter-action potential splits to give the curves α and β, corresponding to a triplet and a singlet collision complex. There is a substantial overlap between β with zero point vibrational energy and α with one molecule vibrationally excited. Thus NO–NO collisions are extremely efficient at inducing vibrational relaxation, whereas NO–Ar collisions cause relaxation at the normal rate predicted by (ix). It may be noted that T is independent of the constant C of equation (iv), increase of which simply shifts the potential curves to greater x, and does not affect the steepness of the potential at the classical turning point.

2.6. It is now required to evaluate the off-diagonal matrix element

$$V_{01} = \int_{-\infty}^{+\infty} \exp{(X/l)}\, \psi_0(X)\, \psi_1(X)\, dX.$$

The wave functions for a diatomic harmonic-oscillator are

$$\left.\begin{aligned}
\psi_0(r) &= (\alpha/\pi)^{\frac{1}{4}} \exp{(-\tfrac{1}{2}\alpha r^2)} \\
\psi_1(r) &= (\alpha/\pi)^{\frac{1}{4}} \sqrt{(2\alpha)}\, r \exp{(-\tfrac{1}{2}\alpha r^2)} \\
\psi_2(r) &= (\alpha/\pi)^{\frac{1}{4}}\, 1/\sqrt{2}\, (2\alpha r^2 - 1) \exp{(-\tfrac{1}{2}\alpha r^2)}
\end{aligned}\right\} \qquad (x)$$

where r is the displacement of the internuclear separation from its equilibrium value, and $\alpha = 2\pi\nu\mu_{BC}/\hbar$. However, we have fixed the centre of gravity of BC and employed the reduced mass of A and BC in the translation wave equations, as required for a gas-phase collision. In the X co-ordinate with B vibrating, the amplitude of vibration is reduced (assuming that the amplitude of oscillation at the periphery of a molecule is the same as that of the nuclei) such that $rM_C/(M_B + M_C) = X - X_e$.

We have equated V_{nn} to unity above, and it is therefore necessary to retain the same form as the normalized functions (X), and simply write $\alpha = 2\pi\nu\mu_{BC}(M_C + M_B)^2/\hbar(M_C)^2$.

A further factor $\exp(X_e/l)$ is common to all matrix elements and will drop out. Then

$$\left.\begin{aligned}
V_{00} &= \exp(\gamma) \simeq 1 \\
V_{01} &= \sqrt{(2\gamma)} \exp(\gamma) \simeq \sqrt{(2\gamma)} \\
V_{02} &= \sqrt{(2)}\gamma \exp(\gamma) \simeq \sqrt{(2)}\gamma \\
V_{11} &= (1+2\gamma) \exp(\gamma) \simeq 1
\end{aligned}\right\} \quad \text{(xi)}$$

where $\gamma = (4\alpha l^2)^{-1}$.

Generally $\gamma \ll 1$ and the approximations given in the right-hand side of equations (xi) are satisfactory. For example, for N_2, $\exp(\gamma) = 1\cdot01$. However, hydrides are abnormal because the amplitude of vibration is comparable with the range of the repulsive forces, and V_{00}/V_{11} differs considerably from unity for low-frequency oscillators. This means that when we solve (vii), $F_0(x)$ and $F_1(x)$ will represent waves scattered from different inter-action potentials which will influence the relaxation rate for the reasons illustrated on Fig. 2. Mies[12] has taken such effects into account, and employed the more precise Morse anharmonic oscillator functions.

In the general case for the harmonic oscillator,

$$(V_{01})^2 = 2\gamma = \hbar M_C / 4\pi\nu l^2 \, M_B(M_B + M_C)$$

We now need to consider collisions with the other end of the molecule for which V_{01} has M_B and M_C interchanged. Adding and dividing by 2,

$$(V_{01})^2 = \frac{M_B^2 + M_C^2}{M_B M_C (M_B + M_C)} \frac{\hbar^2}{4l^2[\Delta E]}. \quad \text{(xii)}$$

The mass function tends to be large for hydrides; for N_2 and HI it is respectively $0\cdot07$ and 1. This is because the large amplitude of vibration of a light atom favours efficient coupling with translation. Collision with the heavy end of the molecule is comparatively ineffective.

The relative probability of a double quantum transition depends on

$$(V_{02}/V_{01})^2 = \gamma$$

which is $\ll 1$ except for hydrides. This also holds true for a Morse oscillator. However, T_{02}/T_{01} will usually prohibit a double quantum change because ΔE occurs in the exponent. (The harmonic oscillator selection rule generally will not hold if there is chemical affinity between A and BC; first, it will no longer be permissible to separate the translational and vibrational wave functions in the manner described; secondly, the interaction potential will not be given by (iv).)

2.7. Schwartz, Slawsky and Herzfeld deduced values of l from the Lennard–Jones parameters ϵ and r_0, which have been tabulated by Hirschfelder, Curtis and Bird,[9] and thereby reduced the theory to numbers and compared with experiment. With Takayanagi,[15] they also deduce factors which take account of unfavourable orientation. A careful and detailed account of the required calculations is given by Herzfeld and Litovitz.[8] The problem is to obtain a satisfactory fit of the exponential interaction potential with the repulsive region of the Lennard–Jones function

$$V = 4\epsilon \left[\left(\frac{r_0}{r} \right)^{12} - \left(\frac{r_0}{r} \right)^{6} \right]$$

which is employed for the analysis of transport phenomena in gases. Thereby l is expressed as a function of ϵ and r_0. Integration of equation (viii) through the velocity distribution and substitution of the experimental l, provides numbers which are in order of magnitude agreement with experiment (on the harmonic oscillator model). However, the extrapolation of collision parameters determined from viscosity data, to the high velocities required for vibrational relaxation, can only lead to an approximate value for l, to which the transition probabilities are extremely sensitive. Mies[16] has calculated relaxation rates for H_2–He collisions, employing a Hartree–Fock potential.

Schwartz and Herzfeld[17] have considered the three-dimensional scattering problem, with a breathing surface replacing the oscillating plate. Except for a numerical factor close to unity, the form of the result is unchanged. Some of the theoretical results are compared with experiment in the sections which follow. However, it is beyond the scope of this chapter to provide a comprehensive review of all major contributions to the theory, the emphasis being on experimental data.

2.8. Finally we need to discuss briefly the separation made in subsection 2.3,

$$\Psi = \sum_n \psi_n(X) f_n(x)$$

This assumes that the vibrational eigenfunctions of the oscillator with A in close proximity are identical with those of the isolated molecule and are not distorted by the collision. This is a good approximation for most molecular systems because the amplitude of the vibration is generally small compared to the range of the intermolecular repulsion. The force exerted by A compresses the vibrator, but since A exerts the same force in the expanded and contracted configurations, the frequency is only slightly affected and the eigenfunctions are essentially unchanged. However, this seems to be a relatively poor approximation for hydrides, which have large vibrational amplitudes; the force exerted by A is much greater in the expanded than in the contracted configurations. The eigenfunctions will be appreciably changed, and the vibrator becomes anharmonic in the inverted sense. The effect increases with vibrational amplitude and therefore decreases V_{nn}/V_{00} and increases the rate of vibrational relaxation; i.e. the potential curves converge. Some approximate calculations indicate that the effect on V_{11}/V_{00} may be as important as the anharmonic effect of Mies. [18]

3. Rotation–translation energy transfer

3.1. Rotational relaxation is generally rapid, and except for H_2 and D_2, requires less than ten collisions. When investigating vibrational and electronic energy transfer, in a system which has been disturbed from equilibrium, one frequently experiences the experimental situation that rotation and translation effectively have the same temperature, as defined by the Boltzmann equation, which is different from that of the other degrees of freedom. The facile interchange of rotational and translational energy may be anticipated from equation (2, ix); the energy separation is small between the low-lying rotational levels which are appreciably populated under normal experimental conditions.

3.2. Ultrasonic dispersion and absorption have been applied extensively to the measurement of the rates T–T, T–R and T–V energy transfer, in diatomic and more complex gases. The velocity of sound (V) is related to the ratio of the specific heats at constant pressure and constant volume (γ), the molecular weight (M) and the absolute temperature (T) by the equation $V^2 = \gamma RT/M$. At high frequencies, excitation and de-excitation of some types of molecular motions may be too slow to contribute to the specific heat, and consequently γ increases and therefore the sound velocity rises. The sound frequency (ω) at the point of inflection of a plot of V^2 against $\log \omega$ is related to the relaxation time (τ) by

$$\omega_{\text{inf.}} = C_0/\tau C_\infty$$

where C_0 and C_∞ refer to the apparent specific heats respectively on the low- and high-frequency extremes of the dispersion region. From the relaxation time, the probability of energy transfer per collision (P or $1/Z$) is easily computed from the simple hard sphere equations relating collision frequency with concentration, or preferably, with a more refined model.[9] In a dispersion region, energy is released out of phase, and the medium becomes very inelastic. Therefore, measurement of the variation of the absorption coefficient with frequency also yields the energy transfer relaxation times. Polyatomic gases show several relaxation regions, at least one for vibration and another, at higher frequencies, when rotation ceases to follow. Experimentally, velocity dispersion and absorption are conveniently measured by setting up a stationary wave system in a cylindrical tube containing the gas; as the length of the column of gas is increased, points of resonance are observed at half-wavelength intervals; the decreasing amplitude with length of the column of gas yields the absorption coefficient.[7]

3.3. Rotational relaxation rates have also been deduced from the structure of shock waves; the shock front is distorted because translational relaxation is more rapid than rotational relaxation. A light beam is partially reflected from the shock front because of the refractive index gradient, and the distortion is recorded directly with a photomultiplier. From the form of the intensity–time curves, the probability of relaxation can be computed.[19,20]

Other methods for rotational relaxation are fluorescence and crossed molecular beams. The first of these requires excitation of a molecule into an excited electronic or vibrational state with monochromatic light, so that only one or two rotational states are populated. If the lifetime of the excited state can be determined by considering the total effect of radiative and radiationless transitions, from the rotational spreading of the fluorescence spectrum with added gases, rotational relaxation can be studied quantitatively. The crossed molecular beam method is becoming increasingly important for kinetic measurements. When two molecules collide in the region of the crossing point of two molecular beams, scattering occurs over all regions of space; however, it is frequently a property of such systems that the intensity of the scattered beam peaks in the direction of the initial relative velocity vector. For elastic scattering, the peak intensity is found in a direction corresponding to the sum of the final relative velocity vectors and the vector representing the motion of the centre of gravity of the system. On the other hand, in the extreme inelastic case where the kinetic energy is converted to internal energy, the beam would peak in the direction of the centre of gravity vector. Measurement of the intensity of the scattered beam permits identification of the energy transfer process and the determination of the cross-section. Thus Blythe [21] has shown that the cross-section for rotational excitation of $ortho$-D_2 by atomic potassium is 1 Å².

3.4. Ultrasonic and shock tube measurements indicate that $Z \simeq 10$ for rotational relaxation of N_2 and O_2. [7, 8] This represents an average over a range of J states, with $J \simeq 7$ having the greatest population at ambient temperature. Holmes, Jones and Lawrence [22] have shown that $Z \simeq 3$ for N_2O and CO_2, by ultrasonic dispersion. Hydrogen relaxes comparatively slowly, with $Z \simeq 350$ at 273 °K, because of the small moment of inertia and consequent large spacing between rotational states. [7, 8] A perturbation treatment for R–T relaxation, similar to that of section 2, predicts that for symmetric and unsymmetric molecules, respectively, $\Delta J = \pm 2$ or ± 1 transitions should be the most probable. [8] However, it is now established that transitions are not restricted to these changes in rotational quantum number, if the rotational quanta are of magnitude comparable with kT.

The most interesting results have been obtained with the fluorescence method. Broida and Carrington [23] observed rotational relaxation of NO in the $A^2\Sigma^+$ state by populating a single rotational level with an atomic line at 2,144 Å. The source was a specially constructed cadmium resonance lamp, excited with an R.F. discharge. With a low NO pressure, and in the absence of foreign gases, essentially only a single line in each rotational branch was observed in emission. Addition of Ar and N_2 produced rotational transitions, observed by a spreading of the branches. The results showed that $\Delta J < \pm 5$ transitions can occur for single collision events and that Ar and N_2 collisions are equally effective. This suggests that R–R transfer is not important. Similar conclusions were drawn from ultrasonic measurements in O_2. [24]

The occurrence of multiple quantum rotational transitions is also proved conclusively in the detailed studies of Brown and Klemperer [25] and Steinfeld and Klemperer [26] on the resonance fluorescence of $I_2 B^3\Pi_{Ou^+}$. Employing either the sodium D lines to populate predominantly $J' = 44$ and 37 in $v' = 15$, or the mercury green line to populate $J' = 34$ in $v' = 25$, they investigated rotational and vibrational energy transfer, and induced pre-dissociation, in the presence of He, Ne, Ar, Kr, Xe, H_2, D_2, N_2, O_2, NO, CO_2, SO_2 and CH_3Cl. Several very interesting features were identified. The efficiency of R–T transfer increases smoothly with increase of reduced mass, except for the polyatomic species (CO_2, SO_2 and CH_3Cl) which are comparatively more efficient with cross-sections up to 70 Å². For all added gases, $\Delta J < 40$ transitions occurred at single collision events. The high efficiency of the polyatomic molecules may arise because of the comparative ease of R–R transfer, resulting from the abundance of available rotational states and the consequent facility of conserving angular momentum for such processes. However, the interpretation is complicated because of the simultaneous occurrence of induced predissociation.

3.5. The NO and I_2 fluorescence experiments demonstrate that although the optical rules do not hold for rotational energy transfer, complete thermalization of rotation does not occur. The results suggest that the smaller the rotational spacing, the greater is the probability of a large ΔJ transition. As ΔJ increases, the decreasing translational overlap (equivalent to equation (2, ix) on a three-dimensional scattering model) will be restrictive at ambient temperature when the energy to be converted to translation exceeds 50 cm⁻¹; transitions with $\Delta E > \sim 200$ cm⁻¹ should have low probability. Three other novel features were demonstrated from the $I_2 B^3\Pi_{Ou^+}$ results.

(a) Some persistence of the initial rotational states was observed, following $\Delta v' = \pm 1$ transitions in He, H_2, HD and D_2, but not in the other gases.

(b) No persistence of the initial rotational states was observed following $\Delta v' = \pm 2$ transitions, which have one-quarter of the probability of $\Delta v' = \pm 1$ transitions. This was attributed to the stronger coupling required for the double quantum vibrational transition.

(c) The HD ($J = 0 \rightarrow 1$) rotational spacing (91 cm^{-1}) is very nearly identical with the energy of the vibrational transition $v' = 15 \rightarrow 14$. HD is not abnormally effective at inducing vibrational relaxation.

Some of these experimental features are brought into the discussion below.

4. Vibration–translation energy transfer

4.1. Vibration–translation relaxation is the only type of energy transfer that is presently understood in any detail, and a large number of simple molecules have been investigated. The main experimental methods are the ultrasonic techniques, which have been considered, and the shock tube. The vibrational temperature is initially lower than the translational and rotational temperatures in a shock-heated gas, and with time the gas density increases in the hot zone as equilibrium is established. The relaxation time can be computed by measuring the rate of density change by Schlieren photography, by following the population of excited levels by absorption or emission spectroscopy, and by direct measurement of the vibrational temperature by atomic line reversal. The latter technique is discussed in section (8) under E–V transfer. The shock tube has been particularly successful for simple molecules, especially diatomic, from 500 to $\sim$ 3,000 °K, and provides most of our information about temperature dependence. [27–29]

Millikan and White [30] have recently reviewed and systematized shock-tube data for diatomic molecules, and developed the empirical equation

$$\log_{10}(P\tau) = 5 \times 10^{-4} \mu^{\frac{1}{2}} \theta^{\frac{4}{3}}[T^{-\frac{1}{3}} - 0\cdot015\,\mu^{\frac{1}{4}}] - 8$$

where P = pressure (atm), τ = relaxation time (s), μ = reduced mass, and $\theta = h\nu/k$ (°K). The equation reproduces the observed relaxation times to within a factor of 2, for systems as diverse as N_2, I_2, O_2—H_2. However, it does not take account of the effect of the large amplitude of vibration of light molecules, and presumably will therefore not be applicable to diatomic hydrides. It has recently been reported that the HI relaxation time is much shorter than the empirical equation predicts. [18] Systematic correlations also have been given by Losev and Osipov. [31]

TABLE 1. *Collision frequencies for vibrational relaxation*

System	ν (cm^{-1})	Temperature (°K)	$Z_{\text{exp.}}$	$Z_{\text{calc.}}$ (A)	$Z_{\text{calc.}}$ (B)
O_2	1,556	288	$2 \cdot 1 \times 10^7$	$6 \cdot 8 \times 10^8$	$7 \cdot 3 \times 10^7$
O_2	1,556	1,372	$3 \cdot 6 \times 10^4$	$8 \cdot 2 \times 10^4$	2×10^4
N_2	2,330	550	$\sim 10^8$	10^{10}	$7 \cdot 6 \times 10^8$
N_2	2,330	3,640	$2 \cdot 4 \times 10^4$	$2 \cdot 8 \times 10^4$	$2 \cdot 3 \times 10^4$
Cl_2	557	288	3×10^4	$5 \cdot 8 \times 10^5$	$1 \cdot 8 \times 10^5$
Cl_2	557	1,000	550	$4 \cdot 9 \times 10^3$	$2 \cdot 1 \times 10^3$

4.2. The data of Table 1 are taken from Herzfeld and Litovitz who compare theory with experiment for the number of gas kinetic collisions required for V–T energy transfer (deactivation). Values of $Z_{\text{calc.}}$ are derived from the S.S.H. theory with the exponential interaction potential fitted to the Lennard–Jones potential at the most probable velocity for relaxation, to determine l in terms of r_0 and ϵ. In method 'A', the two functions are fitted in magnitude and slope at the most probable velocity; in method 'B' they are fitted in magnitude at the potential corresponding to the most probable velocity, and also at the point of zero potential on the Lennard–Jones function. The data have been selected to demonstrate the dependence on frequency and velocity, as required by equation (2, ix). The variation between methods A and B illustrates the extreme sensitivity of $Z_{\text{calc.}}$ on the form of the potential. No account is taken of the non-identity of V_{11} and V_{00} due to anharmonicity.

The data of Table 2, obtained by Millikan, [32] demonstrate the effect of varying the reduced mass. The results were obtained with a novel and ingenious technique. CO gas at 1 atm and flowing at 10 cm s^{-1} was excited by infra-red radiation from a CH_4/O_2 flame.

Resonance fluorescence was observed downstream, corresponding to a vibrational temperature of 993 °K. The probabilities of V–T transfer were determined from the increased rate of decay of the fluorescence in the presence of added gases. According to Millikan, with μ small, Z_{10} is larger than predictions from S.S.H. theory.

T A B L E 2. *The effect of reduced mass on vibration–translation relaxation*

Deactivator	H_2	HD	D_2	He	Ne	CO
Z_{10} (286 °K)	2×10^6	$3 \cdot 4 \times 10^6$	$3 \cdot 8 \times 10^7$	$3 \cdot 2 \times 10^7$	$2 \cdot 3 \times 10^8$	10^9

4.3. Vibration–translation energy transfer of nitric oxide is abnormally fast. [33–35] A direct illustration by flash spectroscopy is shown in Plate 1 (facing p. 152). When nitric oxide is subjected to an intense light pulse, the NO $X^2\Pi$ ground state is excited to $A^2\Sigma^+$, which is deactivated electronically due to self quenching, to populate almost exclusively the first vibrational level of the ground electronic state. [36] (The potential curves are shown in Fig. 3.)

$$\text{NO } X^2\Pi + h\nu \rightarrow \text{NO } A^2\Sigma^+$$

$$\text{NO } A^2\Sigma^+ + \text{NO } X^2\Pi \rightarrow \text{NO } X^2\Pi \ (v = 1) + \text{NO } X^2\Pi \ (v = 0)$$

A large excess of inert gas prevents any significant temperature rise, and pressure broadens the γ bands to provide adequate light absorption. Employing only a single optical pass, the NO partial pressure should preferably be in the range 2–5 mm if

$$\text{NO } X^2\Pi \ (v = 1)$$

is to be observed by absorption of the γ (0, 1) band at 2,370 Å. Since the self-quenching half-pressure is $\sim 0 \cdot 9$ mm, [37] the mechanism of vibrational excitation is as given above, and is not due to fluorescence of NO $A^2\Sigma^+$. It is not understood why the second reaction causes excitation predominantly to $v'' = 1$. Photometry of plates similar to Plate 1 permits the determination of the relaxation rates due to NO collisions, and due to added gases. These experiments are referred to again in the section on V–V transfer.

Nikitin [14] has suggested that the abnormally fast rate of vibrational relaxation of NO is due to splitting between the singlet and triplet states of the $(NO)_2$ complex, which was discussed in section 2.5 and illustrated in Fig. 2 in terms of converging potential curves. Wray [35] has shown that the experimental data at low temperature are interpreted by Nikitin's theory, and at high temperature by the S.S.H. equations.

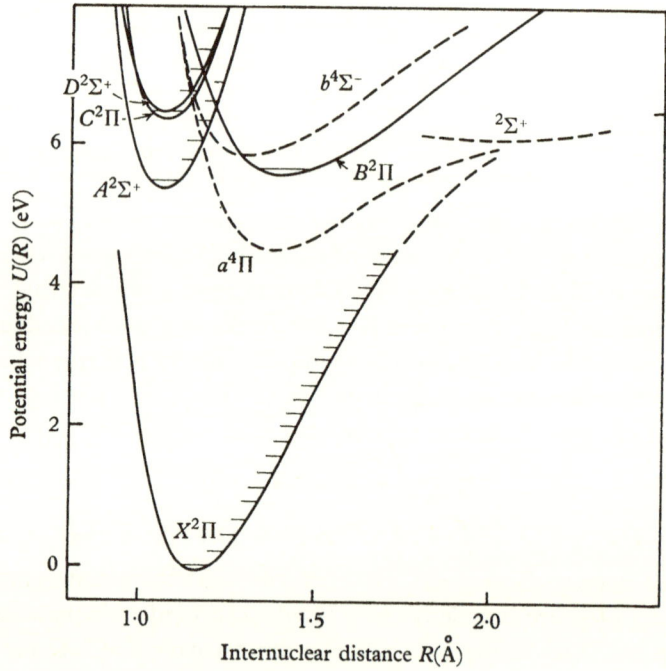

Fig. 3. The potential diagram of nitric oxide.

4.4. *A priori*, we might have anticipated that polyatomic molecules would show several vibrational relaxation times. However, the vast majority show only single vibrational relaxation; dispersion of the sound velocity is characteristic of the entire vibrational specific heat. Lambert and Salter [38] showed that for a given class of compound log Z is proportional to the lowest vibrational frequency ($\nu_{min.}$) of the molecule. They recognized three classes, molecules without H atoms, molecules with one H atom and molecules with more than one H atom. Deuterides behave similarly to hydrides. Their diagram is reproduced in Fig. 4. The increase in log Z with $\nu_{min.}$ is predicted by equation (2, ix). We

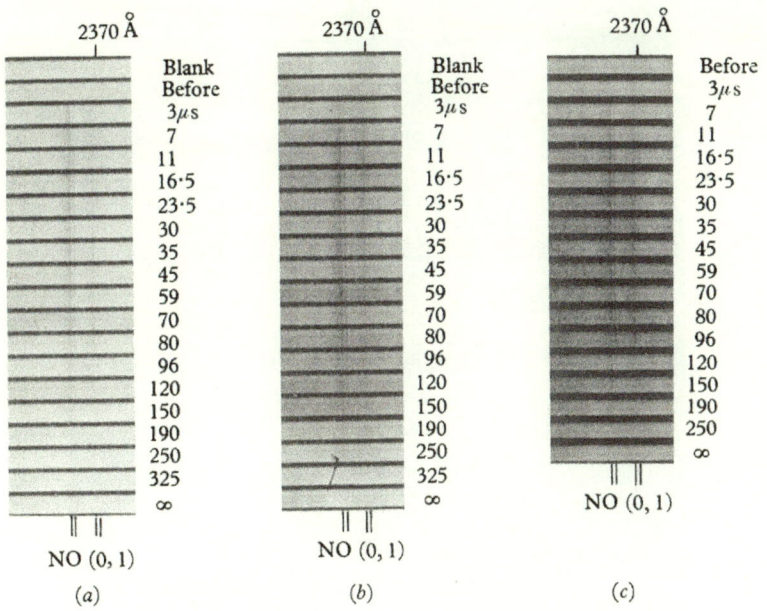

1 Formation and decay of NO²Π ($\nu = 1$) in N₂, Kr and He. 1600 J flash energy. (a) 2·2 mm NO and 430 mm N₂. (b) 2·2 mm NO and 430 mm Kr. (c) 2·25 mm NO and 540 mm He.

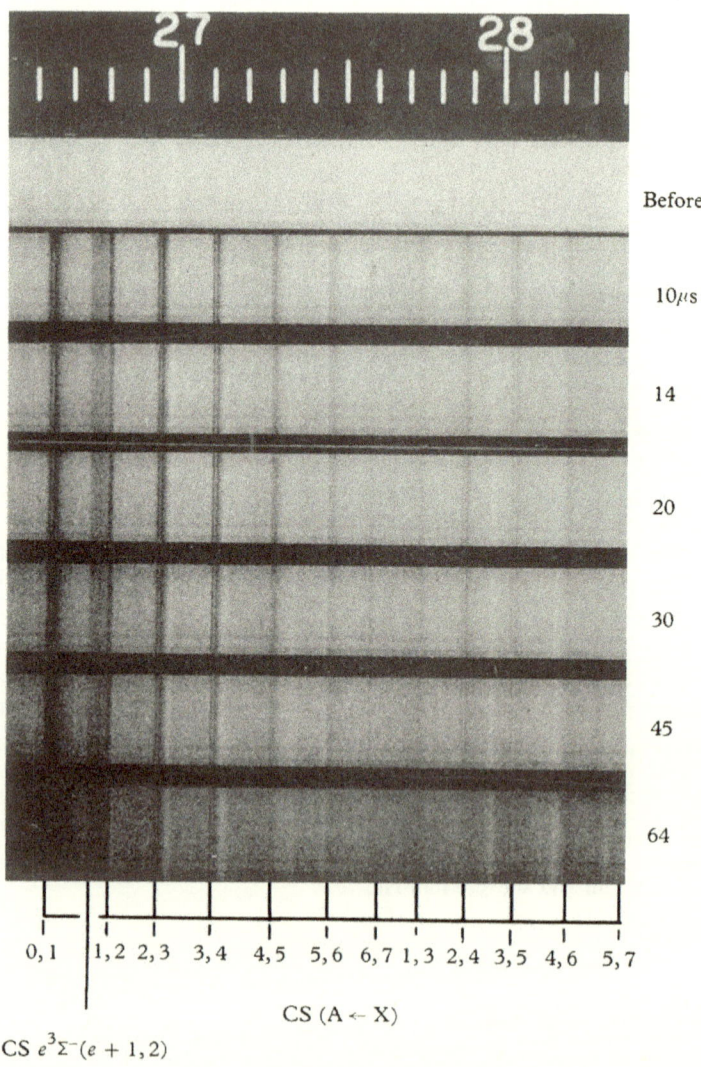

2   Enlargement of the (0, 1) and (0, 2) sequences of CS (A ← X). 0·5 mm CS₂, 50 mm O₂ with 450 mm N₂. 1·5 m path length; 2000 J flash energy.

shall see later that the slope of the non-hydride plot is approximately the same for other types of energy transfer, when log Z is plotted against the energy to be converted to translation.

Lambert and Salter interpreted their empirical correlation in the following way. In most polyatomic molecules, the energy levels are such that the energy corresponding to the lowest vibrational

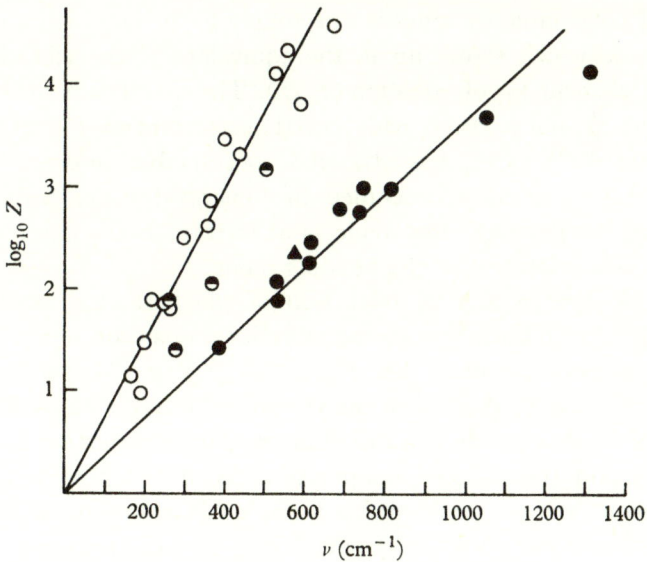

Fig. 4. The Lambert–Salter diagram for vibration translation energy transfer (reproduced by permission of Dr D. J. Lambert). O, No H atoms. ◓, one H atom; ●, two or more H atoms; ▲, two or more D atoms.

frequency is much larger than the energy difference between the lowest and the next higher frequency. Since the rate of energy transfer is controlled by the overlap of the translational wave functions which falls off exponentially as the change in translational energy increases, the rate-determining step for vibrational relaxation is excitation to (or de-excitation from) the first vibrational level. From this level, the energy spreads rapidly on collision. The system therefore behaves as a set of coupled oscillators, and exhibits a single dispersion region.

The vibration levels of a few simple polyatomic-molecules are such that the energy difference between the lowest and the next higher frequencies is comparable with or greater than the lowest frequency. These molecules may show two dispersion regions; examples are SO_2, C_2H_6 and CH_2Cl_2. [39-41]

The S.S.H. theory has been extended by Tanczos[42] to polyatomic molecules, and Stretton[43] has recently shown that numerical calculations for hydrides are in good agreement with experiment. According to Stretton, the comparatively fast relaxation of polyatomic hydrides is due simply to the large vibrational amplitude which shows up in the equivalent of the vibrational matrix element V_{01} of equation (2, xii). The calculation for nonhydrides has not yet been made. Stretton recommends $l = $ 0·18 Å.

Relaxation of CO_2 has attracted considerable interest. [44-46] Witteman[47] recently concluded that energy transfer from the bending to the stretching mode is at least 10 times faster than direct V–T relaxation of the bending mode.

(5) The probability of relaxation of the harmonic oscillator should increase linearly with the vibrational quantum number. It may easily be verified that $(V_{n, n-1}/V_{1,0})^2 = n$. Montroll and Shuler[48] have shown that if the vibrational levels of an oscillator are initially in a Boltzmann distribution, during relaxation due to a change in translational temperature the distribution remains Boltzmann, and the oscillator has a single relaxation time which is identical with that obtained by considering excitation only to $v = 1$.

The selection rules and dependence on v have not yet been adequately explored experimentally. Hooker and Millikan[49] observed the variation with time of the fundamental ($v = 1 \to 0$) and overtone ($v = 2 \to 0$) emissions and showed that the former increased linearly with time, whereas the latter showed an induction period. Decius[50] has shown that on the multistate harmonic oscillator model of Montroll and Shuler with the selection rule $|\Delta v| = 1$, $\log [1 - (I/I_\infty)]$ against time should be linear for the fundamental with a slope equal to the reciprocal relaxation time, whereas for the overtone $\log [1 - (I/I_\infty)^{\frac{1}{2}}]$ should be linear with the same slope. I/I_∞ is the ratio of the intensity of emission to the equilibrium intensity. Thus Hooker and Millikan obtained relaxation times of 172 μs and 190 μs, for $v = 1$ and $v = 2$ respectively, which may be evidence for excitation of $v = 2$ by the step

$$CO\ (v = 1) + CO\ (v = 0) \to CO\ (v = 2) + CO\ (v = 0)$$

However, the relaxation time for the V–V exchange

$$2CO\ (v = 1) \rightarrow CO\ (v = 2) + CO\ (v = 0)$$

is very short at the pressures that they employed, and would appear to 'Boltzmannize' the vibrational distribution irrespective of the detail of the relaxation mechanism.

Recently Chow and Greene [18] have attempted to measure the rate of population of the individual vibrational levels of shock wave heated HI vapour by the change in far ultraviolet absorption. The results appeared to show that the rates of excitation are roughly equal. Z for HI/HI collisions and HI/Ar collisions is several powers of 10 smaller than the S.S.H. prediction. H_2 also relaxes abnormally rapidly. [51] The discrepancy may be partly due to distortion of the oscillator which is not considered on the simple S.S.H. model.

Flash photolysis offers many opportunities of investigating highly vibrationally excited species, though few quantitative measurements have yet been made. Three mechanisms are available for vibrational excitation. The first is electronic excitation and collisional deactivation, as exemplified by the nitric oxide experiments. Secondly, vibrationally hot species may be produced by exothermic reactions, [52–54] e.g.

$$NO_2 + h\nu \rightarrow NO + O, \qquad O + NO_2 \rightarrow O_2^\dagger + NO$$
$$O_3 + h\nu \rightarrow O_2 + O, \qquad O + O_3 \rightarrow 2O_2^\dagger$$

Accurate vibrational distributions have recently been measured for the ozone photolysis, [55] and relaxation was shown to be consistent with single quantum vibrational transitions. However, the ozone photolysis produces $O_2^\dagger$ with 20 or more vibrational quanta, and the general form of the decay is rather insensitive to the magnitude of Δv. The third mechanism is excitation resulting from direct photochemical fragmentation, e.g.

$$NOCl \rightarrow NO^\dagger + Cl \qquad (v \leqslant 10)\ [56]$$
$$CS_2 \rightarrow CS^\dagger + S \qquad (v \leqslant 7)\ [57, 58]$$
$$(CN)_2 \rightarrow 2CN^\dagger \qquad (v \leqslant 7)\ [59]$$
$$CSe_2 \rightarrow CSe^\dagger + Se \qquad (v \leqslant 3)\ [60]$$
$$SO_3 \rightarrow SO^\dagger + O_2 \qquad (v \leqslant 3)\ [138]$$

An attempt at a quantitative study of the CS system indicated that the decay rates are not consistent with the optical rules. The atomic sulphur appears to catalyse relaxation of CS, because its removal by addition of oxygen,

$$S + O_2 \rightarrow SO + O$$

slows down relaxation. Plate 2 shows the formation of vibrationally excited CS in the presence of O_2, and Fig. 5 shows the decay of the

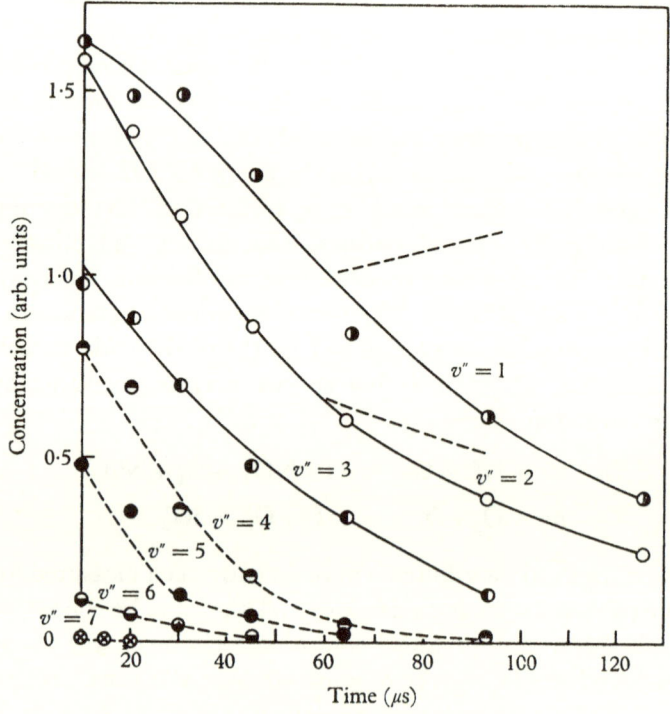

Fig. 5. Decay of vibrationally excited CS. $0\cdot5$ mm CS_2, 50 mm O_2, with 450 mm N_2.

various vibrational states following the flash, which has effectively terminated at 60 μs. The dotted lines are predicted from the optical rules, if $v = 3$ decays at $3k$, $v = 2$ at $2k$ and $v = 1$ at k, where k is determined from the $v = 3$ decay after 60 μs. The cause of the abnormal behaviour is not known, but may possibly be due to collisions of CS with atomic oxygen. Some similar effects can be observed in the CSe_2 photolysis; the relaxation rate of CSe can be *slowed down* by increasing the total pressure, due to an increase in the rate of three-body removal of atomic selenium.

It is reasonably well established from these experiments that atomic S and Se will both induce vibrational relaxation. This may simply be due to chemical affinity, or to V–E–T energy transfer via the low-lying electronic states of the isolated atoms, or to the splitting resulting from loss of degeneracy during the collision. A formal treatment of vibrational relaxation leading to chemical decomposition and based on the optical rules, is yet to find justification by experiment.

5. Vibration–vibration energy transfer

5.1. It has long been recognized that small quantities of impurities can catalyse V–T energy transfer. Some well-known examples, which have been discussed by Herzfeld and Litovitz, are the effect of H_2O on the relaxation of CO_2, the H_2O/O_2 system, and the Cl_2/CO system. Some of these effects are undoubtedly due to hetero-molecular exchange of vibrational energy, which is very much faster than V–T relaxation provided the difference of energy between the vibrational quanta is small compared to the total energy of the vibrational quantum. Thus Tuesday and Boudart [61] have shown that Z for exchange of vibrational quanta between O_2 and H_2O is approximately 2,000 (probably much too high). The vibrational frequencies are respectively 1,556 and 1,593 cm^{-1}. D_2O with the bending frequency of 1,178 cm^{-1} has no marked effect. In other cases, the impurity effect may be due to chemical affinity and complex formation, though no definite example of this is known.

There is very little direct evidence that the acceptor molecule is vibrationally excited in the processes described in this section and interpreted as V–V transfer. In the flash photolysis experiments described below, [62] relaxation of NO by CO is accompanied by vibrational excitation of the CO which can be observed in the vacuum ultraviolet. Unfortunately there are several other processes which may cause excitation. However, there has recently been an interesting development in laser applications, which proves that vibrational exchange occurs. In 1962, Morgan and Schiff [63] reported that N_2 from an electric discharge is highly vibrationally excited, and that relaxation is catalysed by addition of N_2O and CO_2. It was suggested [64] that the energy is transferred to N_2O into ν_3 (2,223·5 cm^{-1}), which then has the opportunity of fast stepwise relaxation via the intermediate vibrational levels. Legay-Sommaire, Henry and Legay [65] recently have been able to stimulate light emission from ν_3 to intermediate levels in N_2O and CO_2, added to an electric discharge in nitrogen, providing an efficient laser system.

5.2. The theory of V–V transfer is similar to that given in section 2, though the square of the off-diagonal vibrational matrix elements of both species appears in the expression for the transition probability. Consequently for harmonic oscillators in the general exchange process

$$AB\,(v = m) + CD\,(v = n) \to AB\,(v = m-1) + CD\,(v = n+1)$$

$P_{n,\,n+1}^{m,\,m-1} = m(n+1)\,P_{0,\,1}^{1,\,0}$ and $\Delta v = \pm 1$ for both species. Z contains the steric factor 3^2, compared to 3 for V–T transfer, because both species require favourable orientation.

For identical molecules (exact-resonance transfer), the theory reduces to the simple form

$$Z_{n,\,n+1}^{m,\,m-1} = 2 \times 10^{-3}\,(M_B + M_C)\left(\frac{M_B M_C}{M_B^2 + M_C^2}\right)^2 \frac{(\nu r_0)^2}{T}\,[m(n+1)]\,e^{-\epsilon/kT}$$

where M_B and M_C are the atomic masses (relative to H), ν is the frequency in cm^{-1} and r_0 (Å) and ϵ are the Lennard-Jones parameters. Thus for exchange between N_2 ($v = 0$) and N_2 ($v = 1$) at 300 °K, $Z = 2,000$. V–T transfer requires 10^{10} collisions. An approximate rate for off-resonance exchange can be obtained by multiplying the above equation by an exponential term to take account of the missmatch of the overlap of the translational wave functions, such as given in equation (2, ix) integrated through the velocity distribution. The mass function favours very fast vibrational exchange between hydrides, and this term is of course due to the large amplitude of vibration being effective for both molecules. The inverted anharmonic effect discussed in section 2.5 may not be quite as important as for V–T relaxation, because V–V transfer occurs at comparatively low velocities.

5.3. The flash photolysis experiments with NO in various pressures of Ar and N_2, described in section (4), show quite clearly that self relaxation is extremely fast and is largely responsible for vibrational relaxation. However, by varying the N_2 pressure, it was possible to show that relaxation of NO by collision with N_2 is a significant though minor effect, and the probability of relaxation was roughly estimated. A more marked effect results from addition of CO (Fig. 6, curve (b)), because the CO vibrational

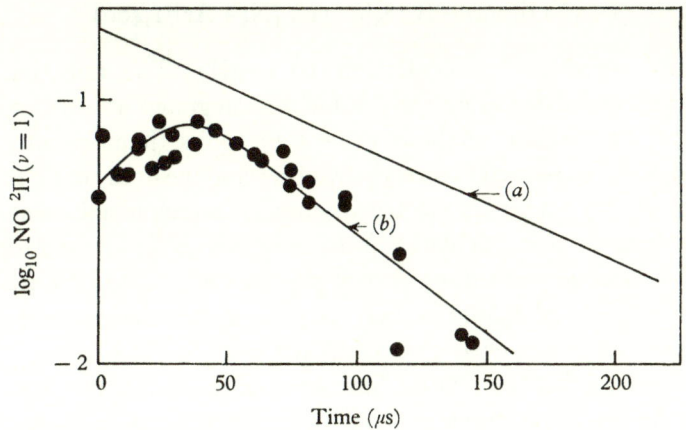

Fig. 6. Effect of carbon monoxide on the decay of vibrationally excited nitric oxide. (a) 2 mm NO with 600 mm N_2. (b) 2 mm NO, 40 mm CO, with 560 mm N_2. Flash energy 1600 J.

frequency is close to that of NO. The decay rate in the absence of CO is indicated by the full line (a). The two results were consistent with theoretical expectation and are listed in Table 3 below.

In an attempt to obtain the rate of V–V transfer for two diatomic molecules at exact resonance, exchange of vibrational quanta between nitric oxide in the $A^2\Sigma^+$ state (Fig. 3) and N_2 was investigated, since the two species have almost identical vibrational frequencies. [66] Nitric oxide was excited to $A^2\Sigma^+$ ($v = 0, 1, 2$ and 3) with light from a high pressure xenon arc, which provides continuous ultraviolet radiation down to 1,800 Å; the technique was to expose a photographic plate to fluorescent light normal to the exciting light, and to measure relative fluorescent intensities by means of plate photometry. To avoid complications due to variations in line width due to changes in pressure broadening, the total pressure was kept constant with added argon. The lifetime of the $A^2\Sigma^+$ state is shorter than the radiative lifetime because of radiationless transitions, and self quenching is especially marked. By investigating quenching of the fluorescence with added CO_2 and constructing a set of Stern–Volmer diagrams, the true quenching half pressures were determined by extrapolation to zero pressures. The lifetime τ is related to the rate of spontaneous radiation k_f by the equation,

$$\tau k_f = (1 + [NO]/0{\cdot}91 + [CO_2]/0{\cdot}31 + [N_2 + Ar]/1{,}400)^{-1}$$

where $[X]$ represents concentration in mmHg. The radiative lifetime was calculated from the f value, or integrated absorption coefficient, by standard methods to be $2{\cdot}2 \times 10^{-7}$ s. Jeunehomme and Duncan [67] measured k_f directly by observing the decay of light emission following a pulsed electric discharge, and found a lifetime twice as long as that calculated from the f value. It is hard to think of serious complications in their experiments, so our energy transfer collision probabilities may require appropriate scaling. Later reports, [67a] however, give a lifetime close to $2{\cdot}2 \times 10^{-7}$ s.

Plate 3 shows that the relative intensities of the four γ progressions in Ar are quite different from those with N_2 as inert gas. However, the enhancement of the γ ($v' = 0$) progression is due not only to vibrational exchange with N_2, but also because of excitation and deactivation of the $C^2\Pi$ ($v = 0$) state to the $A^2\Sigma^+$ state. If the γ bands are isolated by means of filters placed between the xenon arc and the fluorescence cell, the various processes can be fully analysed. Figure 7 shows the resulting Stern–Volmer plots; the probability of vibrational exchange increases with increasing vibrational quantum number (Table 4) and at least 85% of transitions occur with $\Delta v = \pm 1$.

The CO results of Table 3 were obtained by Millikan [32] with the infra-red fluorescence method. The results are presented graphically in Fig. 8. The first two processes were measured in the endothermic direction and the Boltzmann factor has been taken out so that the experimental Z given in Fig. 8 corresponds either to the reverse process, or to the process as measured, provided the collision has sufficient kinetic energy to make up the discrepancy. The non-hydride systems form an approximately linear set

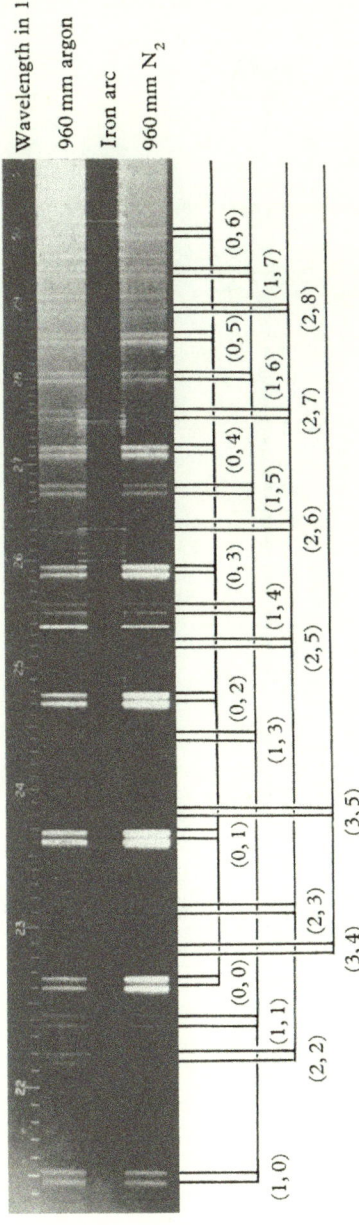

Wavelength in 100 Å

960 mm argon

Iron arc

960 mm N₂

(1,0)

(2,2)
(1,1)
(0,0)

(3,4)
(2,3)
(0,1)

(3,5)
(1,3)
(0,2)

(2,5)
(1,4)
(0,3)

(2,6)
(1,5)
(0,4)

(2,7)
(1,6)
(0,5)

(2,8)
(1,7)
(0,6)

3 Fluorescence of the NO γ-bands in argon and nitrogen. Nitric oxide pressure 1·0 mm.

2967 Å

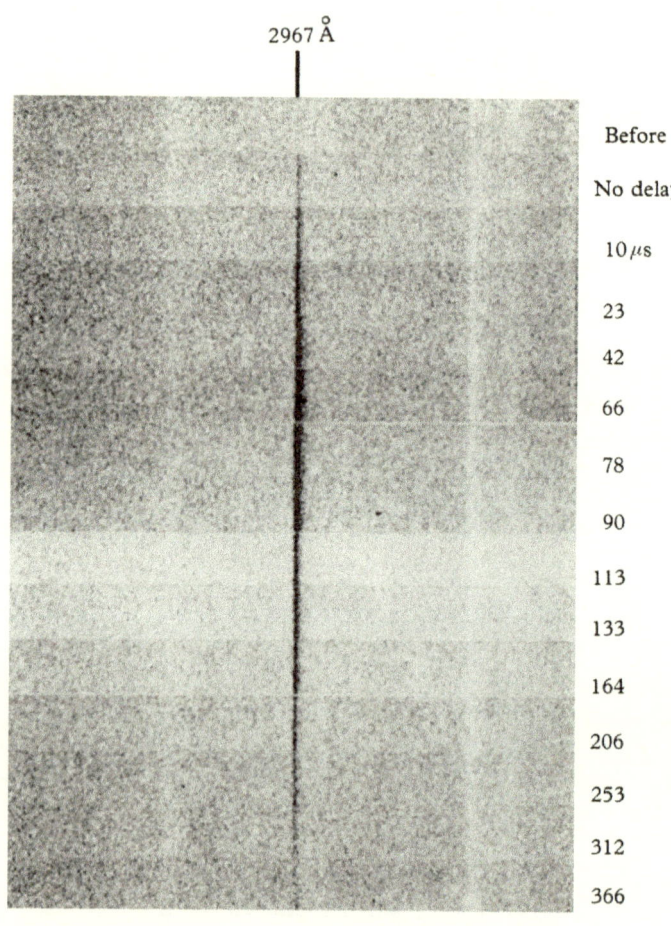

Before

No delay

10 μs

23

42

66

78

90

113

133

164

206

253

312

366

4   Production of Hg (6^3P_0) in flashed mercury/nitrogen mixtures.

(points 7, 8, 9, 10 on curve B) and give a rough illustration of how the cross-section for vibrational exchange falls off as the energy discrepancy increases. Furthermore, the slope of the plot is

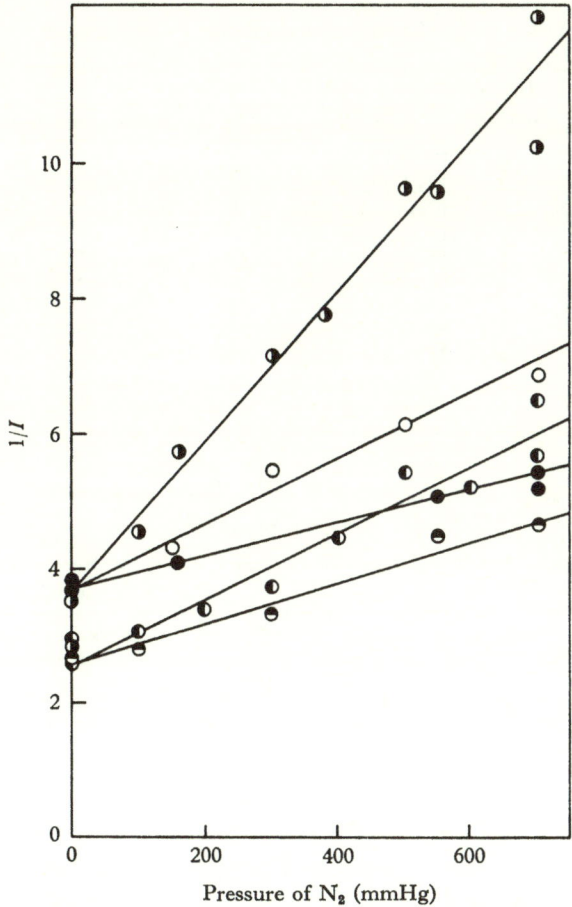

Fig. 7. The reciprocal intensity of the γ-bands as a function of the partial pressure of N_2 in (NO and N_2 + argon mixtures), total pressure 700 mmHg.

- ◐ (3, 4) band ⎫
- ○ (2, 2) band ⎬ 2·0 mmHg NO
- ● (1, 1) band ⎭

- ◑ (2, 2) band ⎫ 1·0 mmHg NO
- ◓ (1, 1) band ⎭

approximately the same as that found by Lambert and Salter (Fig. 4) for vibration translation relaxation of non-hydrides. This was first pointed out by Lambert.[69] The overlap of the translational wave functions, equation (2, ix), dominates the variation of cross-section with change of translational energy (the translational term is in fact identical for the two types of energy transfer).

TABLE 3. *Collision frequencies for V–V transfer*

System	Fig. 11 number	Method	$\Delta\nu$ (cm^{-1})	P_{1-0}	References
NO $X^2\Pi$/N$_2$	9	Flash photolysis	454	4×10^{-7}	[62, 64]
NO $X^2\Pi$/CO	8	Flash photolysis	267	0.25×10^{-4}	[62, 64]
NO $X^2\Pi$/H$_2$O	2	Flash photolysis	281	70×10^{-4}	[62, 64]
NO $A^2\Sigma^+$/N$_2$	7	Fluorescence	12	1.3×10^{-3}	[66, 64]
CO $X^1\Sigma^+$/O$_2$	10	Fluorescence	587	2.2×10^{-7}	[32]
NO $X^2\Pi$/CH$_4$	5	Microwave-pulse	350	9.1×10^{-4}	[68]
CO $X^1\Sigma^+$/CH$_4$	6	Fluorescence	617	3×10^{-5}	[32]
NO $X^2\Pi$/D$_2$O	4	Microwave-pulse	696	1.0×10^{-3}	[68]
NO $X^2\Pi$/H$_2$O	2	Microwave-pulse	281	6.2×10^{-3}	[68]
NO $X^2\Pi$/D$_2$S	1	Microwave-pulse	-15	10×10^{-3}	[68]
NO $X^2\Pi$/H$_2$S	3	Microwave-pulse	586	3.2×10^{-3}	[68]

TABLE 4. *Variation of V–V collision frequencies with quantum number*

$Z_{0,1}^{1,0}$	$Z_{0,1}^{2,1}$	$Z_{0,1}^{3,2}$	$Z_{0,1}^{1,0}$ (calc)
790	440	200	2,000

Curve A of Fig. 8 was calculated from equations (65–9) and (65–14) of Herzfeld and Litovitz for non-resonant exchange for diatomic molecules, molecular weight 30, with the frequency of one oscillator fixed at 1,900 cm^{-1}. It is necessary to interpolate to the calculated value at exact resonance. The agreement with the experimental line B is satisfactory. The more precise theoretical treatment of Rapp and Englander-Golden [70] also is in satisfactory agreement with experiment.

Relaxation of NO $X^2\Pi$ ($v = 1$) by the triatomic hydrides shows another uniform correlation with change in internal energy, indicated by curve D. However, the rates are too fast to be consistent with the standard S.S.H.–Tanczos equations, assuming a 'normal' interaction potential. Corresponding to the points 1–4, are calculated theoretical probabilities given by α, β, γ and δ. The cause of the fast relaxation by the triatomic hydrides is not definitely known, but may be due to hydrogen bonding in the collision complex. The CH_4 results, however, are in excellent agreement with the theoretical probabilities, indicated by curve C.

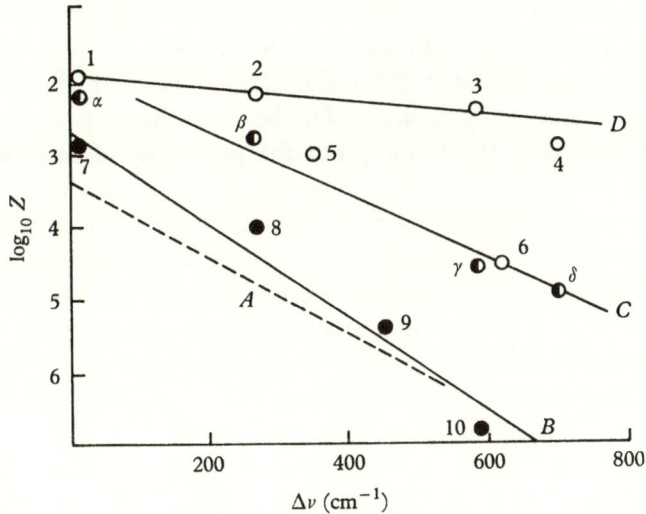

Fig. 8. Diagram for intermolecular exchange of vibrational energy. A, Two non-hydride diatomic molecules (theory); B, two non-hydride diatomic molecules (experiment); C, one diatomic molecule and one polyatomic hydride (theory); D, one diatomic molecule and one triatomic hydride (experiment).

5.4. Lambert, Edwards, Pemberton, Stretton and Parks-Smith [71,72] have measured ultrasonic dispersion in mixtures of pairs of polyatomic gases which have near-resonant vibrational levels. Some mixtures showed a non-linear dependence of reciprocal relaxation time on molar compositions, and the results showed clearly that in a mixture A and B in which B has the lowest vibrational frequency, A is excited from B by intermolecular vibrational exchange. Energy level diagrams were considered for various mixtures, and it was concluded that exchange of vibrational energy occurs at about one in fifty collisions. The high efficiency of exchange results from the lower vibrational frequencies, compared to the diatomic systems, and the rates showed a dependence on the square of the frequency as required by theory. Similar effects occur in shock-heated gas mixtures. For example, in hydrocarbon/O_2 mixtures, O_2 is excited by exchange from the hydrocarbon. [73]

6. Rotation–vibration energy transfer

Some recent experimental observations suggest the possibility of efficient interconversion of vibrational and rotational energy in gas collisions. Millikan and Osburg [74] have shown that *ortho*- and *para*-hydrogen differ significantly in their ability to deactivate vibrationally excited CO, the latter being four times more efficient than the former, and the possibility of rotational excitation of H_2 was discussed. However, there appears to be no obvious energy matching between rotational spacings in *para*-hydrogen and the vibrational frequency of CO. Millikan and White [75] investigated relaxation of O_2 by O_2 and by Ar, and found the former to be five-fold more effective than the latter. Although the difference could be due to transfer of some energy via rotation, it might equally well be due to some other complications; for example, loss of degeneracy as with nitric oxide.

Cottrell and Matheson [76] discovered that the vibrational relaxation time of CH_4 is less than that of CD_4, notwithstanding the higher vibrational frequencies of the former. They suggest that molecules with a small moment of inertia are able to fulfil the Landau–Teller condition for interaction with vibration, because of their high peripheral speeds of rotation. By a semi-classical calculation, a theoretical value for the ratio of the relaxation times of CH_4 and CD_4 was derived in agreement with experiment. Similar results and conclusions resulted from experiments comparing PH_3 to PD_3, SiH_4 to SiD_4 and AsH_3 to AsD_3. [77,78] (The vibrational relaxation time of NH_3 is apparently about one-sixth that of ND_3, and NH_3 relaxes abnormally rapidly compared with predictions from the Lambert–Salter plot. In this case, Cottrell and Matheson [77] put forward the interesting view that the abnormal rate is due to the occurrence of inversion during collision, which has an impulsive effect on the intermolecular repulsion. This is equivalent to a splitting of the potential surface, as in the nitric oxide model.) Stretton [43] has recently demonstrated that the relaxation time of C_2H_2 is greater than that of C_2D_2, which is the normal behaviour shown by the Lambert–Salter diagram. The acetylenes do not have an axis of low moment of inertia and the results therefore support the view that rotation does participate in these processes. Klemperer *et al.* [25,26] (section 3.4) demonstrated the absence of any specific effect of HD for vibrational relaxation of excited I_2, where the rotational spacing is almost identical to the vibrational quantum. However, in this system V–T relaxation is extremely fast and may obliterate any effects due to V–R transfer. These problems can be solved only when rotational excitation has been investigated by direct experimental techniques. In a comparison of experimental rates with the S.S.H.–Tanczos prediction, Stretton [43] suggests that the V–R process enhances the overall V–T transfer by three- to six-fold. This seems to be a reasonable assessment of the present state of the subject.

7. Electronic–vibration and electronic–translation energy transfer with $\Delta E < 1$ eV

7.1. There is presently very little known about E–V and E–T transfer, and it is convenient to lump them together here, from the experimental aspects and also in discussion. They are extremely interesting in relation to photochemistry and reaction kinetics. It is important first to consider two quite different models for E–V transfer:

(*a*) A resonance process in which the potential curves are parallel. In these cases the probability of energy transfer increases as the energy difference between the electronic change and the vibrational quantum decreases, because of the increasing overlap of the translational wave functions. The energy which is transferred in this type of process is generally < 1 eV.

(*b*) A process in which energy transfer occurs because of a crossing or pseudo-crossing of potential curves (see Fig. 2 of section 2 and related discussion). Except in collision with the inert gases, highly electronically excited species appear to show chemical affinity with collisional partners, as evidenced by rough correlations of quenching cross-section with ionization potential, indicating some charge transfer to the vacant orbital in the transition complex. [79] Chemical interaction is expected to be greater for the collision complex with the highest internal energy, and therefore we anticipate many cases in which the potential curves of the initial and final states converge or even cross. We may describe [80] this second class of E–V transfer as 'the formation and fragmentation of a transition complex to produce a low yield of vibrational energy, tending towards a random distribution of the energy among the available degrees of freedom'.

We shall attempt to fit the experimental results into this conceptual framework. Except for very simple systems, it is extremely difficult with the present state of our knowledge to predict an experimental result. It is interesting, and may be stimulating, to group the experimental results together; some of the tentative conclusions may require modification. In this section we discuss cases (*a*), and defer the examples belonging to (*b*) to section 8.

7.2. Some of the simplest of electronic energy-transfer processes involve change of orientation of electron spin with the electronic orbital angular momentum. The internal energy change is quite small, $\ll 1$ eV, compared to a full electronic transition which corresponds to several eV. Spin-orbit relaxation of Na ($3\,^2$P) occurs at every collision with inert gases, e.g.

$$\text{Na} \,(3\,^2\text{P}_{\frac{3}{2}}) + \text{Ar} \rightarrow \text{Na} \,(3\,^2\text{P}_{\frac{1}{2}}) + \text{Ar}$$

The spin-orbit coupling is very weak, corresponding to a splitting of only 17 cm^{-1} (1 eV = 8,068 cm^{-1}), and is disturbed easily even by collision with an inert gas molecule. The change in angular momentum is not restrictive. Measurements can be made by irradiating sodium vapour with one of the D lines isolated with a monochromator. The other D line occurs as fluorescence only if a foreign gas is added. [6] The ground state of NO consists of the $^2\Pi_{\frac{1}{2}}$ and $^2\Pi_{\frac{3}{2}}$ components, with a splitting of 121 cm^{-1}. It has been shown by sound absorption measurements [33] that the probability of relaxation per gas kinetic collision is ~ 0.062 at 298 °K. Compared to the sodium experiments, spin-orbit relaxation is slow for NO because of the increase in the magnitude of the energy to be converted to translation. Spin-orbit relaxation of mercury

$$\text{Hg} \,(6\,^3\text{P}_1) + \text{M} \rightarrow \text{Hg} \,(6\,^3\text{P}_0) + \text{M}$$

has been the subject of much experiment and speculation, because of its importance in mercury photosensitized reactions. The splitting is $1,767$ cm^{-1}, and relaxation by collision with inert gases at ambient temperature is too slow to measure at pressures up to 1 atm. The mercury resonance line at $2,537$ Å is absorbed by mercury vapour to produce Hg $(6\,^3P_1)$, and the quenching cross-sections can be determined by measuring the quenching of the fluorescence by added gases, though the problem is complicated because of reabsorption and imprisonment of the radiation. In the past, there has been confusion as to which molecules quench to the $(6\,^3P_0)$ state and which quench directly to the $(6\,^1S_0)$ ground state. Zemansky [6] has attempted to correlate the quenching cross-sections with the energy discrepancy between the spin-orbit splitting and the fundamental frequency of the quenching molecule and constructed a typical resonance curve with a rapidly diminishing cross-section with increasing resonance defect. Unfortunately, nearly all these molecules quench directly to the $(6\,^1S_0)$ state. The situation has been aggravated by misleading experimental results. Hg $(6\,^3P_0)$ is optically metastable because the matrix elements of the transition moment are zero if J remains zero, and the only state of lower energy is Hg $(6\,^1S_0)$. Thus Hg $(6\,^3P_0)$ is long lived $(\sim 10^{-3}$ s$)$, will diffuse to a surface at low pressures, and has the property of ejecting electrons from a metal electrode. However, with this method of detection, substantial yields of Hg $(6\,^3P_0)$ have been reported in systems where it is not produced, or is present only in very low yield. [81] The method is not specific to mercury metastables and any electronically excited molecules (and possibly free radicals and vibrationally excited species) would be suitable candidates provided they had the required energy.

7.3. Some clarification as to which molecules quench Hg $(6\,^3P_1)$ to the $(^3P_0)$ state has been obtained by flashing mercury vapour with a foreign gas, and by observing the absorption due to Hg $(6\,^3P_0)$ by kinetic spectroscopy. [82,83] The inert-gas discharge tubes, normally employed for flash photolysis, give continuous light and only a very small fraction is absorbed in mercury vapour by the resonance line at $2,537$ Å. To obtain adequate light absorption, it is necessary to broaden the mercury line with a high pressure of an inert gas. If N_2, CO, H_2O or D_2O is added to the inert gas, strong absorption occurs due to Hg $(6\,^3P_0)$. Metastable atoms have not been detected in flashed mercury vapour containing any one of the gases, NO, H_2, O_2, CO_2, N_2O, CH_4, C_2H_6, C_3H_8, C_2H_6, C_2H_4 and NH_3. The formation and decay of Hg ($6\,^3P_0$) in N_2 is shown on Plate 4, by absorption at 2,967 Å ($6\,^3D_1 \leftarrow 6\,^3P_0$).

Scheer and Fine[84] have investigated the formation of $^3P_0$ atoms, in CO and N_2, by observing the rate of ejection of electrons from a silver electrode. They concluded that for equal pressures, CO produces only 1·3 times more metastables than N_2. The interpretation of the results seems to be straightforward, and it follows that CO quenches $^3P_1$ atoms largely to the ground electronic state, because CO quenches $^3P_1$ ten times faster than N_2 does. Matland[85] recently has re-measured the cross-section for quenching Hg($6\,^3P_1$) by N_2, and found 0·42 Å² at 298 °K. It is instructive to divide this value by a Boltzmann factor corresponding to 563 cm^{-1}, since the fundamental of N_2 is 2,330 cm^{-1} and the spin-orbit splitting is 1,767 cm^{-1}. This leads to a cross-section of 6·7 Å² for quenching of $^3P_1$ by N_2 molecules which have sufficient kinetic energy to excite the N_2 fundamental. The equivalent quantity for CO is 3·5 Å² according to Scheer and Fine's results. If it is accepted that quenching causes vibrational excitation, the process

$$N_2\,(v = 1) + Hg\,(6\,^3P_0) \rightarrow N_2\,(v = 0) + Hg\,(6\,^3P_1)$$

occurs twice as rapidly as

$$CO\,(v = 1) + Hg\,(6\,^3P_0) \rightarrow CO\,(v = 0) + Hg\,(6\,^3P_1)$$

Matland's data strongly suggest that vibrational excitation occurs, because the activation energy for quenching does correspond to 563 cm^{-1}. Thus the back reaction, as written above, would have negligible activation energy. According to this interpretation, the process with the largest energy discrepancy is the faster. On Fig. 9 are plotted energy discrepancy against quenching cross-section (Boltzmann factor, or activation factor, taken out), for the four molecules which quench to the $^3P_0$ state. There is no systematic fall in quenching cross-section with increasing energy discrepancy, as proposed by Zemansky.[6]

7.4. The absence of any systematic correlation from the mercury results is due to attractive interactions in the transition complex, referred to above. The magnitude of the interaction and the form of the potential curves will be peculiar to each deactivator, and we have seen, even for V–T relaxation, the extreme sensitivity to the form of the interaction potential. In the hope of discovering systems which do show systematic behaviour, we have recently commenced investigation of low-lying electronic states, which differ from the ground state only in respect of orientation of spin and orbital angular momentum. Atomic selenium (Se $4\,^3P$) was selected for the first experiments, [86] the splitting of the ground state being: $^3P_0$, $\Delta E = 2,534$ cm^{-1}: $^3P_1$, $\Delta E = 1,990$ cm^{-1}: $^3P_2$, $\Delta E = 0$. In

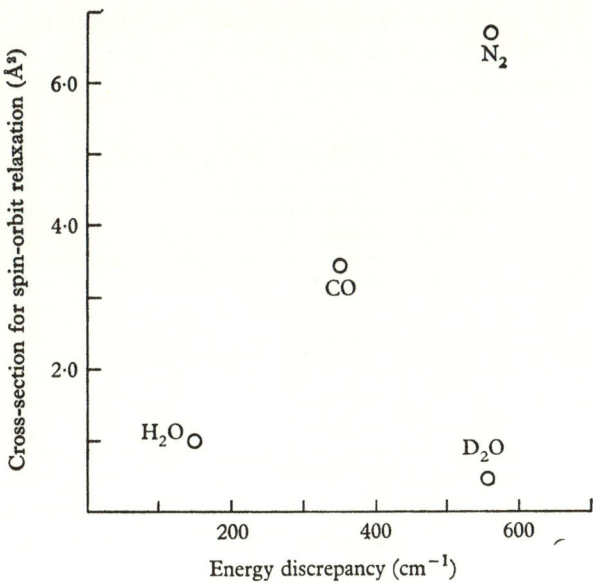

Fig. 9. The effect of energy discrepancy on the relaxation rate. The ordinate is the cross-section for the transition $^3P_1 \rightarrow {}^3P_0$ for collisions which have sufficient energy to excite vibration in the quenching molecule.

principle such a set of energy levels offers opportunities to investigate E–T energy transfer ($^3P_0 \rightarrow {}^3P_1$) and E–V energy transfer ($^3P_0 \rightarrow {}^3P_1$ or $^3P_2$). In practice with the available equipment, it was possible to make a quantitative study only of the rate of decay of Se ($4\,^3P_0$), and consequently the fate and mechanism of relaxation of Se ($4\,^3P_1$) are not understood. Furthermore, it has not yet been established which of the two states are produced by relaxation of Se ($4\,^3P_0$).

Atomic selenium can be produced electronically hot by flash photolysis of CSe_2 which predissociates in the electronically excited state; the CSe persists for some seconds, and the system therefore provides a clean source of atoms, and a large excess of inert gas prevents any temperature change. Plate 5 shows the formation and slow decay of Se $(4\,^3P_0)$ in Ar, and Plate 6 illustrates the fast relaxation induced by added N_2 and O_2. By photometering plates of the type illustrated in Plate 5, the absolute rate constants for deactivation of Se $(4\,^3P_0)$ were measured and are listed in Table 5. The collision numbers were calculated assuming Se $(4\,^3P)$ to have a collision diameter equal to that of krypton, and employing the 6–12 potential model.

TABLE 5. *Deactivation of Se $(4\,^3P_0)$*

Deactivator	Bimolecular rate constant (cm^3 molecule^{-1} s^{-1})	Z	ΔE (cm^{-1})
Ar (1)	2·4 ($\pm$0·3) $\times$ 10^{-14}	9,400	-544
CO (2)	1·1 ($\pm$0·3) $\times$ 10^{-12}	326	-391
O$_2$ (3)	1·5 ($\pm$0·3) $\times$ 10^{-12}	155	-544
N$_2$ (4)	3·0 ($\pm$0·3) $\times$ 10^{-12}	83	-203
H$_2$	3·5 ($\pm$0·7) $\times$ 10^{-10}	1·6	-544
N$_2$O (5)	1·2 ($\pm$0·15) $\times$ 10^{-10}	2·1	$+45$
CO$_2$ (6)	1·4 ($\pm$0·1) $\times$ 10^{-10}	2·0	$+128$

In computing the listed changes in internal energy (ΔE) or energy discrepancies, it is assumed that O_2, H_2 and Ar cause E–T transfer ($^3P_0$–$^3P_1$), that CO_2 and N_2O cause E–V transfer ($^3P_0$–$^3P_1$) by excitation of ν_2, and that N_2 and CO cause E–V transfer ($^3P_0$–$^3P_2$). That is to say, ΔE corresponds to the minimum energy which has to be converted to translation (ignoring rotational excitation) and assuming only single quantum vibrational excitation.[87] A plot of ΔE against Z is shown on Fig. 10, with the H_2 result omitted. The numbering of the points from 1 to 6 correspond to column 1 of Table 5. Point 7 relates to spin-orbit relaxation of NO $X^2\Pi$, and point 8 was derived from a study of the process,

$$\text{Fe } (a^5D_3) \xrightarrow{\text{Ar}} \text{Fe } (a^5D_4),$$

by isothermal flash photolysis of iron carbonyl.[88] Thus all the data of Plate 5 relate to spin-orbit relaxation of ground electronic states.

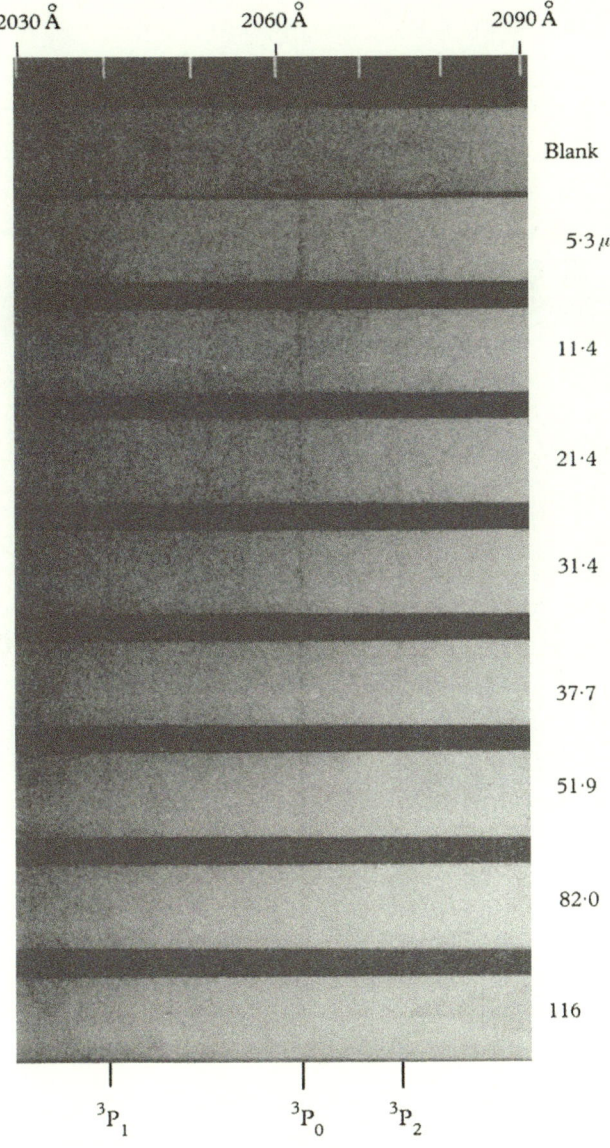

2030 Å   2060 Å   2090 Å

Blank

5·3 μ

11·4

21·4

31·4

37·7

51·9

82·0

116

$^3P_1$ $^3P_0$ $^3P_2$

5   Formation and decay of hot selenium atoms in argon.
0·05 mm CSe_2 + 50 mm Ar. 2025 J.

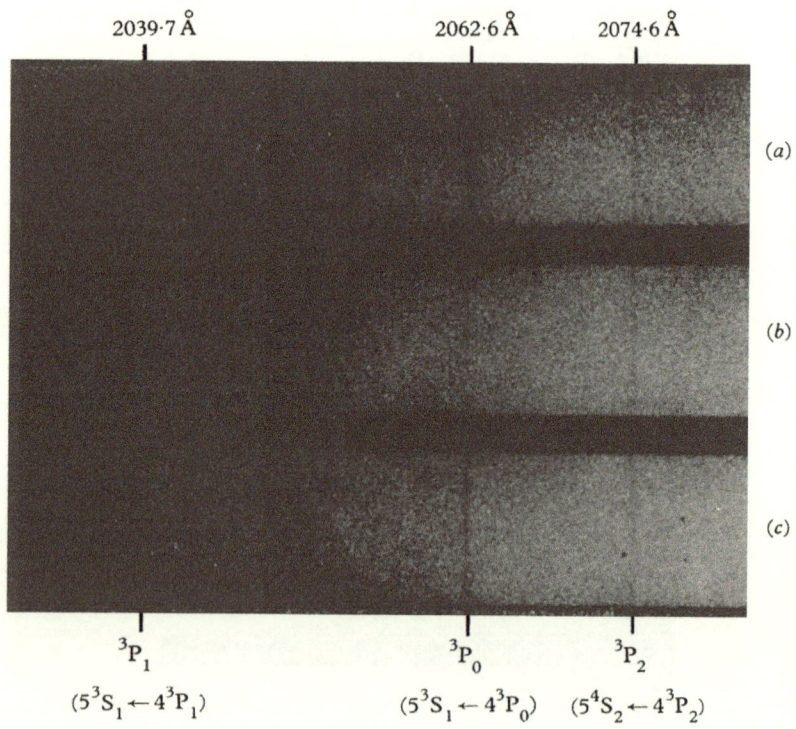

6 Formation and decay of electronically excited atomic selenium. (a) o·1 mm CSe_2 + 50 mm Ar. (b) o·1 mm CSe_2 + 10 mm N_2 + 40 mm Ar. (c) o·1 mm CSe_2 + 10 mm O_2 + 40 mm Ar.

Except for the Se/O_2 result, the date may show an approximate trend corresponding to the slope of the Lambert–Salter plot given by the broken line. Apparently $d \log Z/d\Delta E$ is roughly the same for

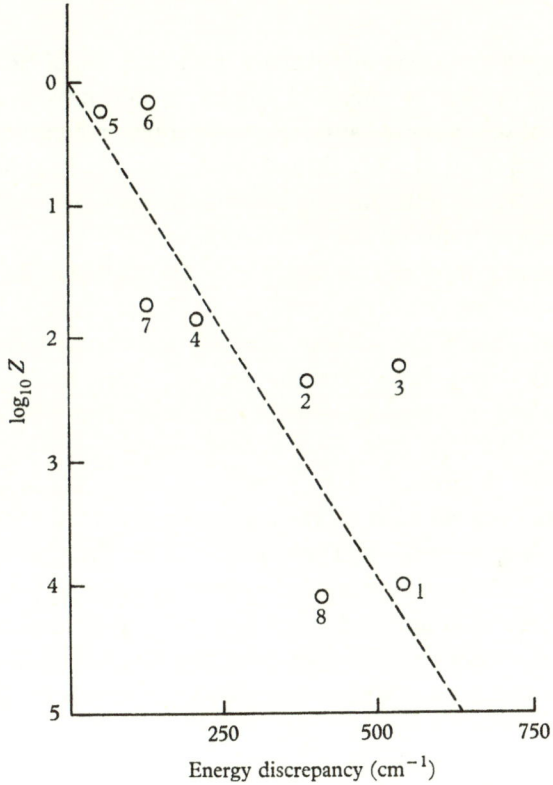

Fig. 10. Spin-orbit relaxation of ground electronic states.
- - - - -, Lambert–Salter slope.

a number of simple V–T, V–V, E–T and E–V energy transfer processes. The occurrence of an approximate law of the form

$$\log Z = A\Delta E + B$$

may arise because the expression for the overlap of the translational wave functions, equation (2, ix), is common to all energy transfer processes in which the intermolecular repulsion is approximately exponential, and in which the quantities equivalent to the diagonal matrix elements V_{nn} are equal. Variations of l, reduced mass, and

V_{nn} (e.g. due to chemical affinity or splitting) scatter the data about the mean slope. However, we must be cautious about the interpretations of the so-called E–V processes until the mechanisms are established, and until many more similar systems have been studied.

Oxygen appears to cause abnormally fast relaxation of Se $(4\,^3P_0)$. This is probably due to chemical affinity, which provides some convergence of the potential surfaces in the collision complex. The extent of splitting of the $^3P_1$ three-fold degeneracy is difficult to assess accurately, and a further complication arises because quintet, triplet and singlet collision complexes can arise. [14]

It is interesting to note that the $J = 0 \to 2$ transition in $p - H_2$ corresponds to 365 cm^{-1}, supplying all but 179 cm^{-1} of the $^3P_0$–$^3P_1$ energy separation, without affecting the change in angular momentum required due to translation away from the centre of gravity. This may evolve as a general feature of electronic energy transfer.

7.5. Several theoretical papers on spin-orbit relaxation are largely directed at a solution of the 'high energy' problem which is applicable only if the spin-orbit coupling is weak. Thus Moskowitz and Thorson [89,90] have calculated the rates of spin-orbit relaxation of the alkali metal atoms and compared them with experiment. Interaction of translation with spin-orbit coupling occurs because, in the collision complex, the angular momentum couples to and precesses round the internuclear axis. If the relative velocity is high, the orbital angular momentum is wrenched away from the electron spin and may reorientate after collision. The calculations of Moskowitz and Thorson confirm earlier work of Bates, [91] and are in satisfactory agreement with the somewhat fragmentory experimental data. Nikitin [92] has developed the theory, and considered the effects of loss of degeneracy during collision. Bichovskii and Nikitin [93] had previously emphasized the importance of loss of degeneracy in the E–V processes

$$\text{Hg}\,(6\,^3P_1) + AB\,(v = 0) \to \text{Hg}\,(6\,^3P_0) + AB\,(v = 1)$$

and thereby interpreted the random variation of cross-section with the AB vibrational frequencies. However, chemical interaction or charge transfer may be more important than loss of degeneracy, though the latter obviously is another factor which may scatter the data of Fig. 10 about the Lambert–Salter line. In the case of a substantial change in internal energy, numerical calculations have not yet been carried out.

7.6. Donovan and Husain [94–96] have recently measured the rate of spin-orbit relaxation of atomic iodine, by flash photolysis of I_2, CH_3I, and CF_3I, and kinetic absorption spectroscopy in the vacuum ultraviolet. The excitation energy is 7,603 cm^{-1}. The results are very interesting, and are listed in Table 6. It is difficult to understand these data in relation to the theoretical models discussed above.

TABLE 6. *Deactivation of* $I\,(5\,^2P_{\frac{1}{2}})$ *to the* $I\,(5\,^2P_{\frac{3}{2}})$
ground state (300 °K)†

Deactivator	Rate constant (cm³ molecule⁻¹ s⁻¹)
I_2	5×10^{-12}
H_2	9×10^{-14}
D_2	$1\cdot1 \times 10^{-13}$
HI	$1\cdot5 \times 10^{-13}$
Ethyl iodide	$3\cdot7 \times 10^{-13}$
Propane	$5\cdot6 \times 10^{-14}$
CO_2	$1\cdot3 \times 10^{-16}$
SF_6	$3\cdot1 \times 10^{-15}$
N_2	2×10^{-16}
Argon	Very small
Helium	Very small
Xenon	Very small

The very slow rate of deactivation of $I\,(5\,^2P_{\frac{1}{2}})$ by Ar and He may turn out to be consistent with Fig. 10, and the relevant discussion. Deactivation by I_2 is very fast, and corresponds to $Z \simeq 15$; the efficiency of I_2 may be comparable with the rate of chemical exchange.

$$I^* + I_2 \rightarrow I_2 + I$$

However, the relaxation rate by N_2, $Z \sim 10^6$, is too fast to be consistent with a model in which the potential surfaces are approximately parallel and with the $\Delta v = \pm 1$ restriction. The rate predicted by such a model would be $Z \sim 10^{40}$ and the disparity is

† The author is greatly indebted to Dr D. Husain and Mr R. J. Donovan for generously communicating results prior to publication.

surely too large to be accounted for on the grounds of convergence of potential surfaces at the ambient-temperature turning point, due to chemical affinity, loss of degeneracy, or change of spatial electron configuration; i.e. it is difficult to justify postulation of crossing or near crossing of potential curves for deactivation of $I (5\,{}^2P_{\frac{1}{2}})$ by N_2. (The activation energy for formation of NI is $\sim$ 200 kcal/mole.) The most reasonable reaction is

$$I (5\,{}^2P_{\frac{1}{2}}) + N_2 (v = 0) \rightarrow I (5\,{}^2P_{\frac{3}{2}}) + N_2 (v = 3 \text{ only}) \qquad [97]$$

for which $\Delta E = 697$ cm^{-1}. If the effects of anharmonicity on V_{03} and the diagonal elements are taken into account, together with an interaction potential minimum of 2 kcal mole^{-1}, theory and experiment match reasonably well. It is now of some interest to establish experimentally the extent of vibrational excitation. More amenable for study is the equivalent process with CO, where vibrational excitation may be observed directly by infra-red emission. (A parallel problem in V–V transfer,

$$AB (v = 2) + CD (v = 0) \rightarrow AB (v = 0) + CD (v = 1)$$

is virtually unexplored. We have not observed any special effect of nitrogen ($\nu_{1-0} = 2{,}330$ cm^{-1}) on vibrational relaxation of CS ($\nu_{2-0} = 2{,}331$ cm^{-1}) or CSe ($\nu_{2-0} = 2{,}033$ cm^{-1}), though the experiments were complicated because of relaxation by transient species, and were difficult to analyse.)

Relaxation of $I (5\,{}^2P_{\frac{1}{2}})$ by the hydrogens and by hydrides may be complicated because of chemical affinity. With H_2, the activation energy for formation of the IHH transition complex should be $\sim$ 10 kcal mole^{-1}. This type of reaction is presently under investigation. [97]

7.7. Yamazaki and Cvetanović [98] have shown that of the inert gases, only Xe efficiently deactivates O ($2\,{}^1D$) to O ($2\,{}^3P$). The attractive interaction potential must be strong enough to cause convergence of the initial and final potential curves, so that Xe, with the largest polarizability, is the most effective. Xe may also be providing external spin-orbit coupling, because of its high atomic weight. CO_2 also is an efficient deactivator for excited atomic oxygen. Unlike spin-orbit relaxation, the initial and final states will not correspond to parallel potential curves.

8. Electronic–vibration and electronic–translation energy transfer with $\Delta E > 1$ eV

8.1. It is of great interest, especially in photochemistry, to discover how much energy is converted to vibration, accompanying transfer of a substantial quantity of electronic energy, corresponding either to a change in the orbital angular momentum of an atom, or to a change of principle quantum number. Such highly energetic species are usually deactivated very efficiently by any polyatomic gas, and the quenching cross-sections show some approximate correlation (especially with members of homologous series), with polarizability and ionization potential. Energy transfer has a high probability because of chemical complex formation, and consequent near-crossing of potential surfaces. [80]

The quenching and excitation of Na $(3\,^2P)$ has been superficially investigated,

$$AB + \text{Na} \,(3\,^2P) \rightarrow AB + \text{Na} \,(3\,^2S)$$

Inert gases have very small cross-sections for quenching Na $(3\,^2P)$, whereas for all the polyatomic molecules which have been studied, the quenching cross-sections are of the same order of magnitude as gas-kinetic cross-sections. It is suggested later than this is not because a polyatomic molecule can accept most of the energy internally. Table 7 lists quenching cross-sections for a number of diatomic molecules, and has been modified from that given by Pringsheim. [5] In each case, for the given vibrational level the quenching is exothermic, the cross-sections being too large to correspond to vibrational levels of higher energy than the electronic energy of the sodium. Oxygen is exceptional in having electronically excited states of lower energy than Na $(3\,^2P)$. There is no relationship between quenching cross-section and the minimum energy that cannot be converted to vibration.

TABLE 7. *Cross-sections for deactivation of Na $(3\,^2P)$*

Quenching gas	N_2	NO	O_2	CO	H_2
Quenching cross-section (Å^2)	24·9	31·6	52·2	13·4	15·7
Vibrational level	7	9	11	8	4
Energy discrepancy (eV)	0·15	0·09	0·12	0·07	0·2

8.2. Recent experiments of Gaydon and co-workers [28] show conclusively that vibrationally excited CO and N_2 will excite Na $(3\,^2P)$, and conversely vibrationally excited species must be produced in the quenching process. The temperature of a flame can be measured by focusing a black-body source through it onto the slit of a spectrograph. If the light source and flame subtend the same angle to the detector, when they have the same temperature the flame does not affect the light flux falling from the black-body on to the detector. If reversal of a strong atomic resonance line is observed, for example the sodium D lines, the appropriate temperature is defined by the populations in the $(3\,^2S)$ and $(3\,^2P)$ states. The electronic temperature of the sodium may be higher or lower than the translational temperature. If the excited state is populated either by chemical reaction [99] or by absorption of radiation, [100] in addition to excitation by collision (thermal), the electronic temperature is higher than the translational temperature. Clouston, Gaydon and Glass [101] discovered that, in shock-wave heated N_2, the electronic temperature of sodium is below the translational temperature immediately behind the shock front, and from the relaxation time it was evident that the electronic temperature is related to the vibrational temperature of the N_2. The same phenomenon was later observed in CO, and also in both gases with chromium atom reversal. [28] Evidently the process

$$Na\,(3\,^2S) + N_2\,(v = n) \rightarrow Na\,(3\,^2P) + N_2\,(v = o)$$

occurs rapidly, though the shock-wave experiments have not given any information about the magnitude of n.

8.3. Karl and Polanyi [102] have described a very interesting attempt to measure directly the yield of vibrational energy in a quenching molecule. Highly vibrationally excited CO was observed by infra-red emission from a mixture of mercury vapour and CO on irradiation at 2,537 Å. Certainly either one or the other and probably both of the reactions

$$Hg\,(6\,^3P_{1\,or\,0}) + CO\,(v = o) \rightarrow CO\,(v = n) + Hg\,(6\,^1S_0)$$

produces vibrationally excited CO. The CO pressure was 0·135 mmHg, and since the radiative lifetime of the fundamental is 0·033 s and the diffusivity is about 10^4 cm^2 s^{-1} it would appear that the time resolution is determined by wall deactivation and not by spontaneous radiation. Wall removal should be diffusion controlled at 10^{-1} mmHg. The relaxation time for vibrational exchange is 10^{-3} s, which is about the same as the time between wall collision. Therefore, the initial distribution should not be destroyed entirely. The possibility of freezing the initial distribution by wall destruction at low pressure now seems to be recognized by Polanyi. [103] Recent results for the CO/Hg show that some 20 % of the electronic energy is converted to vibration. There is a fairly smooth distribution into the lower vibrational levels up to $v = 8$, with a very small yield into $v = 9$, and negligible population into $v = 10$ or higher levels. Excitation to $v = 10$ corresponds to 0·5 of the total electronic energy.

8.4. Dickens, Linnett and Sovers [104] have given a D.W.A. calculation of the cross-section for electronic quenching of an atom by a diatomic molecule. An *ultra-simplified* model was chosen in which the electronic eigenfunctions are spherically symmetrical in both electronic states. Despite the crudity of the model, the conclusions are important, and probably have some general validity. The quenching cross-sections increase with decreasing energy discrepancy but are extremely small because of the vanishingly small vibrational matrix elements for multiple quantum vibrational transitions. Therefore, 'resonance quenching' can only result in the excitation of a single vibrational quantum in the quenching molecule. Dickens *et al.* concluded that for efficient quenching there must occur either a crossing or near-crossing of potential curves. They suggest that the diatomic species take up only a single quantum of vibration. It is unlikely that this always will be true, and of course, multiple-quantum vibrational excitation has now been demonstrated by Polanyi's group. Chemical interaction generally will be required for efficient quenching, in order that the potential curves of the initial and final states may approach. Because of the change in electronic configuration, in the transition complex the separated species lose their identity and selection rules for the separated species will not be important. Excitation to a range of vibrational states occurs, depending on the bond length of the quenching molecule in the transition complex and the form of the potential surfaces. This type of model has been discussed by Laidler [105] and more recently by Polanyi. [106] Two classes of electronic-vibrational energy transfer were defined at the beginning of this section. According to the views expressed here, collisional quenching, when several eV of electronic energy is converted to other forms, belongs to the second type. Polyatomic molecules quench electronically excited atoms more efficiently than the inert gases, because the former are chemically more reactive than the latter.

8.5. At first sight it is difficult to see how the observations of Gaydon *et al.* [28] can be consistent with a quenching mechanism described by 'fragmentation of a transition complex to produce a low yield of vibrational energy, tending towards a random distribution of the energy among the available degrees of freedom'. In fact, experiment and theory are quite consistent. Consider a simple scheme whereby the quenching of a metal atom (M) produces molecules only in the zeroth and first vibrational levels. In a shock-heated diatomic gas at about 3,000 °K and 1 atm pressure, the relaxation time for electronic excitation is extremely short ($\sim 10^{-10}$ s) compared to the vibrational relaxation time (10^{-4} s). Therefore the fraction of atoms in the electronically excited state is simply the ratio of the rate of excitation to the rate of deactivation, each per unit concentration of M and M^*, respectively. Deactivation should occur at approximately every collision with a diatomic molecule (AB), irrespective of its vibrational content, at a rate $k(AB)(M^*)$. The possible steps are

$$M + AB\,(v = 0) \underset{k}{\overset{k_1}{\rightleftharpoons}} M^* + AB\,(v = 0)$$

$$M + AB\,(v = 0) \underset{k}{\overset{k_2}{\rightleftharpoons}} M^* + AB\,(v = 1)$$

$$M + AB\,(v = 1) \underset{k}{\overset{k_3}{\rightleftharpoons}} M^* + AB\,(v = 0)$$

$$M + AB\,(v = 1) \underset{k}{\overset{k_4}{\rightleftharpoons}} M^* + AB\,(v = 1)$$

Because of the need for detailed balancing,

$$k_1 = k e^{-E/kT}, \qquad k_2 = k e^{-(E+h\nu)/kT}$$

$$k_3 = k e^{-(E-h\nu)/kT}, \qquad k_4 = k e^{-E/kT}$$

if M and M^* have equal weight, E is the energy of the electronic transition, ν is the vibrational frequency, and T is the translational temperature. If we now write down the ratio of the rate of excitation of M^* to the rate of deactivation of M^* with the equilibrium condition

$$AB\,(v = 1)/AB\,(v = 0) = e^{-h\nu/kT}$$

it may be verified that $M^*/M = e^{-E/kT}$. However, suppose that the concentration of $AB\,(v = 1)$ is zero due to lack of vibrational excitation. The simple scheme then reduces to

$$M + AB\,(v = 0) \underset{k}{\overset{k_1}{\rightleftharpoons}} M^* + AB\,(v = 0)$$

$$M + AB\,(v = 0) \overset{k_2}{\rightharpoonup} M^* + AB\,(v = 1)$$

$$M + AB\,(v = 1) \underset{k}{\leftharpoonup} M^* + AB\,(v = 0)$$

$$\frac{\text{Excitation}}{\text{De-excitation}} = \frac{AB(v = 0)\,[k_1 + k_2]\,M}{2kM^*AB(v = 0)} = 1$$

$$\frac{M^*}{M} = \frac{ke^{-E/kT} + ke^{-(E+h\nu)kT}}{2k} = \frac{(1 + e^{-h\nu/kT})\,e^{-E/kT}}{2}$$

Therefore the electronic temperature of a metal atom will be intermediate between the translational temperature and the vibrational temperature, if the quenching reaction gives a finite yield into any one of the vibrationally excited levels of a diatomic gas. From the above model it can be shown that

$$\frac{\left(\dfrac{M^*}{M}\right)_{t=\infty} - \left(\dfrac{M^*}{M}\right)_t}{\left(\dfrac{AB(v = 1)}{AB(v = 0)}\right)_{t=\infty} - \left(\dfrac{AB(v = 1)}{AB(v = 0)}\right)_t} \simeq \frac{e^{-E/kT}\,(e^{h\nu/kT} - 1)}{2}$$

where $t = $ time. Therefore the apparent electronic relaxation time and the vibrational relaxation time will be very nearly identical, even though the vibrational and electronic temperatures are quite different.

A better approximation can be achieved when the yields into the various vibrational levels are known, so that the k's for the deactivation reactions can be suitably weighted. Some recent experimental data of Hurle [107] are too scattered to fit to the theory though the 'electronic temperature' at the shock front was shown to be intermediate between the vibrational and translational temperatures.

9. Electronic–electronic energy transfer

9.1. *Exchange between atoms.* Although it is established that E–E transfer may occur between atoms, the results are largely qualitative and are only superficially understood. The formal theory of the exchange of electronic energy between atoms has been reviewed by Massey and Burhop [108] and Mott and Massey. [109] If the change in internal energy (ΔE) is very small and the transitions are optically allowed for both atoms, long-range resonance interaction should result in a large cross-section for excitation transfer, which may approach 5×10^{-14} cm^2. The theory predicts a sharp decrease of cross-section with increasing energy discrepancy. If the transitions of each atom are associated with a quadrupole, the cross-section for $\Delta E = 0$ falls to 10^{-15} cm^2, which is approximately the gas kinetic cross-section. Qualitatively, cross-sections for electronic energy transfer should be less than gas kinetic if $l|\Delta E|/hv \gg 1$, where v is the relative speed, l is the length characteristic of the interaction potential, and ΔE is the change in internal energy. It may be supposed that this is the ratio of the duration of the collision (l/v) to a time interval characteristic of the electronic motion ($h/|\Delta E|$). At ambient temperature, the mean molecular speed is about 5×10^4 cm/s, and with $l = 10^{-8}$ cm, the above ratio is equal to unity when $\Delta E = 133$ cm^{-1}. Therefore, the probability of exchange of electronic energy per gas-kinetic collision should fall below unity if $\Delta E > 200$ cm^{-1}. The same conclusion appears in equation (2, ix). For the few experimental systems which have been investigated quantitatively, the cross-sections are all smaller than gas kinetic, and therefore the theory has not been tested. If the exchange of electronic excitation occurs because of a pseudo-crossing of potential curves, a substantial fraction of the energy can be converted to translation, though the probability per collision of course must be below unity. A particularly favourable case exists when the crossing occurs at the turning point of the potential curve of the state with the higher internal energy. [110] However, as yet no process has been investigated for which the potential curves are known. Given a choice, if the opportunity exists for an excitation transfer with $\Delta E \sim 0$, this will be the most probable process because of 'translational overlap'. Nearly all experimental results are consistent with this rule.

9.2. Pringsheim[5] has reviewed some of the early experiments. Beutler and Josephy[111] examined the fluorescence of sodium sensitized by mercury, and concluded that the process with the largest cross-section is

$$Hg\,(6^3P_1) + Na\,(3\,^2S_{\frac{1}{2}}) \rightarrow Hg\,(6^1S_0) + Na\,(9\,^2S_{\frac{1}{2}})$$

for which $\Delta E = +162$ cm^{-1}. The absolute cross-section is not known. The intensities of the sharp and diffuse series of lines, measured by plate photometry, indicated that

$$8S\,(\Delta E = -444\;\text{cm}^{-1}), \quad 7D\,(\Delta E = -212\;\text{cm}^{-1})$$

$$\text{and} \quad 8D\,(\Delta E = +317\;\text{cm}^{-1})$$

are also populated directly by excitation transfer from the mercury, the intensities of the various lines in emission being about one-fifth that of the $9\,^2S \rightarrow 3\,^2P$ line. (The intensity of a single line is not, of course, a direct measure of the rate of population of a state. One needs to know the rate of radiation at this wavelength, and the rates of all other radiative and radiationless transitions. [112]) Beutler and Josephy did not record emission from the P states of sodium. The transitions to Na $(3\,^2S_{\frac{1}{2}})$ (principal series) are largely reversed out, and presumably the 7,000–8,000 Å region, for detection of $n\,^2P \rightarrow 4\,^2S$ transitions, was inaccessible. Recently, Rautian and Sobelman[113] postulated that the sharing between the $9\,^2S$ and $8\,^2P$ states should be according to the scheme,

$$Hg(6^3P_1) + Na(3^2S) \begin{array}{c} \nearrow \quad 1/6\;Na(9^2S) + Hg(6^1S_0) \\[1em] \searrow \quad 5/6\;Na(8^2P) + Hg(6^1S_0) \end{array}$$

This apparently is based on the assumption that the cross-section depends on $e^{-(\Delta E/kT)}$. However, Frish and Bochkova[114] claim to have demonstrated that the 8P state is produced only in very low yield. They observed the emission from excited sodium atoms in an electric discharge in He, Hg and Na mixtures at very low total pressures. The observed intensities were divided by the Einstein A coefficients to obtain the relative stationary concentrations in the various states. The concentrations in the P levels were determined from the principal series, and reversal was allowed for, though this is a difficult and hazardous calculation. It is not clear why the populations of the P and F states were not measured from the infra-red transitions. Some of their results are given in Table 8. The second column gives the change in the stationary concentrations of the excited states of sodium, on adding mercury to the discharge. The data appear to show that the cross-section for excitation of the 8P state, the process with the smallest energy discrepancy and for which optical transitions of both atoms are allowed, is smaller than the cross-section into the 9S state, which is optically forbidden for the sodium. The high concentration in the 7S, 6P and 7D states apparently is due to excitation transfer from Hg (6^3P_0), and the changes in internal energy are given in brackets. Apparently the P state is preferred in this case, though the mercury transition is mildly forbidden. A detailed investigation of this energy-transfer system would be worthwhile. (Decomposition of a set of Σ and Π diatomic intermediates would give S, P, D, F, etc., states in the ratio $1:1\cdot7:1:0\cdot7$, etc., if the $\Delta\Lambda = 0$ rule dominates. [110])

TABLE 8. *Concentrations of excited sodium atoms produced by mercury sensitization*

State	Concentration (atom/cm³ × 10^{-6})	ΔE (cm⁻¹)
6P	37	$-2,116\ (-349)$
7S	27	$-1,400\ (+367)$
6D	11	$-1,025\ (+742)$
6F	Not measured	$-1,012$
6H	Not measured	$-1,069$
7P	$1\cdot1$	-871
8S	$1\cdot8$	-444
7D	$5\cdot3$	-212
7F	Not measured	-203
8P	$1\cdot2$	-113
9S	$5\cdot9$	$+162$
8D	$3\cdot3$	$+316$

Excitation transfer from Hg (6^3P_1) to indium is again consistent with the rule that the exchange is accompanied by a small change in internal energy. [5] Photographic examination of the sensitized fluorescence indicates that at 900 °C, the processes

$$Hg~(6^3P_1) + In~(5\,^2P) \rightarrow Hg~(6\,^1S_0) + In~(7\,^2P)...\Delta E~(cm^{-1})$$
$$= -551 \quad and \quad -440$$

$$Hg~(6^3P_1) + In~(5\,^2P) \rightarrow Hg~(6\,^1S_0) + In~(6\,^2D)...\Delta E~(cm^{-1})$$
$$= -364 \quad and \quad -314$$

have approximately equal cross-sections. The first is optically forbidden and the second allowed.

The mercury photosensitized fluorescence of thallium vapour appears to provide an exception to the rule, because the reaction with the largest cross-section corresponds to a substantial change in internal energy, notwithstanding the opportunity of electronic energy transfer to a state of almost identical internal energy. At 900 °C, the reaction

$$Hg~(6\,^3P_1) + Tl~(5\,^2P_\frac{1}{2}) \rightarrow Hg~(6\,^1S_0) + Tl~(6\,^2D)...\Delta E~(cm^{-1})$$
$$= -3{,}218 \quad and \quad -3{,}294$$

is apparently preferred to

$$Hg~(6^3P_1) + Tl~(5\,^2P_\frac{1}{2}) \rightarrow Hg~(6\,^1S_0) + Tl~(8\,^2S)...\Delta E~(cm^{-1})$$
$$= -666$$

Pringsheim [5] discusses a number of complications, and suggests that excitation of $Tl~(6\,^2D)$ may occur by species other than $Hg~(6^3P_1)$.

9.3. Recent research in relation to optical masers provides our only quantitative information about electronic excitation transfer between atoms. In an electric discharge containing a trace of neon, energy transfer occurs from metastable helium atoms $He~(2\,^3S_1)$ to neon and excites $Ne~(2s)$, to produce population inversion with respect to the $(2p)$ and $(1s)$ states. In pure helium, Phelps and Brown [115] have measured the rate of decay of $He~(2\,^1S_0)$ and $He_2 a^3\Sigma_u^+$, following a pulsed d.c. discharge. The metastable species were detected photoelectrically, by observing reversal of suitably isolated lines from a helium discharge lamp. The production and decay of $He~(2\,^3S_1)$ in an electric discharge, obtained by the microwave pulse method [116] is illustrated by Plate 7(a). Javan, Bennett and Herriott [117] employed the method of Phelps and Brown to observe the increased rate of decay of He (2^3S_1) in the presence of neon, and recorded a cross-section of $3\cdot7 \pm 0\cdot5 \times 10^{-17}$ cm². This value was confirmed by the microwave pulse measurements (Plate 7(b)). Javan et al. proved that transfer occurred to the neon, by observing the $2s \to 2p$ emission lines during the afterglow; the decay of the emission was simultaneous with the He (2^2S_1) decay. Apparently all four of the $2s$ levels are populated directly, though the individual cross-sections have not yet been reported. The internal energy changes for the processes

$$\text{He } (2^3S_1) + \text{Ne } (2^1S_0) \to \text{He } (1^1S_0) + \text{Ne } (2s_{2,3,4 \text{ or } 5})$$

are, respectively, -314, -469, $-1{,}053$ and $-1{,}247$ cm^{-1}. The transitions are optically forbidden for the helium, and the second and fourth are forbidden for the neon. Benton, Matson, Ferguson and Roberts [118] employed the same technique to measure excitation transfer from He (2^1S_0) to neon. Transfer occurs predominantly to the $3s_2$ state. [119]

$$\text{He } (2^1S_0) + \text{Ne } (2^1S_0) \to \text{He } (1^1S_0) + \text{Ne } (3s_2)$$

The changes in internal energy for transfer to $3s_2$, $3s_3$, $3s_4$ and $3s_5$ are, respectively, $+387$, $+337$, -357 and -441. The transition is optically forbidden for the helium and allowed for the neon. The cross-section was recorded as 4×10^{-16} cm². If this value is correct, it represents very efficient energy transfer because at 300 °K, only one collision in 6·8 has sufficient kinetic energy to supply the internal energy increase. He (2^1S_0) is short lived even in pure helium, and it is difficult to measure its decay rate accurately.

Stepp and Anderson [120] have suggested recently that there may be partial conservation of electronic angular momentum accompanying energy transfer between atoms, and interpreted experiments on mercury fluorescence by means of the steps:

$$\text{Hg } (6^3P_1) + \text{Hg } (6^3P_0) \to \text{Hg } (6^1S_0) + \text{Hg } (6^3D_1)$$

$$\text{Hg } (6^3P_1) + \text{Hg } (6^3P_2) \to \text{Hg } (6^1S_0) + \text{Hg } (6^3D_3)$$

9.4. The above experimental results make up the bulk of our

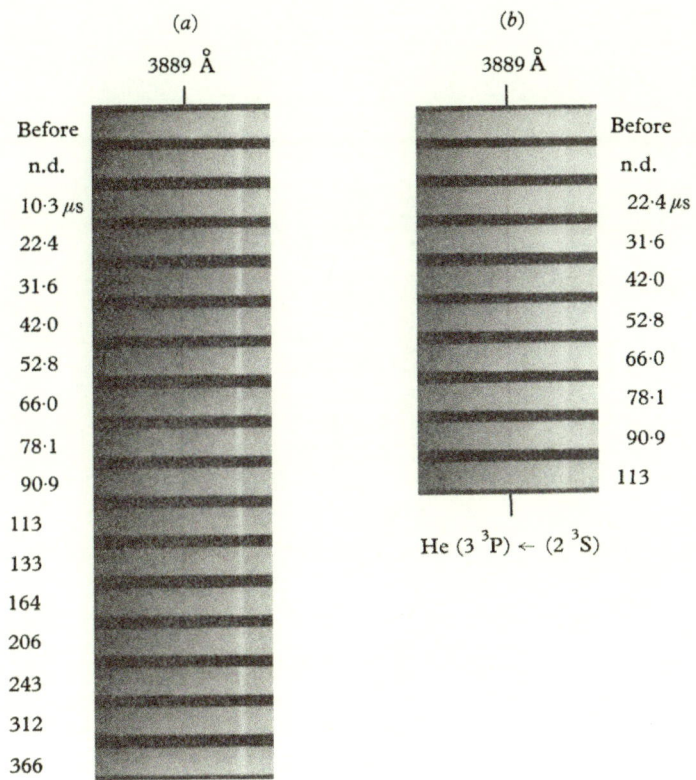

7 Production and decay of He (2^3S_1) in a microwave pulse. (a) Formation and decay of He (2^3S_1) in 5 mm of He. (b) Decay of He (2^3S_1) in 5 mm He + 1·2 × 10⁻² mm Ne.

knowledge of the exchange of electronic energy between atoms, without ionization, and we can summarize as follows:

(a) The opportunity exists for excitation transfer with $\Delta E \sim 0$ in all systems which have been investigated. To a first approximation, transfer does occur to minimize ΔE.

(b) If there are several opportunities with $\Delta E < \sim 1,000$ cm^{-1} the precise course of the energy transfer cannot be predicted. A linear relationship between $\log P$ and ΔE has not been established.

(c) The optical rules for the separated atoms have no obvious influence on electronic energy transfer.

(d) Cross-sections have been measured only for transfer from helium to neon.

Long-range resonance interaction is unimportant in all the atomic systems which have been investigated so far. Excitation transfer occurs at short range, and will depend on the interaction between the sets of potential curves corresponding to the initial and final states. Clearly more quantitative experiments would be of great value (see Table 9).

TABLE 9. *Summary of excitation transfer between atoms*

Donor		Acceptor		ΔE	Cross-section	Comments†
Initial	Final	Initial	Final	(cm^{-1})	(Å^2)	
Hg ($6\,^3P_1$)	($6\,^1S_0$)	Na ($3\,^2S_{\frac{1}{2}}$)	($9\,^2S_{\frac{1}{2}}$)	$+162$	—	F preferred to A with $\Delta E = -113$ cm^{-1}
Hg ($6\,^3P_1$)	($6\,^1S_0$)	In ($5\,^2P$)	($7\,^2P$)	-440	—	F } ~equally probable
Hg ($6\,^3P_1$)	($6\,^1S_0$)	In ($5\,^2P$)	($6\,^2D$)	-314	—	A }
Hg ($6\,^3P_1$)	($6\,^1S_0$)	Tl ($5\,^2P$)	($6\,^2D$)	$-3{,}218$	—	A
He ($2\,^3S_1$)	($1\,^1S_0$)	Ne ($2\,^1S_0$)	($2s_2$)	-314 }		FHe ANe }
He ($2\,^3S_1$)	($1\,^1S_0$)	Ne ($2\,^1S_0$)	($2s_3$)	-469 }	0.37	FHe FNe } ~equally probable
He ($2\,^3S_1$)	($1\,^1S_0$)	Ne ($2\,^1S_0$)	($2s_4$)	$-1{,}053$ }		FHe ANe }
He ($2\,^3S_1$)	($1\,^1S_0$)	Ne ($2\,^1S_0$)	($2s_5$)	$-1{,}247$ }		FHe FNe }
He ($2\,^1S_0$)	($1\,^1S_0$)	Ne ($2\,^1S_0$)	($3s_2$)	$+387$	~4	FHe ANe

† F = Optically forbidden transition; A = optically allowed transition. The Hg transition is mildly forbidden.

9.5. *Exchange between an atom and a complex molecule.* According to Massey and Burhop, [108] electronic energy transfer can occur efficiently from an atom to a polyatomic molecule, even if there is a considerable electronic energy mismatch, because of the opportunity for the energy defect to be converted to vibrational energy. This is based on the fact that Hg (6^3P_1) is quenched with approximately unit probability per gas-kinetic collision by NO, O_2, olefins, N_2O, etc. (Table 10). However, this view is not particularly compelling. The observations simply reflect the circumstances that in all the known experimental systems of atom transfer, the opportunity exists to exchange practically the entire energy as electronic, whereas this is not the case for the polyatomic molecules which have been investigated, because they have a lower density of electronic levels in the region of 5 eV. The quenching of Hg (6^3P_1) provides practically our entire knowledge of the exchange of electronic energy from an atom to complex molecules. Strausz and Gunning [123] have shown recently that quenching cross-sections tend to increase with the electron-donating properties of the quenching species. Strong attractive interaction permits crossing between surfaces in the transition complex. The quenching molecule loses its identity as a separate chemical species, and optical rules relating to the separated molecules should not be appropriate to the overall transition. Laidler [105] has suggested some interpretations of the quenching of Hg (6^3P_1) based on potential hypersurfaces. In general, we might anticipate that the electronic energy discrepancy will not be converted efficiently to vibration. Experimental techniques are now closing in on this problem.

TABLE 10. *Cross-sections for deactivation of Hg (6^3P_1) and Hg (6^3P_0)* [83,95,121,122]

Gas	$\sigma^2(\text{Å}^2)^3P_1$	$\sigma^2(\text{Å}^2)^3P_0$
N_2	0·42	9×10^{-6}
H_2	6·01	0·018
O_2	13·9	0·093
CO	4·07	0·028
CO_2	2·48	0·0014
H_2O	1·0	0·0066
D_2O	0·46	0·0048
N_2O	12·6	0·51
NO	24·7	0·34
CH_4	0·16	0·007
C_2H_6	0·11	0·011
C_3H_8	1·6	0·16
C_2H_4	22	0·6
NH_3	2·94	0·0033

9.6. Quenching cross-sections in Table 11 are a selection from Steacie, [124] and Strausz and Gunning. [126] The breakdown of the transition complex frequently produces free radicals, though inter-action of C_2H_4, NO and O_2 with Hg (6^3P_1) appears to result in electronic energy transfer. [125-127] The yield of vibrational energy is not known in any case.

TABLE 11. *Comparison of Hg (6^3P_0) cross-sections to those of other excited atoms*

Quenching gas	Cd $(5\,^3P_1)$ (3·78 eV)	Na $(3\,^2P)$ (2·09 eV)	Hg $(6\,^3P_1)$ (4·88 eV)	Hg $(6\,^3P_0)$ (4·66 eV)
CH_4	0·012	0·1	0·06	0·007
C_2H_6	0·024	0·17	0·11	0·011
C_3H_8	0·012 (?)	0·2	1·3	0·16
C_2H_4	24·9	44	26	0·6
NO	—	31·6	25	0·35
HCl	—	—	33	—
NH_3	—	—	3·0	0·0048
PH_3	—	—	26	—
C_6H_6	28·4	75	60	—

It is curious that the cross-sections for deactivation of Hg (6^3P_0) are all very small compared to those for Hg (6^3P_1). Hg (6^3P_0) is optically metastable (because J cannot remain zero) and also is metastable with respect to collisional deactivation. However, it is extremely unlikely that the very small dipole moment integral is responsible for the small deactivation cross-sections. First, we have examined various experimental results and shown that the optical rules have no obvious influence on excitation transfer between atoms. Secondly, the cross-sections correspond to a low prob-ability of deactivation per collision, and therefore long-range inductive resonance is not likely to be important. Thirdly, if as suggested above, the mercury atom interacts strongly with a quenching molecule, selection rules for the separated species would be totally unrelated to the problem of transitions between potential surfaces in the transition complex. It might be supposed, for example in deactivation of Hg (6^3P_1) by NO, that a linear Π transition state is produced and that the initial electronic configura-tions only allow the formation of certain multiplets. However, the selection rules still permit radiationless transitions in all cases. It is surprising, therefore, that the cross-sections for deactivation of Hg (6^3P_0) are consistently very much smaller than for Hg (6^3P_1) and at present there is no satisfactory explanation. Brennen and Kistiakowsky[128] recently have reported preliminary observations on energy transfer between $N_2A^3\Sigma_u^+$ and transition metal atoms; transfer of electronic energy occurs indiscriminately. Bennett, Faust, McFarlane and Patel[129] have investigated energy transfer processes in the Ne/O_2 laser. They have shown that in its reactions with O_2, Ne (3^3P_2) (lowest excited state) is less reactive than the $^3P_1$ and $^3P_0$ multiplets by a factor of 2. It would appear that there is very little difference in reactivity of the different J components for Ne (3^3P), provided collisional population of the multiplets is slow. (However, Ne (3^3P_0) may be somewhat metastable to collisional deactivation.)[130]

Table 11 shows quite clearly that the quenching cross-sections cannot be related directly to the excitation energy of the atom. If a quenching gas had an electronically excited state between 4·88 eV and 4·66 eV, a substantial difference in deactivation cross-section between Hg (6^3P_1) and Hg (6^3P_0) would be expected simply from energetic considerations. A possible example is NO, because the $a^4\Pi$ state lies within 4·5 ± 0·2 eV above the ground state.[131] However, this situation does not occur generally, and comparison of the Na (3^2P) cross-sections with those for the excited mercury atoms does suggest that there is some specific reason why Hg (6^3P_0) resists collisional deactivation. A possible approach to this problem, following the suggestion of Bykhovskii and Nikitin[93] could be to consider the effect of removal of the degeneracy during collision. If three transition states are produced, the probability of a transition would be favoured simply because of the multiplicity of the potential surfaces, and the increased probability that the symmetry requirements can be fulfilled. In addition, the coupling of the atomic angular momentum to the molecular electric-field may provide long-range attraction for $^3P_1$ energy transfer. Wilkinson[132] has recently reviewed energy transfer between complex organic molecules.

Callear and Smith[133] and Sagert and Thrush[134] have recently investigated electronic energy transfer between diatomic molecules:

$$NO\ C^2\Pi + N_2\ X^1\Sigma_g^+ \rightarrow NO\ X^2\Pi + N_2\ A^3\Sigma_u^+$$
$$N_2\ A^3\Sigma_u^+ + NO\ X^2\Pi \rightarrow N_2\ X^1\Sigma_g^+ + NO\ A^2\Sigma^+$$

It was suggested that N_2, formed in the second step, is probably highly vibrationally excited.

10. General applications and discussion

10.1. The last few years have shown an increasing interest in the chemistry of individual quantum states. The principal experimental techniques have been flash photolysis, afterglows in discharge tubes (especially examination of infra-red emission), and crossed molecular beams. One of the main problems is to determine how the heat of a chemical reaction is transferred to and distributed amongst the various degrees of freedom in the product species. A similar problem in photochemistry is to determine the distribution of energy accompanying photolytic dissociation. This article is devoted almost entirely to non-reactive energy transfer processes, which comprise some of the simplest of all kinetic systems. This subject aids our understanding of reactive energy transfer in two ways. Examples in sections 8 and 9 have theoretical and empirical features in common with chemical reactions and photochemistry. From the theoretical discussion, it is anticipated on classical and wave-mechanical grounds that a reaction in which there is a contraction in molecular dimensions will tend to retain the chemical energy as internal energy, usually as vibration; i.e. there will be no recoil forces to separate the product fragments. [106] As well as providing an introduction to reactive energy transfer, an understanding of non-reactive processes is required for the analysis of experimental data from reacting systems. Under what conditions can the initial distributions be observed, and what are the main processes which establish equilibrium? The feasibility of observing an initial distribution with a particular experimental technique depends on what the distribution is. If a large number of quantum states are populated, because relaxation of the higher levels is very rapid, it is difficult to achieve adequate time resolution. An initial distribution cannot be derived from a 'stationary' distribution unless all energy processes have been independently established and measured. It is comparatively easy to establish an initial distribution, if only a few low vibrational states are populated, especially if the vibrational frequency is high and relaxation is slow. A good example is the reaction of S ($3\,^2P$)

with O_2, which produces SO in the few lowest vibrational states. [58] In setting up experimental conditions for analysis of these problems, all possible processes need to be considered. A frequent complication seems to be relaxation by collision with transient species. Thus far, very few experimental systems have been adequately analysed, the most sophisticated attempts being measurements of the infra-red emission by J. C. Polanyi and his collaborators.

10.2. Much of the pioneering work on energy transfer in chemical reactions, was carried out by Professor Norrish with various co-workers. [135] Lipscomb, Norrish and Thrush [52] showed that vibrationally excited oxygen, produced by the reaction

$$O + NO_2 \rightarrow O_2^{\ddagger} + NO$$

can be observed by kinetic absorption spectroscopy. However, the nitric oxide is not highly vibrationally excited, and these experiments and similar studies with the $O + ClO_2$, $O + O_3$, and $O + RH$ reactions, indicated that the nascent bond is rich in vibrational energy, which is what one would expect intuitively. Recent semi-quantitative measurements show that quite a small fraction of the total exothermicity appears as vibration in these reactions. [136,137] It would be of considerable value to obtain precise rate constants into each quantum state. Norrish and Oldershaw [138] made some extremely interesting observations of vibrational excitation of SO ($v \leqslant 3$) by photolysis of SO_3, P_2 ($v \leqslant 7$) from phosphine, and PN ($v \leqslant 1$) from PH_3 and NH_3 mixtures. [139] The last two provide rare (as yet) examples of vibrationally excited products of radical–radical reactions, e.g.

$$2PH \rightarrow P_2^{\ddagger} + H_2$$

A problem which is even more difficult is to investigate the yield of vibrational energy accompanying three-body combination of atoms. The combination rate constants tend to be small compared to the fast (and second-order) vibrational relaxation. Thus the stationary concentration of the diatomic product usually corresponds very closely to an ambient Boltzmann distribution. Callear [58] observed S_2 from S atom combination, but concluded that the relaxation was too fast for any meaningful quantitative measurements to be made.

10.3. Polanyi [106] and co-workers have described a detailed investigation of the reaction of H with Cl_2 in a flow system:

$$H + Cl_2 \rightarrow HCl^\dagger + Cl$$

From the infra-red emission, a non-Boltzmann vibrational distribution was observed in the HCl product, and by analysing the various relaxation processes, the absolute rates into each quantum state were investigated. Charters and Polanyi [140] employed total pressures of $\sim 10^{-2}$ mmHg, with HCl partial pressures of $\sim 10^{-4}$ mmHg, to avoid 'Boltzmannization' by V–V transfer. According to the equations of section 6, $Z_{0,1}^{1,0}$ is 9 at 400 °K, corresponding to a relaxation time of 3×10^{-2} s for the process

$$HCl\ (v = 2) + HCl\ (v = 0) \rightarrow 2HCl\ (v = 1)$$

under their conditions. However, Findlay and Polyani [141] have demonstrated that Z for the exchange

$$HCl\ (v = 6) + HCl\ (v = 0) \rightarrow HCl\ (v = 5) + HCl\ (v = 1)$$

is roughly 5×10^3. The process is slowed down because there is an internal energy change of 620 cm^{-1} due to anharmonicity. This is an advantageous situation which protects the initial distribution in the higher vibrational levels. The time resolution should be determined not only by the rate of spontaneous radiation ($\sim 10^{-2}$ s) and pumping rate, but also by wall deactivation. Under the conditions of Charters and Polanyi, the mean time for diffusion to the wall is $\sim 10^{-4}$ s. At 10^{-2} mmHg total pressure, before returning to the bulk phase, a molecule will make $\sim 10^4$ collisions with the wall, and therefore we expect wall removal to be diffusion controlled and an efficient trap for vibrationally excited species. The effect of wall deactivation is not yet fully analysed but, under suitable conditions, a technique of 'killing' the products on the wall should permit direct observation of the initial distribution. From the $H + Cl_2$ reaction, the yield of vibrational energy is less than 10 % of the heat of reaction.

10.4. Formation of vibrationally excited species by photochemical methods also has been pioneered by Norrish and co-workers; the CS_2 and CSe_2 systems have already been discussed. Photochemical dissociation of molecules invariably produces excited fragments though this is not always easy to demonstrate by kinetic spectroscopy. Polyatomic fragments relax extremely rapidly through low-frequency bending modes, and even vibrationally excited diatomic hydrides are difficult to detect because of rapid V–V transfer, and V–V transfer to the parent hydride. One of the most spectacular of the photochemical reactions is the formation of NO with $v \geqslant 11$, from dissociation of NOCl. [56] Some V–V exchanges were investigated. Production of vibrationally excited CN radicals from photolysis of $(CN)_2$ and CNBr also has been studied in some detail. [59]

10.5. The most attractive and direct technique of studying reactive energy transfer is the method of crossed molecular beams [142–145] in which reactants and products are completely isolated from destructive collision processes. However, it does involve rather complex engineering and very low pressures. The concentration of products in the region of the crossing point is $\sim 10^{-9}$ mmHg, and the total yield of product would correspond to about one monolayer per month. Suitable sensitivity has so far been achieved only for alkali metal atoms and alkali metal compounds, by means of surface ionization detectors. Detectors may soon be available for the halogens. [146] If the yield of translational energy is small, the products are thrown forward in the direction of the initial centre of gravity vector, and considerations of angular momentum conservation show that the product beam must peak at certain specific angles, which has been verified experimentally. This technique has been confined largely to reactions of alkali metal atoms with alkyl halides, and in every case, almost the entire heat of reaction is retained as internal energy. Presumably there is little recoil or repulsive force at the point of the broken bond, in this type of reaction.

10.6. As yet only V–T and V–V energy transfer have been investigated experimentally in any detail. Except for some hydrides, the theory interprets the orders of magnitude of the observed data. Other types of energy transfer have been only superficially investigated, both experimentally and theoretically. The initial studies indicate that for some simple processes in which the mechanism is established with reasonable certainty, there is a very approximate general law for the transfer of energy, such that for given masses and temperature,

$$\log Z = A\Delta E + B$$

ΔE is the energy which is converted to translation, and A is independent of the type of process. Such a law appears to hold because the 'overlap' of the translational wave functions is restrictive in all energy transfer processes. Marked deviations from this approximate rule are anticipated if the potential surfaces converge, or diverge, in the collision complex.

The author is indebted to Mr J. F. Wilson for correcting the manuscript for this chapter.

REFERENCES

1   Oldenberg, O. *Phys. Rev.* 1931, **37**, 194.
2   Landau, L. and Teller, E. *Phys. Z. Sowjetunion*, 1936, **10**, 34.
3   Zener, C. *Phys. Rev.* 1931, **37**, 556.
4   Schwartz, R. N., Slawsky, Z. I. and Herzfeld, K. F. *J. chem. Phys.* 1952, **20**, 1591.
5   Pringsheim, P. *Fluorescence and Phosphorescence*. Interscience, New York, 1949.
6   Mitchell, A. C. G. and Zemansky, M. W. *Resonance Radiation and Excited Atoms*. Cambridge University Press, 1934.
7   Cottrell, T. L. and McCoubrey, J. C. *Molecular Energy Transfer in Gases*. Butterworths, London, 1961.
8   Herzfeld, K. F. and Litovitz, T. A. *Absorption and Dispersion of Ultrasonic Waves*. Academic Press, New York, 1959.
9   Hirschfelder, J. O., Curtis, C. F. and Bird, R. B. *Molecular Theory of Gases and Liquids*. Wiley, New York, 1954.
10 Townes, C. H. and Schawlow, A. L. *Microwave Spectroscopy*. McGraw-Hill, New York, 1955.
11 Jackson, J. M. and Mott, N. F. *Proc. Roy. Soc.* A, 1932, **137**, 703.
12 Mies, F. H. *J. chem. Phys.* 1964, **40**, 523.
13 Mies, F. H. *J. chem. Phys.* 1964, **41**, 903.
14 Nikitin, E. E. *Optika Spektrosk.*, 1960, **9**, 16; 1960, **9**, 8.
15 Takayanagi, K. *Prog. Theoret. Phys.* 1952, **8**, 497.
16 Mies, F. H. *J. chem. Phys.* 1965, **42**, 2709.
17 Schwartz, R. N. and Herzfeld, K. F. *J. chem. Phys.* 1954, **22**, 767.
18 Chow, C. C. and Greene, E. F. *J. chem. Phys.* 1965, **43**, 324.
19 Cowan, G. C. and Hornig, D. F. *J. chem. Phys.* 1950, **18**, 1008.
20 Levitt, B. P. and Hornig, D. F. *J. chem. Phys.* 1962, **36**, 219.
21 Blythe, A. R. *J. chem. Phys.* 1964, **41**, 1917.
22 Holmes, R., Jones, G. R. and Lawrence, R. *J. chem. Phys.* 1964, **41**, 2955.

23 Broida, H. P. and Carrington, T. *J. chem. Phys.* 1963, **38**, 136.
24 Holmes, R., Jones, G. R., Pusat, N. and Tempest, W. *Trans. Faraday Soc.* 1962, **58**, 2342.
25 Brown, R. L. and Klemperer, W. *J. chem. Phys.* 1964, **41**, 3072.
26 Steinfeld, J. I. and Klemperer, W. *J. chem. Phys.* 1965, **42**, 3475.
27 Bradley, J. N. *Shock Waves in Chemistry and Physics.* Methuen, London, 1962.
28 Gaydon, A. G. and Hurle, I. R. *The Shock Tube in High Temperature Chemistry and Physics.* Chapman Hall, London, 1963.
29 Greene, E. F. and Toennies, J. P. *Chemical Reactions in Shock Waves.* Arnold, London, 1964.
30 Millikan, R. C. and White, D. R. *J. chem. Phys.* 1963, **39**, 3209.
31 Losev, S. A. and Osipov, A. I. *Uspekhi Fiz. Nauk*, 1961, **74**, 393.
32 Millikan, R. C. *J. chem. Phys.* 1963, **38**, 2855; 1965, **43**, 1439.
33 Bauer, H. J., Kneser, H. O. and Sittig, E. *J. chem. Phys.* 1959, **30**, 1119.
34 Basco, N., Callear, A. B. and Norrish, R. G. W. *Proc. Roy. Soc.* A, 1961, **260**, 459.
35 Wray, W. L. *J. chem. Phys.* 1962, **36**, 2597.
36 Callear, A. B. *Disc. Faraday Soc.* 1962, **33**, 28.
37 Callear, A. B. and Smith, I. W. M. *Trans. Faraday Soc.* 1963, **59**, 1720.
38 Lambert, J. D. and Salter, R. *Proc. Roy. Soc.* A, 1957, **253**, 277.
39 Sette, D., Busala, A. and Hubbard, H. C. *J. chem. Phys.* 1955, **23**, 787.
40 Lambert, J. D. and Salter, R. *Proc. Roy. Soc.* A, 1957, **243**, 78.
41 McCoubrey, J. C., Milward, R. C. and Ubbelohde, A. R. *Proc. Roy. Soc.* A, 1961, **264**, 299.
42 Tanczos, F. *J. chem. Phys.* 1956, **25**, 439.
43 Stretton, J. L. *Trans. Faraday Soc.* 1965, **61**, 1053.
44 Daen, J. and de Boer, P. C. T. *J. chem. Phys.* 1962, **36**, 1222.
45 Winter, T. G. *J. chem. Phys.* 1963, **38**, 2761.
46 Herzfeld, K. F. *Disc. Faraday Soc.* 1962, **33**, 22.
47 Witteman, W. J. *J. chem. Phys.* 1962, **37**, 655.
48 Montroll, E. W. and Shuler, K. E. *J. chem. Phys.* 1957, **26**, 454.
49 Hooker, W. J. and Millikan, R. C. *J. chem. Phys.* 1963, **38**, 214.
50 Decius, J. C. *J. chem. Phys.* 1960, **32**, 1262.
51 White, D. R. *J. chem. Phys.* 1965, **42**, 447.
52 Lipscomb, F. J., Norrish, R. G. W. and Thrush, B. A. *Proc. Roy. Soc.* A, 1956, **233**, 455.
53 McGrath, W. D. and Norrish, R. G. W. *Z. phys. Chem.* 1958, **15**, 245.
54 Basco, N. and Norrish, R. G. W. *Disc. Faraday Soc.* 1962, **33**, 99.
55 Fitzsimmons, R. V. and Bair, E. J. *J. chem. Phys.* 1964, **40**, 451.
56 Basco, N. and Norrish, R. G. W. *Proc. Roy. Soc.* A, 1962, **268**, 291.
57 Callear, A. B. and Norrish, R. G. W. *Nature, Lond.*, 1960, **188**, 53.
58 Callear, A. B. *Proc. Roy. Soc.* A, 1963, **276**, 401.

59 Basco, N., Nicholas, J. E., Norrish, R. G. W. and Vickers, W. H. J. *Proc. Roy. Soc.* A, 1963, **272**, 147.
60 Callear, A. B. and Tyerman, W. J. R. *Trans. Faraday Soc.* 1965, **61**, 2395.
61 Tuesday, C. S. and Boudart, M. *Tech. Note 4, Contract AP*33(038)-23976. Princeton University, 1965.
62 Basco, N., Callear, A. B. and Norrish, R. G. W. *Proc. Roy. Soc.* A, 1963, **269**, 180.
63 Morgan, J. E. and Schiff, H. I. *Can. J. Chem.* 1963, **41**, 903.
64 Callear, A. B. *Disc. Faraday Soc.* 1962, **33**, 283.
65 Legay-Sommaire, N., Henry, L. and Legay, F. *C. r. Acad. Sci.*, 1965, **260**, 3339.
66 Callear, A. B. and Smith, I. W. M. *Trans. Faraday Soc.* 1963, **59**, 1735.
67 Jeunehomme, M. and Duncan, A. B. F. *J. chem. Phys.* 1964, **41**, 1692.
67a Jeunehomme, M. *J. chem. Phys.* 1966, **45**, 4433.
68 Callear, A. B. and Williams, G. J. *Trans. Faraday Soc.* 1966, **62**, 2030.
69 Lambert, J. D. *Disc. Faraday Soc.* 1962, **33**, 93.
70 Rapp, D. and Englander-Golden, P. *J. chem. Phys.* 1964, **40**, 573, 3120; Rapp, D. *ibid.* 1965, **43**, 316.
71 Lambert, J. D., Edwards, A. J., Pemberton, D. and Stretton, J. L. *Disc. Faraday Soc.* 1962, **33**, 61.
72 Lambert, J. D., Parks-Smith, D. G. and Stretton, J. L. *Proc. Roy. Soc.* A, 1964, **282**, 380.
73 White, D. R. *J. chem. Phys.* 1965, **42**, 2028.
74 Millikan, R. C. and Osburg, C. A. *J. chem. Phys.* 1964, **41**, 2196.
75 Millikan, R. C. and White, D. R. *J. chem. Phys.* 1963, **39**, 98, 3209.
76 Cottrell, T. L. and Matheson, A. J. *Trans. Faraday Soc.* 1962, **58**, 2336.
77 Cottrell, T. L. and Matheson, A. J. *Trans. Faraday Soc.* 1963, **59**, 824.
78 Cottrell, T. L., Dobbie, R. C., McClain, J. and Read, A. W. *Trans. Faraday Soc.* 1964, **60**, 241.
79 Callear, A. B. *Chemical Society Annual Reports*, 1964, **61**, 48.
80 Callear, A. B. *Applied Optics* (Suppl. 2), 1965, p. 145.
81 Darwent, B. de B. and Hurtubise, F. G. *J. chem. Phys.* 1952, **20**, 1684.
82 Callear, A. B. and Norrish, R. G. W. *Proc. Roy. Soc.* A, 1962, **266**, 299.
83 Callear, A. B. and Williams, G. J. *Trans. Faraday Soc.* 1964, **60**, 2158.
84 Scheer, M. D. and Fine, J. *J. chem. Phys.* 1962, **36**, 1264.
85 Matland, C. G. *Phys. Rev.* 1953, **92**, 637.
86 Callear, A. B. and Tyerman, W. J. R. *Nature, Lond.*, 1964, **202**, 1326.
87 Callear, A. B. and Tyerman, W. J. R. *Trans. Faraday Soc.* 1966, **62**, 2760.
88 Callear, A. B. and Oldman, R. J. *Nature, Lond.*, 1966, **210**, 730.
89 Thorson, W. R. *J. chem. Phys.* 1961, **34**, 1744.

90 Thorson, W. R. and Moskowitz, J. W. *J. chem. Phys.* 1963, **38**, 1848.
91 Bates, D. R. *Proc. Phys. Soc.* 1959, **73**, 227.
92 Nikitin, E. E. *J. chem. Phys.* 1965, **43**, 744.
93 Bykhovskii, V. K. and Nikitin, E. E. *Optika. Spektrosk.*, 1964, **16**, 201; *Opt. Spectry*, 1964, **16**, 111.
94 Donovan, R. J. and Husain, D. *Nature, Lond.*, 1965, **206**, 171.
95 Donovan, R. J. and Husain, D. *Trans. Faraday Soc.* 1966, **62**, 1050.
96 Donovan, R. J. and Husain, D. Unpublished results.
97 Callear, A. B. and Wilson, J. F. *Nature, Lond.*, 1966, **211**, 517.
98 Yamazaki, H. and Cvetanović, R. J. *J. chem. Phys.* 1964, **41**, 3703.
99 Padley, P. J. and Sugden, T. M. *Proc. Roy. Soc.* A, 1958, **248**, 248.
100 Carrington, T. *J. chem. Phys.* 1959, **31**, 1243.
101 Clouston, J. G., Gaydon, A. G. and Glass, I. I. *Proc. Roy. Soc.* A, 1958, **248**, 429.
102 Karl, G. and Polanyi, J. C. *J. chem. Phys.* 1963, **38**, 271.
103 Polanyi, J. C. Private communication quoting unpublished results.
104 Dickens, P. G., Linnett, J. W. and Sovers, O. *Disc. Faraday Soc.* 1962, **33**, 52.
105 Laidler, K. J. *J. chem. Phys.* 1947, **15**, 712.
106 Polanyi, J. C. *J. Quant. Spectry. Radiative Transfer*, 1963, **3**, 471.
107 Hurle, I. R. *J. chem. Phys.* 1964, **41**, 3911.
108 Massey, H. S. W. and Burhop, E. H. S. *Electronic and Ionic Impact Phenomena*. Clarendon Press, Oxford, 1952.
109 Mott, N. F. and Massey, H. S. W. *The Theory of Atomic Collisions*. Oxford University Press, 1949.
110 Herzberg, G. *Spectra of Diatomic Molecules*. Van Nostrand, New York, 1950.
111 Beutler, B. and Josephy, H. *Z. Physik.* 1929, **53**, 747.
112 Bennett, W. R., Jr. *Appl. Opt.* Suppl. 1, 1962, 24.
113 Rautian, S. G. and Sobelman, I. I. *Zh. éksp. teor. Fiz.* 1960, **39**, 217.
114 Frish, S. E. and Bochkova, O. P. *Zh. éksp. teor. Fiz.* 1962, **43**, 1831.
115 Phelps, A. V. and Brown, S. C. *Phys. Rev.* 1952, **86**, 102; 1955, **99**, 1307.
116 Callear, A. B., Green, J. A. and Williams, G. J. *Trans. Faraday Soc.* 1965, **61**, 1831.
117 Javan, A., Bennett, W. R., Jr. and Herriott, D. R. *Phys. Rev. Lett.* 1961, **6**, 106.
118 Benton, E. E., Matson, F. A., Ferguson, E. E. and Roberts, W. W. *Phys. Rev.* 1962, **128**, 206.
119 White, A. B. and Gordon, E. I. *Appl. Phys. Lett.* 1963, **3**, 197.
120 Stepp, E. E. and Anderson, R. A. *J. Opt. Soc. Am.* 1965, **55**, 31.
121 Noyes, W. A. and Leighton, P. A. *The Photochemistry of Gases*. Reinhold, New York, 1941.
122 Cvetanović, R. J. *J. chem. Phys.* 1955, **23**, 1208.
123 Strausz, O. P. and Gunning, H. E. *Advances in Photochemistry*, vol. 1. Interscience, New York, 1963.

124 Steacie, E. W. R. *Atomic and Free Radical Reactions*. Reinhold, New York, 1954.
125 Callear, A. B. and Cvetanović, R. J. *J. chem. Phys.* 1956, **24**, 873; Setser, D. W., Rabinowitch, B. S. and Placzek, D. W. *J. Am. Chem. Soc.* 1963, **85**, 862.
126 Strausz, O. P. and Gunning, H. E. *Can. J. Chem.* 1961, **39**, 2549.
127 Volman, D. H. *Advances in Photochemistry*, vol. I. Interscience, New York, 1963.
128 Brennen, W. R. and Kistiakowsky, G. B. *J. chem. Phys.* 1966, **44**, 2695.
129 Bennett, W. R., Jr., Faust, W. L., McFarlane, R. A. and Patel, C. K. M. *Phys. Rev. Lett.* 1962, **8**, 470.
130 Phelps, A. V. *Phys. Rev.* 1959, **114**, 1011.
131 Vanderslice, J. T., Mason, E. A. and Maisch, W. G. *J. chem. Phys.* 1959, **31**, 738; Frosch, R. P. and Robinson, G. W. *J. chem. Phys.* 1964, **41**, 367.
132 Wilkinson, F. *Advances in Photochemistry*, vol. III, 1965.
133 Callear, A. B. and Smith, I. W. M. *Trans. Faraday Soc.* 1965, **61**, 2383.
134 Sagert, N. H. and Thrush, B. A. *Disc. Faraday Soc.* 1964, **37**, 223.
135 Norrish, R. G. W. *The Solvay Institute, 12th Discussion in Chemistry*, 1962, p. 99.
136 Kane, R. A., McGarvey, J. J. and McGrath, W. D. *J. chem. Phys.* 1963, **39**, 840.
137 Bass, A. M. and Garvin, D. *J. chem. Phys.* 1964, **40**, 1772.
138 Norrish, R. G. W. and Oldershaw, G. A. *Proc. Roy. Soc.* A, 1958, **249**, 498.
139 Norrish, R. G. W. and Oldershaw, G. A. *Proc. Roy. Soc.* A, 1961, **262**, 1.
140 Charters, P. E. and Polanyi, J. C. *Disc. Faraday Soc.* 1962, **33**, 107.
141 Findlay, F. D. and Polanyi, J. C. *Can. J. Chem.* 1964, **42**, 2176.
142 Polanyi, M. *Atomic Reactions*. Williams and Norgate, London, 1932.
143 Bull, T. H. and Moon, P. B. *Disc. Faraday Soc.* 1954, **17**, 54.
144 Datz, S. and Taylor, E. H. *J. chem. Phys.* 1955, **23**, 1711.
145 Herschbach, D. R. *Appl. Optics* (Suppl.), 1965, **2**, 128.
146 Scheer, M. D. Private communication.

8

POLYMER CHEMISTRY

J. C. BEVINGTON

1. Introduction

It is doubtful whether Norrish would describe himself as primarily a polymer chemist and yet his studies of polymerizations have earned him recognition as one of the leaders in this field. He has not been involved in the collection of great quantities of information concerning, for example, monomer reactivity ratios or transfer constants. He has not confined himself to well-defined and restricted areas, but he has ranged widely, choosing topics which have subsequently proved to be of great interest and importance.

In his polymer work, as in his other research, Norrish has never adopted the procedure whereby teams of research workers are set to examine slightly different aspects of the same problem and are committed to repetitive measurements. Each of his students has been assigned a problem original in character and has been allowed to develop it. Research in universities has two functions; one of them is to give training in the methods of research and the other is to advance knowledge. The functions sometimes seem to be in conflict, but Norrish has always been able to strike the very delicate balance between them. He has been particularly successful in developing originality and thoroughness in his students, not only in the design of experiments but also in the interpretation of results.

Many of Norrish's papers on polymerization have been published in the *Proceedings of the Royal Society*, and the high standards of that journal have always been maintained. Norrish has not been remote from 'plastics' and many of the topics studied in his laboratories have had real industrial significance, but he has allowed them to develop and to become academic studies of reaction kinetics and reaction mechanisms.

This chapter has been divided into sections, each of which is concerned with papers on particular aspects of polymerization. It is evident that Norrish's pioneering work has covered a very wide range of topics; it is also very plain that, in many cases, a project has made important contributions to several parts of the whole subject.

2. Initiation of radical polymerization

One of Norrish's earliest and most important general contributions to polymerization kinetics was the demonstration that study of the so-called uncatalysed reactions can be unrewarding and even misleading. He realized that, in many cases, it is virtually impossible to achieve absolute purification of the reactants and that results of much greater significance can be obtained by examination of the more reproducible reactions promoted by the deliberate addition of controlled amounts of suitable catalysts. Norrish has been responsible for several detailed and significant studies of the behaviour of particular initiators.

Nicholson and Norrish[1] investigated the decomposition of benzoyl peroxide in solution at temperatures between 60 and 80 °C for pressures up to 3,000 kg cm^{-2}; the work was performed in connection with the polymerization of styrene at high pressures (see section 4). Apart from the special procedures required for work under these conditions, the experimental technique was simple; the residual peroxide in a solution was determined by an iodometric method. The net effect of pressure is to increase the overall rate of decomposition, but this result can be misleading because the peroxide is consumed in two distinct processes, a unimolecular dissociation to radicals and a radical-induced chain decomposition. Critical examination of the results showed that increasing the pressure causes the velocity constant for the unimolecular dissociation to fall and that for the radical-induced decomposition to rise (see Fig. 1). The activated complex in a unimolecular dissociation most probably has a volume larger than that of the original molecule so that dissociation is expected to be suppressed at high pressures. On the other hand, the induced decomposition involves bimolecular processes for which the complexes are expected to be smaller than the starting molecules; it is, therefore, reasonable that this decomposition should be accelerated by the application of high pressures.

Norrish has not published work involving the use of isotopically labelled initiators but tracer techniques would probably have been particularly useful in his studies of the decomposition of benzoyl peroxide and of the use of this substance as an initiator of radical polymerizations. The thermal dissociation of the peroxide gives benzoyloxy radicals almost all of which, in the absence of reactive substances, dissociate further [2]

$$C_6H_5CO \cdot O \cdot \rightarrow C_6H_5 \cdot + CO_2$$

In the main, the photolysis of the peroxide follows a similar course. If suitably labelled peroxide is used, the decomposition can be followed by isotope dilution analysis for the carbon dioxide. The

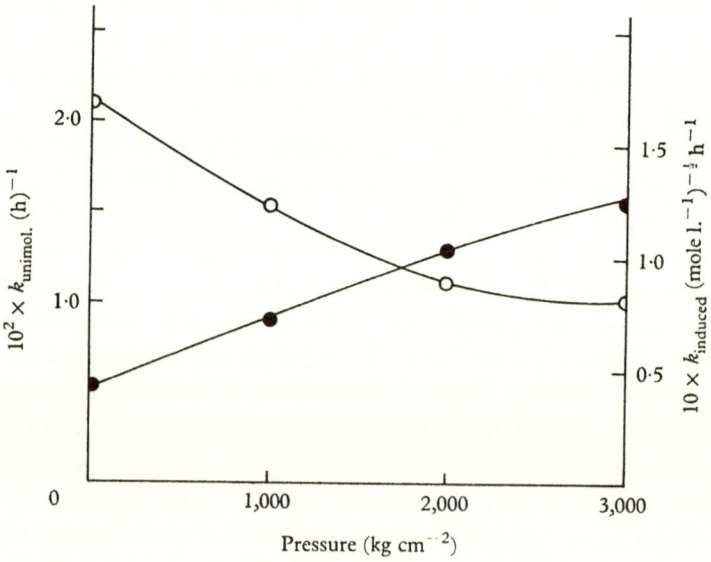

Fig. 1. Variation with pressure of the velocity constant for the decomposition of benzoyl peroxide in carbon tetrachloride at 70 °C. O Unimolecular decomposition; ● radical-induced decomposition.

sensitivity achievable in this procedure makes it possible to use very dilute solutions of peroxide. When monomer is present, there is competition between the decomposition of the benzoyloxy radical and its reaction with monomer

$$C_6H_5CO \cdot O \cdot + CH_2{=}CHX \rightarrow C_6H_5CO \cdot OCH_2CHX \cdot$$

The resulting polymer molecules contain both benzoyloxy and

phenyl end-groups; the relative numbers of these end-groups can be determined quite easily and accurately by tracer techniques. The proportions of the two types of end-group depend upon the ratio of the velocity constants for the competing reactions and upon the concentration of monomer. At high pressures, it is likely that the velocity constant for the dissociation of the radical is decreased while that of the competing initiation is increased. It is probable, therefore, that the ratio of the numbers of the two types of end-group varies significantly with pressure in the sense that benzoyloxy end-groups are favoured at high pressure. The labelled peroxide could also be used for measurement of the rate of initiation and for study of transfer to initiator; such information could be of considerable significance in connection with the detailed examination of polymerizations at high pressures.

In 1937, Norrish published on the principle of primary recombination; [3] the paper was mainly concerned with the photolysis of aldehydes and ketones but it contained also more general points. He indicated that the cage recombination of radicals would become more significant at high pressures. Such an effect could be very important for substances, such as benzoyl peroxide and azoisobutyronitrile, used as the sources of free radicals to initiate polymerizations. In the case of the azonitrile, the primary dissociation is almost certainly according to the equation

$$(CH_3)_2C(CN)N:NC(CN)(CH_3)_2 \rightarrow 2(CH_3)_2C(CN)\cdot + N_2$$

Under normal conditions of pressure, only about 60% of the radicals react with monomer and the remainder undergo geminate recombination giving tetramethylsuccinodinitrile as the main product. [2] It would be interesting to know if the efficiency of initiation depends upon the applied pressure. This problem is another for which the most obvious approach would involve the use of isotopically labelled initiators.

Norrish's studies [4,5] of the photochemical decompositions of *tert.*-butyl hydroperoxide, dicumyl peroxide and cumene hydroperoxide were not directly linked with polymer work but they are very significant in connection with the use of these and related compounds as initiators for radical polymerizations and as agents for promoting cross-linking of polymers. Product analyses confirmed that for these peroxides the initial fission occurs at the oxygen–oxygen bond and that the alkoxy radicals so formed can readily dissociate to a ketone and an alkyl radical, e.g.

$$C_6H_5C(CH_3)_2O \cdot \rightarrow C_6H_5COCH_3 + CH_3 \cdot$$

$$(CH_3)_3CO \cdot \rightarrow (CH_3)_2CO + CH_3 \cdot$$

In polymerizing systems, initiation may occur through the agency of the alkoxy or the derived alkyl radicals. In the case of di-*tert.*-butyl peroxide, tracer studies have shown the presence in polymer of end-groups of the expected types. [2] The competition between the dissociation of the alkoxy radical and its reaction with monomer can be investigated. As in the case of the benzoyloxy and other aroyloxy radicals, it is possible to compare the reactivities of monomers towards small reference radicals.

In a number of cases, Norrish has applied his photo-chemical expertise　problems associated with the initiation of polymerization; one of these concerned the use of cerous ions. Irradiation with light of wavelength 2,537 A of acidified aqueous solutions of cerous salts causes fluorescence and liberation of hydrogen. [6,7] A scheme involving electron transfer satisfactorily accounts for the observations; it can be represented simply thus:

$$Ce^{3+} \cdot H_2O + h\nu \qquad \rightarrow (Ce^{4+} \cdot H \cdot OH^-)^*$$

$$(Ce^{4+} \cdot H \cdot OH^-)^* \rightarrow Ce^{3+} \cdot H_2O + h\nu'$$

$$(Ce^{4+} \cdot H \cdot OH^-)^* + H_3O^+ \rightarrow Ce^{4+} + H \cdot + 2H_2O$$

There is, therefore, a competition between the alternative reactions involving $(Ce^{4+} \cdot H \cdot OH^-)^*$. If a monomer is present in the system, an additional process might be considered, viz.

$$(Ce^{4+} \cdot H \cdot OH^-)^* + H_3O^+ + M \rightarrow Ce^{4+} + 2H_2O + MH \cdot$$

where M represents the monomer and the formation of the radical MH· is regarded as the initiation of a polymer chain. It was shown that polymerizations of acrylonitrile, methacrylic acid and methyl acrylate can indeed be thus initiated and that the initiation competes with the fluorescence. It is possible that a radical-ion containing cerium, rather than the hydrogen atom, might be responsible for initiation of polymerization; this possibility seems remote, however, since no cerium was detected even in polymers prepared under conditions favouring the formation of products of comparatively low molecular weight.

Hussain and Norrish[7] set up the kinetic equations for competing reactions according to the scheme outlined above. They assumed that a steady state is established for the excited ceric ion complex. They showed that, in general, the derived equations are obeyed. Photochemical electron-transfer processes of the general forms

$$M^{z+} \cdot H_2O \rightarrow M^{(z+1)+} + OH^- + H \cdot$$

$$A^{z-} \cdot H_2O \rightarrow A^{(z-1)-} + OH^- + H \cdot$$

$$A^{z-} \cdot H_3O^+ \rightarrow A^{(z-1)-} + H_2O + H \cdot$$

have been considered in detail by Dainton and James.[8] The hydrogen atoms can initiate polymerizations but in many cases there are complications arising from formation of complexes between the monomer and the ion.

In another photochemical study,[9,10] anthracene and other aromatic hydrocarbons were examined as photosensitizers for the polymerization of styrene. The work included the investigation of primary processes by the technique of flash photolysis. It was concluded that the triplet states of the sensitizers are quenched by styrene and that it is the vinyl double bond in that molecule which is responsible for the effect. The triplet is thought to act as a diradical; the polymerizations, however, have the characteristics of monoradical processes so that the actual initiating step may be according to the equation

$$C_6H_5CH{=}CH_2 + {}^3A \rightarrow C_6H_5CH{=}CH \cdot + HA \cdot$$

where ^3A represents triplet anthracene. This possibility was discounted and it was suggested that an intermediate diradical is formed thus

$$^3A + M \rightarrow (AM)^*$$

and that this may either dissociate and be wasted

$$(AM)^* \rightarrow A + M$$

or react with another molecule of sensitizer

$$(AM)^* + A \rightarrow C_6H_5CH{=}CHA \cdot + HA \cdot$$

to give a pair of monoradicals. This scheme accounts for the variation of efficiency of initiation with concentration of sensitizer and agrees with a scheme proposed for the photo-oxidation of anthracene in carbon disulphide.

Dihydroanthracene (DHA) reacts with triplet anthracene so that, if a solution of anthracene containing both styrene and DHA is irradiated, there is competition for the triplet anthracene. The existence of this effect was confirmed by study of the effects of DHA upon the rate of polymerization and the rate of initiation. The latter quantity was calculated from rates of polymerization and molecular weights of polymers after corrections had been made for DHA acting as a transfer agent during the polymerization.

Norrish and Simons[10] concluded that initiation in the unsensitized photopolymerization of styrene proceeds through the self-quenching of the triplet excited monomer. They suggested that the unsensitized thermal polymerization of the monomer might be initiated in a similar way through an intermediate complex. The unsensitized reactions have the kinetic characteristics of monoradical polymerizations and these ideas concerning initiation show that there is a feasible mechanism for the production of the necessary monoradicals.

The photolysis of a polymer containing pendant ketone groups was studied by Guillet and Norrish.[11,12] One of the modes of decomposition leads to the production of radicals capable of initiating polymerization; this work is considered in detail in the section concerned with co-polymerization.

3. Acceleration during polymerization

Norrish has been involved in a number of studies of acceleration during polymerization. In many of them, the kinetic behaviour can be correlated with physical characteristics associated with systems containing polymer, such as the high viscosities of solutions.

It was found[12] that, in the later stages of the polymerization of methyl methacrylate, there is a steady rise in the rate of reaction; at about 95 % conversion, however, the rate falls off rapidly. The acceleration was attributed to self-heating of the reaction mixture; the heating was thought to be due to a reduction in the rate at which heat is dissipated by convection as the viscosity increases.

A comparison of the temperature near the centre of the reaction vessel with that in the surrounding thermostat confirmed that there can be substantial rises in the temperature of the reaction mixture (see Fig. 2).

The problem of acceleration during the polymerization of methyl methacrylate was discussed again by Norrish and Smith[13] in a short but very significant paper (see Fig. 3). They supported the

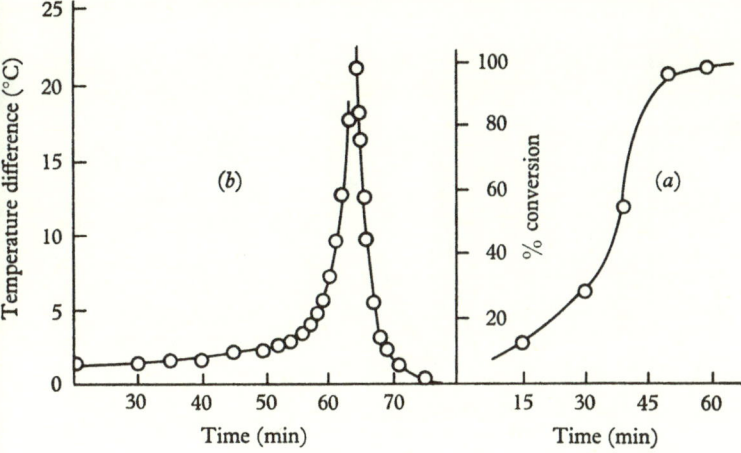

Fig. 2. (a) Conversion/time plot for polymerization at 90 °C of methyl methacrylate containing 0·020 mole% of benzoyl peroxide. (b) Variation with time of difference between temperature of thermostat (90 °C) and temperature inside reaction vessel for methyl methacrylate containing 0·025 mole % of benzoyl peroxide.

view of Schulz and Blaschke[14] that the rise in temperature results from an increase in the rate of polymerization rather than the reverse; they rejected the suggestion, however, that the acceleration is caused by an increase in the rate of initiation. The most telling argument used by Norrish and Smith[13] was based upon a consideration of the molecular weights of polymers produced at various stages in the reaction. Simple kinetic reasoning shows that a rise in the rate of initiation would produce a decrease in the average molecular weight of the polymer, whereas they demonstrated that polymer produced in the later stages of the reaction actually possesses a molecular weight higher than that of material formed early in the reaction. They postulated that, at a certain stage in the polymerization when the viscosity becomes high, the rate of termination of chains decreases. If the characteristics of the propagation process are largely unaffected, there are increases in both the overall rate of polymerization and the chain length at termination.

Norrish was also involved in the study of another system where the rate of termination of polymerization chains is depressed by

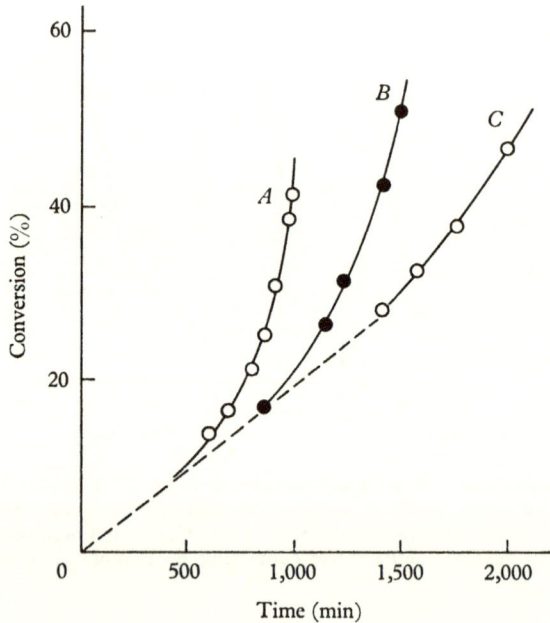

Fig. 3. Conversion/time plots for polymerization of methyl methacrylate. *A*, Pure monomer; *B*, 40% solution of monomer in amyl acetate; *C*, 40% solution of monomer in benzene.

increasing viscosity. His work with Merrett[15] and Nicholson[16] on the polymerization of styrene at high pressures led to the conclusion, confirmed by experiment, that there is an inverse relationship between the velocity constant for termination and the viscosity of the reaction mixture.

Confirmation of the ideas of Norrish and Smith[13] came from polymerizations performed in the presence of diluents. Those liquids which are good solvents for the polymer suppress the acceleration by keeping the viscosity low and enabling the large polymer radicals to diffuse comparatively readily so that they can interact normally. Liquids which cause precipitation of the polymer produce very marked acceleration; the polymer radicals are trapped in the gel so that their interaction is very largely suppressed although their reaction with the freely diffusing monomer is almost unaffected. A similar explanation was advanced later [17] in connection with acceleration during the polymerization of acrylonitrile when the conditions are such that there is precipitation of polymer; definite evidence was found for the presence in the polymer of trapped radicals. Norrish and Smith [13] also showed that conditions likely to produce a reaction mixture of low viscosity reduce the tendency for acceleration to occur. They showed, for example, that polymerizations performed at high temperature or with high concentrations of initiator begin to accelerate only at comparatively late stages.

Later work by Bengough [18] in Melville's laboratory showed that diffusion control of the termination step in radical polymerizations is quite common. An elegant thermal method was developed for examination of the non-stationary state in photo-sensitized polymerizations. It is necessary only to follow the reaction for quite a short period and this can be done at any stage even when the mixture has gelled. The velocity constants for termination and propagation and the activation energies for these processes were determined. It was shown that in very viscous systems there can be a reduction in the velocity constant for propagation. Benson, [19] North and Reed [20] and Allen and Patrick [21] also have made notable contributions to the study of the diffusion control of the elementary steps in radical polymerizations.

Acceleration generally occurs in those systems which involve precipitation of polymer as it is formed. In this connection, Bengough and Norrish studied the polymerizations of vinyl and vinylidene chlorides [22,23] and the co-polymerization of these monomers. [24] Even in homogeneous systems, the polymerizations of these monomers are commonly characterized by a period of inhibition followed by one of acceleration. The inhibition can be eliminated by rigorous purification of the reactants but acceleration remains if the system is one involving precipitation of polymer.

Bengough and Norrish [22] used a simple but very effective procedure to confirm that the acceleration does not result from the gradual consumption of an impurity which acts as a retarder. Limb *A* of the twin dilatometer (see Fig. 4) was charged with benzoyl peroxide and monomer; benzoyl peroxide only was introduced into limb *B*. After some polymerization had occurred in the limb *A*, the seal between *A* and *B* was broken and the unreacted monomer was distilled into limb *B*. The second stage of the polymerization in *B* exhibited exactly the same kinetic characteristics as the first stage in *A* (see Fig. 5).

In another experiment of Bengough and Norrish, [22] the residual monomer was removed from a reaction mixture leaving the precipitated polymer and unused initiator; fresh pure monomer was then added. Polymerization was resumed at the same rate as that being observed when the first stage of the experiment was stopped. These and other observations indicate that the precipitated polymer can act as a catalyst for the polymerization but only if

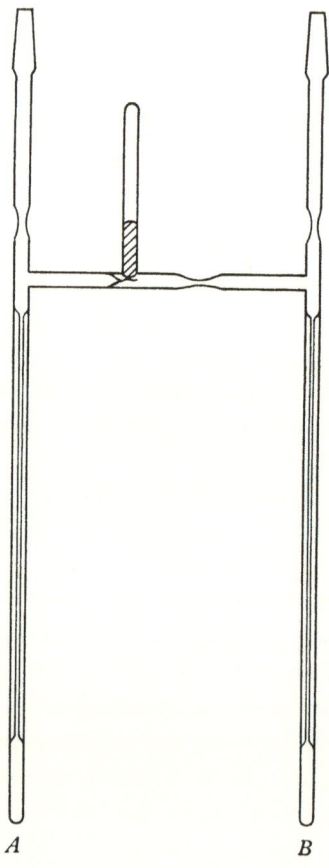

A   *B*

Fig. 4. Twin-dilatometer used by Bengough and Norrish.

reactive centres are simultaneously being generated in the liquid phase. The polymer can, therefore, be described as a co-catalyst.

The precipitated polymer particles are thought to contain immobile radicals able to engage readily in growth reactions but not in termination processes. This view is generally accepted as applicable to systems where polymer is precipitated as it is formed. An important difference between the systems considered by Bengough and Norrish and the case of polyacrylonitrile, for example, is in the nature of the process by which polymer radicals accumulate in the precipitate. Undoubtedly some growing polymer radicals must be trapped in the polymer particles, but Bengough and Norrish [22] concluded that a significant number of radicals are actually produced on the surfaces of the polymer particles by a transfer process involving radicals which may be

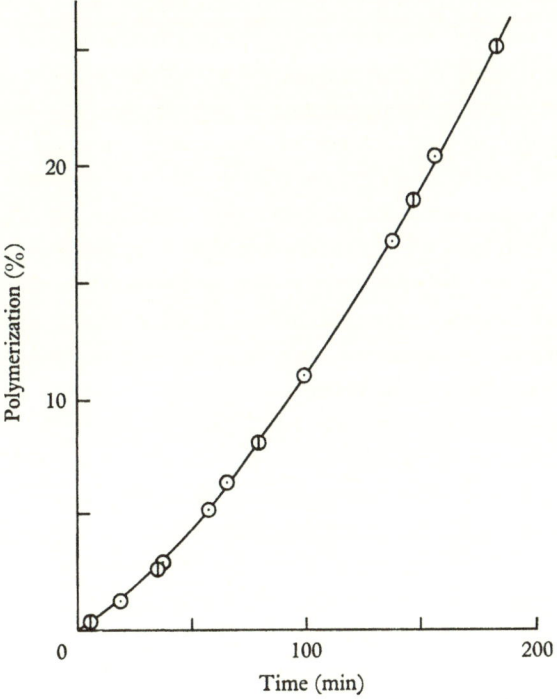

Fig. 5. Conversion/time plots for polymerization of vinyl chloride in twin-dilatometer. ⊙, Polymerization in arm A; ⊕ polymerization in arm B. A refers to the first stage and B to the second stage (the curve is shifted laterally for clarity).

those derived directly from the initiator or short-chain polymer radicals. Their interpretation of the acceleration observed in certain of the co-polymerizations is discussed in another section; it differs somewhat from the scheme just discussed because of the different physical form of the precipitate.

4. Polymerization at high pressures

Norrish began a study of polymerization at very high pressures at about the same time as he started a programme involving the use of very intense sources of light for photochemical reactions. The high pressure work was performed with Merrett[15] and later with Nicholson.[1,16] The first problem, the development of specialized apparatus and techniques new to the laboratory, was successfully overcome. It was decided to examine the polymerization of styrene since the behaviour of that monomer at ordinary pressures was reasonably well understood; further, it was decided to confine the investigation to the early stages of the reaction and so to avoid complications, such as diffusion control of termination, arising from the accumulation in the reaction mixture of excessive quantities of polymer. The progress of polymerization was followed by two techniques — by sampling, and by following the small reduction in pressure arising when the polymerization was conducted at constant volume. Viscometric and osmotic measurements were made on the recovered polymers.

Systematic studies were made of: (i) the effect of the concentration of initiator (benzoyl peroxide) upon the rate of polymerization; (ii) the effect of pressure upon the overall rate of polymerization; (iii) the effect of pressure upon the degree of polymerization of the product. Over the whole range between 0 and 5,000 kg cm^{-2}, the order of the reaction with respect to benzoyl peroxide is not very different from 0·5. For a given concentration of initiator, the rate of polymerization increases continuously with pressure over the whole range examined; the molecular weight of the product increases with pressure up to about 3,000 kg cm^{-2} but is almost independent of it at higher pressures (see Fig. 6). It was recognized that each of the elementary steps in the polymerization might be affected by changes in pressure; various suggestions were made and they were, in the main, substantiated by later work.

Nicholson and Norrish[1,16] presented two papers at a discussion in 1956 in Glasgow of the Faraday Society on reactions at high pressure. An examination[1] of the thermal decomposition of benzoyl peroxide in carbon tetrachloride over a wide range of pressures showed that the direct rate of dissociation of the peroxide to radicals is reduced by increasing the pressure; the rate of initiation of polymerization should be reduced correspondingly and so the increase in the rate of polymerization with pressure cannot be attributed to a rise in the rate of initiation.

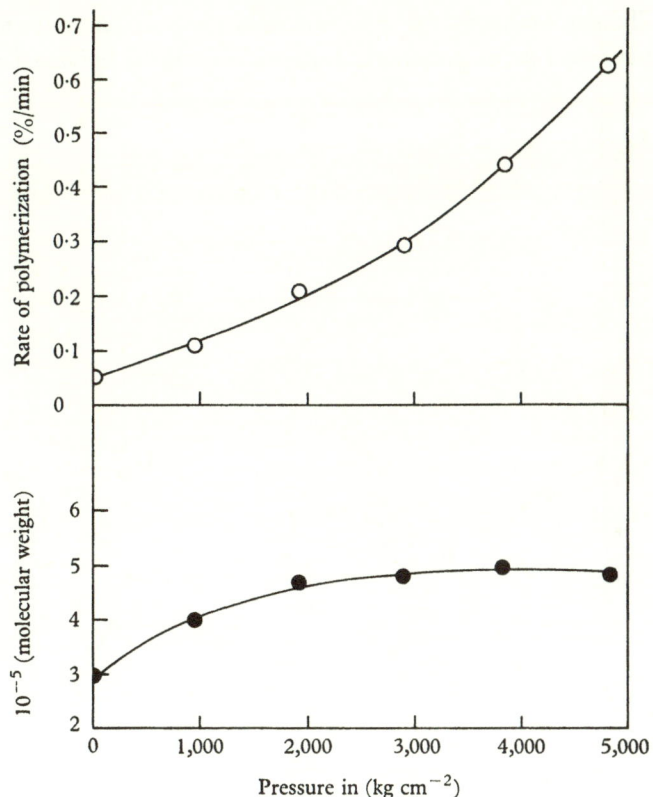

Fig. 6. Effects of pressure on polymerization of styrene. Reactions performed at 60 °C with benzoyl peroxide as sensitizer.

The velocity constants for propagation (k_p) and termination (k_t) were determined [16] for photo-sensitized polymerizations at pressures between 0 and 3,000 kg cm^{-2}; the sector technique was used in conjunction with molecular weight measurements for calculations of kinetic chain lengths and rates of initiation. The value of k_p rises exponentially with pressure as is expected for a bimolecular process of association. On the other hand, k_t falls with

14-2

increasing pressure; the decrease is marked between o and 1,000 kg cm^{-2} but is less pronounced at higher pressures (see Fig. 7). The fall in k_t with pressure was attributed to an increase in the viscosity of the system and the consequent reduction in the rate at which large polymer radicals can come together for inter-action. The simultaneous rise in k_p and fall in k_t would account for the increases in rate of polymerization and degree of polymeriza-

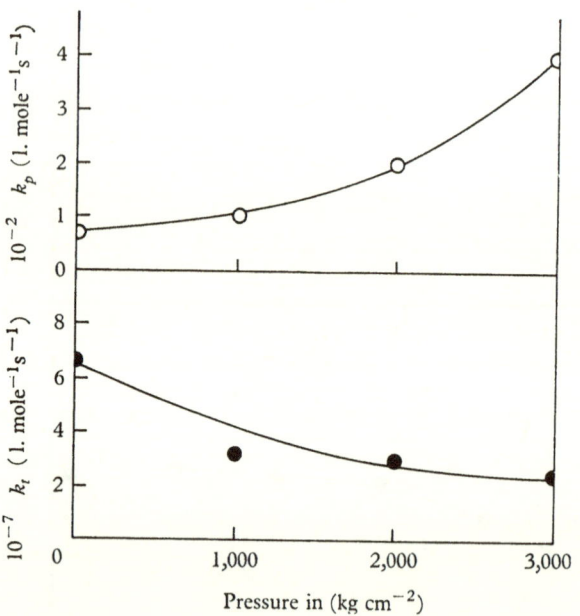

Fig. 7. Variation with pressure of velocity constants for propagation (k_p) and termination (k_t) in polymerization of styrene at 30 °C.

tion. At higher pressures, it is thought that the main control of molecular weight comes from transfer processes; these reactions would be accelerated by rises in pressure, and it is believed that molecular weight remains more or less steady because the effects of pressure upon the rates of propagation and transfer are comparable.

5. Co-polymerization

Some of the very first quantitative studies of co-polymeriza-tion were performed in Norrish's laboratories. He was quick to appreciate that examination of the behaviour of mixtures of monomers could, in many cases, yield information concerning the processes occurring during the polymerizations of single monomers.

Norrish and Brookman [25] described co-polymerizations of styrene and methyl methacrylate with various divinyl compounds such as divinyl benzene. They prepared a series of co-polymers having various densities of cross-linking. They reported that the overall rate of polymerization of styrene or methyl methacrylate can be considerably depressed by the presence in the reaction mixture of comparatively small amounts of the polyfunctional monomer. They suggested that the behaviour of a divinyl compound of general formula $(CH_2=CH)_2X$ is largely governed by the electron-attracting or electron-repelling characteristics of the group X. They further suggested that the characteristics in polymerization of a monomer of formula $CH_2=CHR$ are controlled by electronic effects originating in the group R. The effects of substituents upon the reactivity of the carbon–carbon double bond have been extensively considered by many polymer chemists. The Q and e scheme of Alfrey and Price, [26] and the later scheme of Bamford, Jenkins and Johnston [27] have been successful in correlating data concerning the behaviour of monomers, polymer radicals and transfer agents; in both cases, the treatment is based on the idea that polar effects have profound influences upon the rates of the elementary processes in radical polymerizations.

One of the earliest accounts of systematic study of the kinetics of radical co-polymerization was given by Norrish and Brookman. [12] It described work, involving methyl methacrylate and styrene, which had been undertaken as a contribution to the general study of the mechanism of polymerization processes. The authors put forward various ideas which were shown, in the fullness of time, to be unacceptable but the paper contains mentions of topics which have proved to be of outstanding interest.

Norrish and Brookman [12] did not consider the composition of the co-polymer and its dependence upon the composition of the feed. They directed attention to the more difficult problem of the variation of the rate of polymerization with the composition of the feed, but in so doing they clearly set out the stationary state equation which is the key to the solution of the problem of co-polymer composition, namely that the rates of the two reactions

(polymer radical of type A) + (monomer of type B)

→ (polymer radical of type B)

(polymer radical of type B) + (monomer of type A)

→ (polymer radical of type A)

are equal. Further, they recognized the existence of four distinct growth reactions, each with its characteristic velocity constant. It is known now that only in certain special binary co-polymerizations is it necessary to consider more than four growth reactions; this situation arises when the reactivity of a polymer radical is not governed entirely by the nature of the terminal monomer unit.

It was observed that the rate of polymerization of a mixture of styrene and methyl methacrylate depends to a marked extent upon the composition of the mixture. This general feature of many radical co-polymerizations has attracted much attention; for a while, there appeared to be a satisfactory explanation based upon the so-called ϕ factor and the belief that polymer radicals of one type react more readily with polymer radicals of the other type than with their own kind. Closer examination has revealed that this explanation is of doubtful validity. [2]

Bengough and Norrish[24] also entered the field of co-polymerization; they extended their investigation of the polymerizations of vinyl and vinylidene chlorides to examine the radical co-polymerization of these monomers under conditions which led to precipitation of the product. At about that time, Bamford and co-workers[17] were engaged in their extensive study of another monomer, acrylonitrile, which gives a relatively insoluble polymer but there was a general reluctance in academic laboratories to examine systems having the added complication of precipitation of polymer as it is formed.

An ingenious procedure was used for continuous monitoring of the composition of the mixture of vinyl and vinylidene chlorides during polymerization. [24] The vapour pressure of the mixture of monomers was measured and used to deduce the relative amounts of the two components unreacted; the co-polymer, being insoluble in the reaction mixture, did not interfere. The information derived from measurements of vapour pressure was combined with the results of dilatometry to give not only the total rate of polymerization but also the rates at which the two monomers separately were consumed. They were thus able to find the compositions of the co-polymers and to avoid the substantial errors which would have arisen if these compositions had been derived from rather uncertain analyses for chlorine in the products.

Auto-acceleration was observed in the co-polymerization as in the similar polymerizations of the separate monomers. The acceleration was less marked for systems giving rise to co-polymers containing between about 10 and 50 % of vinylidene chloride; such systems gave a gel-like precipitate instead of a powdery one. The acceleration observed in 'powdery' systems was attributed to the co-catalytic effect considered in connection with the homo-polymerizations; it was concluded, however, that in the 'gel-like' systems there was a decrease in the rate of termination resulting from an effect of the type described by Norrish and Smith. [13]

The papers so far referred to in this section were concerned with co-polymers in which the various monomer units were distributed more or less at random along the polymer chains. Guillet and Norrish [11] investigated the photolysis of polymethylvinylketone in solution and showed that it could be used in connection with the formation of co-polymers of a special type, namely, graft co-polymers, in which the monomer units of the one type form comparatively long branches attached to main chains consisting of monomer units of the other type. There has been considerable interest in the preparation of co-polymers of special types such as this, mainly because they might well have rather unusual physical properties.

The photolysis of polymethylvinylketone generally resembles the photolysis of aliphatic ketones of low molecular weight, a process which received thorough attention in Norrish's laboratory. Product analysis, chemical tests on the residual polymer and examination of its molecular weight, confirmed that two types of primary reaction are possible during the photolysis of the polymer (see Fig. 8). Type I decomposition involves the production of a pair of free radicals by scission of a bond adjacent to a carbonyl group;

the main chain of the polymer molecule is unaffected and a pendant group is detached. Decomposition of Type II, however, involves a split of a carbon–carbon bond α–β to the carbonyl group; free radicals are not liberated but there is degradation of the

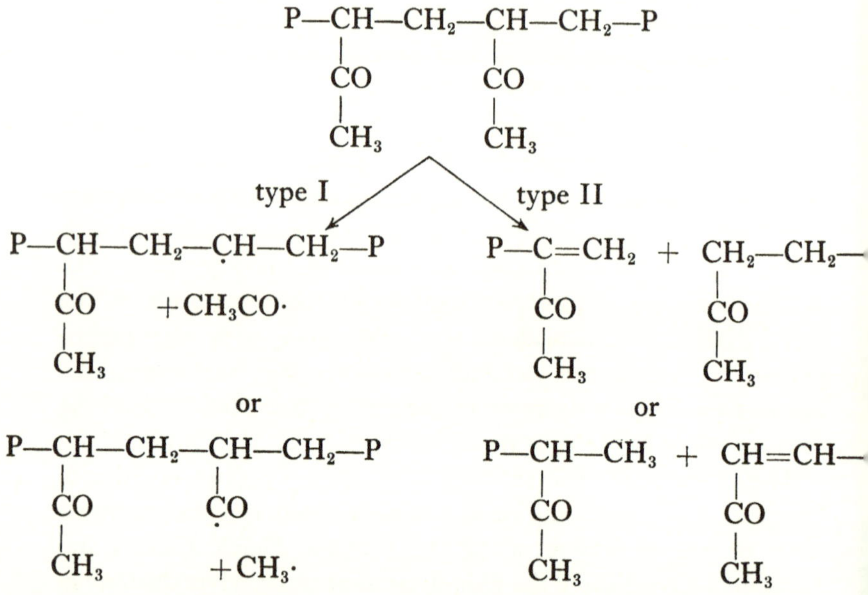

$$P—CH—CH_2—CH—CH_2—P$$
$$\underset{\displaystyle CH_3}{\overset{\displaystyle CO}{|}} \qquad \underset{\displaystyle CH_3}{\overset{\displaystyle CO}{|}}$$

type I $\qquad$ type II

$$P—CH—CH_2—\overset{\cdot}{C}H—CH_2—P \qquad P—C=CH_2 \ + \ CH_2—CH_2—$$
$$\underset{CH_3}{\overset{CO}{|}} \quad +CH_3CO\cdot \qquad\qquad \underset{CH_3}{\overset{CO}{|}} \qquad \underset{CH_3}{\overset{CO}{|}}$$

or $\qquad\qquad\qquad\qquad$ or

$$P—CH—CH_2—CH—CH_2—P \qquad P—CH—CH_3 \ + \ CH=CH—$$
$$\underset{CH_3}{\overset{CO}{|}} \quad \underset{\ \cdot}{\overset{CO}{|}} \quad +CH_3\cdot \qquad \underset{CH_3}{\overset{CO}{|}} \qquad \underset{CH_3}{\overset{CO}{|}}$$

Fig. 8. Reaction scheme for photolysis of polymethylvinylketone.

polymer and terminal unsaturation is introduced. It was concluded that the unsaturated polymer molecules produced in the Type II decomposition can react with radicals liberated in the Type I decomposition to give polymer radicals as in the equation

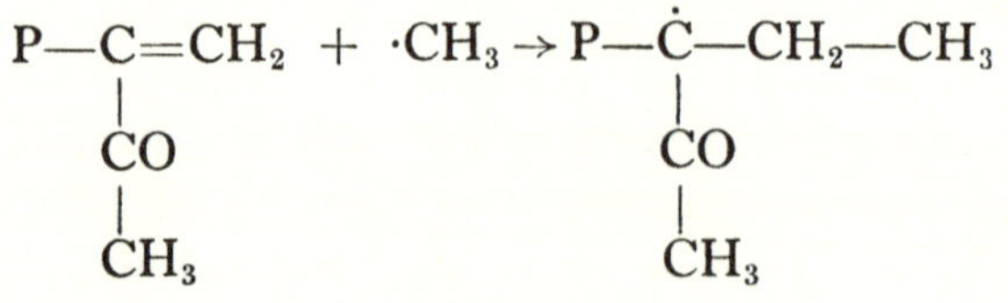

$$P—C=CH_2 \ + \ \cdot CH_3 \rightarrow P—\overset{\cdot}{C}—CH_2—CH_3$$
$$\underset{CH_3}{\overset{CO}{|}} \qquad\qquad\qquad \underset{CH_3}{\overset{CO}{|}}$$

There are, therefore, two distinct routes by which macromolecular radicals can be produced as a result of the photolysis of poly-methylvinylketone.

Photolysis of the polymer in the presence of a polymerizable monomer would be expected to lead to graft co-polymer. This belief was confirmed by experiments reported at the Milan–Turin meeting on polymers in 1954; the first account referred to acrylonitrile but subsequently graft co-polymers containing methyl methacrylate or vinyl acetate were prepared also. Fractionations showed that the formation of the co-polymer was accompanied by the production of homo-polymer, e.g. polyacrylonitrile, and that the final product also contained some polymethylvinylketone. In this respect, the method has the same limitations as so many of the other procedures for preparing graft co-polymers, namely, that the products are mixed with homo-polymers and can be isolated only by using some method of fractionation.

The need for fractionation arose, for example, in the work of Merrett, [28] who grafted methyl methacrylate or styrene chains on to natural rubber. In his work, the reactive points on the main polymer chains were produced by a transfer process mainly involving benzoyloxy or phenyl radicals formed by the dissociation of benzoyl peroxide. The necessity for fractionation was also one of the experimental difficulties in a study of the formation of branched molecules during the polymerization of styrene and vinyl acetate. [29] The growth of branches originated from transfer to polymer; its importance was assessed by means of a tracer technique. In one procedure unlabelled branches were attached to labelled polymer; in another, labelled monomer was used to provide the branches attached to unlabelled polymer.

Fractionation for separation of the products of different types is also important in connection with block co-polymers; these materials generally resemble graft co-polymer but they are composed of essentially linear molecules having the various types of monomer unit segregated in long sequences. Block co-polymers are generally made by techniques involving the re-activation of end-groups of pre-formed polymers; the methods for preparing graft co-polymers, including that of Guillett and Norrish, [11] involve a process which can be described as re-activation of monomer units of pre-formed polymer.

Guillet and Norrish [11] recognized that cross-linked products might be formed during the grafting process if the growing side-chains interact by combination. In the case of methyl methacrylate, a very detailed analysis of the results indicated that at 25 °C about 45 % of the interactions of polymer radicals occur by combination and that the remainder occur by disproportionation. This conclusion agrees with one derived by an entirely different procedure, namely, determination of the average number of initiator fragments incorporated in a polymer molecule produced in a sensitized polymerization.[30]

The work of Andersen, Norrish and Simons[9,10] was not primarily concerned with co-polymerization but it gave important information about another example of a co-polymerization of an unusual type; a substance, unable to polymerize on its own, can enter co-polymers with comparative readiness. It was shown that anthracene can act as a photo-sensitizer for the polymerization of styrene but that it can also engage in co-polymerization; both triplet anthracene and the unexcited singlet anthracene can be involved in the process. The evidence for the occurrence of co-polymerization was based on a comparison of the number of anthracene molecules being removed and the number of polymer molecules being formed in the system; it was clear that, in many cases, an average polymer molecule contained 10 molecules of anthracene. Photo-sensitization of the polymerization of styrene could be brought about by pyrene and chrysene but no evidence for co-polymerization involving these hydrocarbons was found.

Careful interpretation of experimental results made it possible to characterize the various steps involved in the co-polymerization of styrene and anthracene. The values of the velocity constants for the growth reactions

$$P \cdot + {}^3A \rightarrow P \cdot A \cdot$$

$$P \cdot A \cdot + M \rightarrow P \cdot$$

where $P \cdot$ and M represent a polystyrene radical and a molecule of monomeric styrene respectively, were found to be approximately $1 \cdot 5 \times 10^{11}$ l. mole^{-1} s^{-1} and $0 \cdot 8$ l. mole^{-1} s^{-1} respectively. It was confirmed that the molecule of anthracene is most probably incorporated in the 9,10 position. The velocity constant for the growth reaction in the polymerization of styrene at the same temperature is about $1 \cdot 2 \times 10^2$ l. mole^{-1} s^{-1}. The low reactivity of the styrene–anthracyl radical towards monomeric styrene was attributed to the marked resonance stabilization of the radical.

6. Cross-linking

Work on the co-polymerization of styrene and methyl methacrylate with divinyl compounds [25] of general formula $(CH_2{=}CH)_2X$ was mentioned in section 5. A whole range of cross-linked products was obtained; their properties depended upon the precise nature of the divinyl component and upon its concentration in the monomer mixture. Some of the co-polymers were insoluble and infusible materials completely unaffected by liquids; at the other extreme, there were materials which became highly swollen in certain liquids. This type of cross-linking occurred during polymerization and another example of such a process was mentioned in connection with the preparation of graft co-polymers using polymethylvinylketone as stock. [11] An alternative general method of cross-linking involves the reactions of dead polymer molecules; Norrish was concerned in an examination of a process of this type also.

The chlorination of polyvinyl chloride in solution was examined since it appeared that it was possible to prepare a useful material with properties significantly different from those of the starting material. The subject proved to be interesting and, some twenty years later, to be of real industrial significance; Norrish, however, did not publish an account of the work. In the course of the research, it was discovered that polyvinyl chloride, its chlorinated derivatives and similar polymers can be cross-linked by the action of ferric chloride, stannic chloride, and other Friedel–Crafts catalysts. [31] The process of cross-linking is accompanied by a gradual darkening of the polymer accounted for by the introduction into the chain of unsaturation and conjugation.

A simplified reaction scheme was proposed; it was based on the belief that an intermolecular elimination of hydrogen chloride causes the one effect and an intramolecular elimination the other (see Fig. 9).

It was demonstrated that cross-linking and discoloration occur in a similar way for polyvinyl acetate; in this case, acetic acid is eliminated. Chlorinated polystyrenes do not undergo cross-linking with Friedel–Crafts catalysts if the chlorine atoms are confined to the benzene rings; an examination of the effects of the catalysts upon such polymers can be used simply but reliably to give information on the positions of the chlorine atoms.

The rates of cross-linking and discoloration vary according to the nature of the solvent. Liquids, such as nitrobenzene, having high dielectric constants are associated with rapid reactions; this observation fits with accepted ideas concerning the general mechanism of Friedel–Crafts reactions.

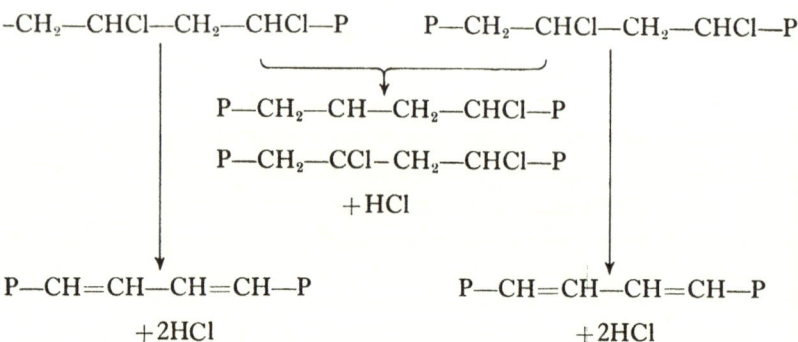

$-CH_2$—$CHCl$—CH_2—$CHCl$—P $\qquad$ P—CH_2—$CHCl$—CH_2—$CHCl$—P

P—CH_2—CH—CH_2—$CHCl$—P

P—CH_2—CCl-CH_2—$CHCl$—P

$+HCl$

P—CH=CH—CH=CH—P $\qquad$ P—CH=CH—CH=CH—P

$+2HCl$ $\qquad\qquad\qquad\qquad$ $+2HCl$

Fig. 9. Reaction scheme for introduction of cross-linking and unsaturation into polyvinyl chloride by the action of Friedel–Crafts catalysts.

The work on catalysed cross-linking was qualitative in nature but this does not reduce its significance. It placed on a proper chemical basis some observations, familiar to plastic technologists, concerning the deleterious effects produced in polyvinyl chloride by certain substances. Further, it gave more support to the view that high polymers are not exceptional in their chemical behaviour and undergo the reactions appropriate to their functional groups. In this connection, reference may be made to the work on the photolysis of polymethylvinylketone;[11] the pattern of behaviour matched that for aliphatic ketones of low molecular weight.

7. Polymerization of aldehydes

One of Norrish's first papers concerned with polymerization was presented at the meeting of the Faraday Society at Cambridge in 1935 and it dealt with the polymerization of gaseous formaldehyde. [32] The paper must be considered as playing an important part in the development of kinetic studies of polymerization.

Previous workers, examining the polymerization of supposedly pure gaseous formaldehyde, had encountered problems of irreproducibility; the kinetics of the process appeared to be very complicated. Carruthers and Norrish [32] overcame these difficulties by using their discovery that formic acid is a powerful promoter of the reaction. They obtained consistent results which led to the derivation of a simple kinetic scheme which could be matched with a plausible mechanism.

The reaction scheme was based upon a set of elementary reactions of a type common in kinetic analysis of chain processes. There was clear distinction between initiation, propagation, branching and termination, as indicated here:

initiation

$$\underset{O}{\overset{H \quad OH}{\underset{\|}{C}}} + CH_2O \rightarrow \underset{O}{\overset{H \quad OCH_2OH}{\underset{\|}{C}}}$$

propagation $P—CH_2OH + CH_2O \rightarrow P—CH_2OCH_2OH$

branching $P—CH_2OH + HCOOH \rightarrow P—CH_2OCH(OH)_2$

termination $P—CH_2OH + CH_2O \rightarrow P—CH_2CHO + H_2O$

The OH group on a polymer molecule is regarded as the active site for reaction. The proposed branching step gives rise to a branched chain polymer as well as to a multiplication of centres. The rate of consumption of formaldehyde was given by the expression

$$\text{Rate} = \frac{k' p_{CH_2O} \, p_{HCOOH}}{k'' p_{CH_2O} - k''' p_{HCOOH}}$$

Perhaps one of the most significant features to emerge from the study of the polymerization of formaldehyde was the observation that the reaction can be suppressed if the temperature of the system is high enough. Dainton and Ivin [33] were able to explain this effect, which is not confined to gaseous formaldehyde, in both kinetic and thermodynamic terms; they introduced the concept of 'ceiling temperature'.

It should be recalled that Carruthers and Norrish described a simple mercury seal to protect greased stopcocks in glass apparatus to be used at elevated temperatures. The devise is so simple and effective that it is worthy of wider use.

Further study in Norrish's laboratory [34] of the polymerization of gaseous formaldehyde showed that hydrogen chloride, boron trifluoride and stannic chloride are even more effective than formic acid in promoting the reaction. The possibility of the overall rate of polymerization being governed by the rate of diffusion of reactants through the reaction vessel was recognized. It was apparent that the original reaction scheme needed modification but the necessary alterations did not destroy the essential simplicity.

Hydrogen chloride cannot engage in a branching reaction of the type envisaged for formic acid and a process of a different type had to be formulated. It was proposed that the catalyst attacks polymer chains at points along their length causing them to split into fragments capable of engaging in further polymerization, as in the equation

$$P-CH_2OCH_2O-P-CH_2OH + HCl$$

$$\rightarrow \quad P-CH_2OH + ClCH_2O-P-CH_2OH$$

The reaction, therefore, is one which leads to kinetic but not structural branching. The proposal was supported by the observation that hydrogen chloride can also catalyse the depolymerization of the polymers at high temperatures; evidently, under these conditions, the reactive centres tend to shed molecules of monomer instead of to react with them.

Carruthers and Norrish [32] had proposed that the reactive centres in the polymerization of formaldehyde are hydroxyl groups at the ends of polymer molecules. The equation suggested for the branching involving hydrogen chloride is based on this assumption also, but the later paper was published at a time when the general features of ionic polymerizations had been recognized and it was suggested that the reactive centres actually are charged bodies.

At the meeting where Carruthers and Norrish presented their paper, Travers [35] gave a short account of a peculiar phenomenon involving acetaldehyde. It was found that the frozen aldehyde does not always melt to a limpid liquid but sometimes becomes very viscous or even gel-like. Staudinger [36] pointed out that there is probably polymerization to a long chain substituted polyoxymethylene having the repeating unit

$$-CH(CH_3)-O-$$

Shortly afterwards, Letort [37] also reported preparation of the polymer.

Subsequent work in Norrish's laboratory [38] contributed to the solution of this peculiar reaction which appears to occur with great readiness at or near the freezing point (-120 °C) of the aldehyde. It was plain that gross irreproducibility would be a great handicap to any thorough study of the system, and so a technique was developed whereby monomer could be frozen in a controlled manner. Aldehyde vapour was passed at a measured rate from a reservoir into a receiver cooled in liquid nitrogen. Controlled amounts of additives could be mixed into the vapour stream and it was shown that some of them have profound effects upon the course of reaction. Acetic acid increases the yield of polymer but reduces its average molecular weight very markedly; water inhibits the polymerization.

The polymer of acetaldehyde was found to be unstable even at 25 °C; degradation and depolymerization could be accelerated by acids. It was proposed that acetic acid exerts its catalytic effects upon the polymerization and depolymerization by a chain scission process of a type similar to that believed to occur in the polymerization of formaldehyde. [34] Suggestions were made that formation of polymer proceeds by an ionic mechanism and that a non-stationary state might be observed in the system. A proposal was made concerning the part played by crystallization of the monomer in the formation of polymer. This last point has been considered in great detail by Letort and his various co-workers [39] who have been able to correlate the crystal structure of the solid aldehyde with the occurrence of polymerization. Later workers have shown that acetaldehyde can be polymerized in processes of other types also and that it is possible to prepare stereoregulated polymers having physical properties quite different from those of the rubbery polymer prepared by the crystallization technique.

As early as 1928, Norrish and Griffiths[40] mentioned that glyoxal can be converted into a solid product. The possibility of polymerizing this substance has been largely ignored until recently, but it has been shown[41] that a polymer can be obtained and that its structure can be represented thus:

$$-\begin{bmatrix} O-C \\ \| \\ CH \\ | \\ OH \end{bmatrix}_x - \begin{bmatrix} O-CH-O-C- \\ | \quad \| \\ CH \quad CH \\ \| \quad / \\ O \ldots HO \end{bmatrix}_y$$

The comparatively small amount of published work concerning polymers of aldehydes and related compounds was reviewed[42] in 1952. At that time, there seemed little likelihood that the polymers could ever be of commercial significance but the picture changed dramatically when materials such as Delrin and Celcon were developed. Very useful polymers having formaldehyde as the main component are now known; some of them are actually prepared from trioxan, the cyclic trimer of formaldehyde, and many of them are actually co-polymers. The possibility of commercial exploitation of polyacetaldehyde has been mentioned on several occasions but it has yet to be realized. The experience with polyformaldehyde, however, suggests that it would be foolish to suppose that polyacetaldehyde will never be made in a useful form. Norrish's work on the polymerization of aldehydes did not contribute directly to the development of the commercial materials but it did provide most useful general information and it contributed significantly to the study of polymers having backbones not consisting entirely of carbon atoms.

8. Cationic and Ziegler–Natta polymerizations

Most of Norrish's work on polymerizations has been concerned with radical reactions but he has ventured into other fields also. Cationic polymerizations and those promoted by mixed organo-metallic catalysts are experimentally very difficult for exact study; after the report of the initial discoveries, the most pressing need has been to assemble sufficient quantitative data to lead to a full and proper understanding of the detailed mechanisms. Norrish and his co-workers attempted to devise comparatively simple experimental systems and to use them to obtain fundamental information; it was hoped that it would then be possible to discriminate between the various mechanisms proposed on the basis of qualitative or semi-quantitative observations. Norrish's work on the Polymerization of aldehydes was essentially an exercise in ionic Polymerization but here it is considered separately (see section 7).

The polymerization of isobutene can proceed very rapidly to give polymers of very high molecular weight even at temperatures below -80 °C. The reaction is promoted by catalysts such as boron trifluoride and titanium tetrachloride and is clearly of an ionic type. Very thorough work at Manchester[43] showed that it is essential that the reaction mixture should contain a third compoment, referred to as a co-catalyst. Several substances, including water, can function in this way.

Norrish and Russell[44] examined the polymerization of iso-outene using stannic chloride as catalyst. This particular catalyst was selected for detailed study because it is less active than boron trifluoride so that there seemed to be reasonable prospects of performing quantitative kinetic studies of the polymerization. The first task was to establish whether or not a co-catalyst is essential in this system also. Polymerizations were performed in dilatometers; the reactants were measured out by means of their partial pressures in a bulb of fixed volume. Extreme precautions were taken eliminate adsorbed water from the vacuum line; the dilatometer has baked under high vacuum at 300 °C and the whole of the glass opparatus was coated with a hydrophobic monolayer.[45] Even after stringent precautions and rigorous purification of materials, there was still an appreciable rate of polymerization even although there had been no deliberate addition of water. It seemed that the triest reaction mixtures still contained roughly 3×10^{-3} mole % of after; this quantity corresponded to only about 0·01 mg in the eilatometer. Subsequently, systematic studies were made of the facts of the concentrations of catalyst and co-catalyst upon the rate of polymerization and upon the molecular weight of the product.

Ziegler and Natta and their co-workers discovered that mixed organo-metallic catalysts, such as those formed by the interaction of an aluminium alkyl and a halide of a transition metal, have remarkable effects upon the polymerizations of ethylene and α-olefins. With the aid of such catalysts, ethylene can be converted to a high polymer at ordinary pressures; the polymer is of high crystallinity, this being due to the absence of the branches which are quite numerous in polyethylene made at high pressures by a radical reaction. The polymers made from the α-olefins by these catalysts are stereo-regulated and therefore of high crystallinity.

Most of the studies of the Ziegler–Natta polymerizations have involved the use of a hydrocarbon diluent, but Lipman and Norrish [46] decided to eliminate one of the components of the reaction mixture and thereby to effect a simplification. They examined the polymerization of ethylene in the gas-phase.

A method of working was developed so that a mixture of gaseous ethylene and the vapour of aluminium trimethyl was made in one bulb, and a similar mixture of ethylene and titanium tetrachloride was made in a second bulb. The two mixtures were admitted to a reaction vessel; after a short induction period, a 'smoke' appeared in the vessel and subsequently was deposited on the walls. The progress of polymerization was followed manometrically. Polymers were recovered from the vessel and examined viscometrically; it was found that generally the polymers had molecular weights between about 5×10^4 and 5×10^5.

The rate of consumption of ethylene fell during the course of a reaction; if the pressure of ethylene was restored to its original value by the addition of fresh monomer, the original rate of polymerization was not restored. It appears, therefore, that the effective amount of catalyst in the system falls during the course of polymerization. It was concluded that this effect is not due to ageing of the catalyst or to aggregation of the catalyst particles. Detailed consideration was given to the possibility that the apparent removal of catalyst was due to its surface becoming covered by polymer. The authors favoured a mechanism for the growth of polymer in which propagation occurs by the successive insertion of monomer units between the chain and the active site. The steric considerations governing such an insertion could well account for the stereoregulation in the chain structure. It was concluded that the deposited polymer has a structure open enough to allow the free passage of monomer through it. Several groups of workers have considered the structures of the active catalysts in the Ziegler–Natta polymerizations, and the natures of the interactions between the two components of the catalyst system. Lipman and Norrish [46] considered the suggestions and were able to show that several of them were consistent with the observed kinetics.

REFERENCES

1   Nicholson, A. E. and Norrish, R. G. W. *Disc. Faraday Soc.* 1956, **22**, 97.
2   Bevington, J. C. *Radical Polymerization.* Academic Press, London, 1961.
3   Norrish, R. G. W. *Trans. Faraday Soc.* 1937, **33**, 1521.
4   Martin, J. T. and Norrish, R. G. W. *Proc. Roy. Soc.* A, 1953, **220**, 322.
5   Norrish, R. G. W. and Searby, M. H. *Proc. Roy. Soc.* A, 1956, **237**, 464.
6   Edgecombe, F. H. C. and Norrish, R. G. W. *Nature, Lond.*, 1963, **197**, 282.
7   Hussain, F. and Norrish, R. G. W. *Proc. Roy. Soc.* A, 1963, **275**, 161.
8   Dainton, F. S. and James, D. G. L. *Trans. Faraday Soc.* 1958, **54**, 649.
9   Andersen, V. S. and Norrish, R. G. W. *Proc. Roy. Soc.* A, 1959, **251**, 1.
10   Norrish, R. G. W. and Simons, J. P. *Proc. Roy. Soc.* A, 1959, **251**, 4.
11   Guillet, J. E. and Norrish, R. G. W. *Proc. Roy. Soc.* A, 1955, **233**, 153, 172; *Nature, Lond.*, 1954, **173**, 625.
12   Norrish, R. G. W. and Brookman, E. F. *Proc. Roy. Soc.* A, 1939, **171**, 147.
13   Norrish, R. G. W. and Smith, R. R. *Nature, Lond.*, 1942, **150**, 336.
14   Schulz, G. V. and Blaschke, F. *Z. phys. Chem.* B, 1941, **50**, 305.
15   Merrett, F. M. and Norrish, R. G. W. *Proc. Roy. Soc.* A, 1951, **206**, 309.
16   Nicholson, A. E. and Norrish, R. G. W. *Disc. Faraday Soc.* 1956, **22**, 104.
17   Bamford, C. H., Barb, W. G., Jenkins, A. D. and Onyon, P. F. *The Kinetics of Vinyl Polymerizations by Radical Mechanisms.* Butterworth, London, 1958.
18   Bengough, W. I. and Melville, H. W. *Proc. Roy. Soc.* A, 1954, **225**, 330.

19 Benson, S. W. and North, A. M. *J. Am. Chem. Soc.* 1959, **81**, 1339.

20 North, A. M. and Reed, G. A. *Trans. Faraday Soc.* 1961, **57**, 859.

21 Allen, P. E. M. and Patrick, C. R. *Makromol. Chem.* 1964, **72**, 106.

22 Bengough, W. I. and Norrish, R. G. W. *Proc. Roy. Soc.* A, 1950, **200**, 301; *Nature, Lond.*, 1949, **163**, 325.

23 Bengough, W. I. and Norrish, R. G. W. *Proc. Roy. Soc.* A, 1953, **218**, 149.

24 Bengough, W. I. and Norrish, R. G. W. *Proc. Roy. Soc.* A, 1953, **218**, 155.

25 Norrish, R. G. W. and Brookman, E. F. *Proc. Roy. Soc.* A, 1937, **163**, 205.

26 Alfrey, T. and Price, C. C. *J. Polymer Sci.* 1947, **2**, 101.

27 Bamford, C. H., Jenkins, A. D. and Johnston, R. *Trans. Faraday Soc.* 1959, **55**, 418.

28 Merrett, F. M. *Trans. Faraday Soc.* 1954, **50**, 759.

29 Bevington, J. C., Guzman, G. M. and Melville, H. W. *Proc. Roy. Soc.* A, 1954, **221**, 437, 453.

30 Bevington, J. C., Melville, H. W. and Taylor, R. P. *J. Polymer Sci.* 1954, **12**, 449.

31 Bevington, J. C. and Norrish, R. G. W. *J. chem. Soc.* 1948, p. 771; 1949, p. 482.

32 Carruthers, J. E. and Norrish, R. G. W. *Trans. Faraday Soc.* 1936, **32**, 195.

33 Dainton, F. S. and Ivin, K. J. *Q. Rev.* 1958, **12**, 61.

34 Bevington, J. C. and Norrish, R. G. W. *Proc. Roy. Soc.* A, 1951, **205**, 516.

35 Travers, M. W. *Trans. Faraday Soc.* 1936, **32**, 246.

36 Staudinger, H. *Trans. Faraday Soc.* 1936, **32**, 249.

37 Letort, M. *C. r. Acad. Sci.* 1936, **202**, 767.

38 Bevington, J. C. and Norrish, R. G. W. *Proc. Roy. Soc.* A, 1949, **196**, 363.

39 Letort, M. and Richard, A.-J. *J. chim. Phys.* 1960, p. 752.

40 Norrish, R. G. W. and Griffiths, J. G. A. *J. chem. Soc.* 1928, p. 2829.

41 Brady, W. T. and O'Neal, H. R. *J. Polymer Sci.* A, 1965, **3**, 2337.

42 Bevington, J. C. *Q. Rev.* 1952, **6**, 141.

43 Plesch, P. H. *Nature, Lond.*, 1947, **160**, 868.

44 Norrish, R. G. W. and Russell, K. E. *Trans. Faraday Soc.* 1952, **48**, 91.

45 Norrish, R. G. W. and Russell, K. E. *Nature, Lond.*, 1947, **160**, 543.

46 Lipman, R. D. A. and Norrish, R. G. W. *Proc. Soc. Roy.* A, 1963, **275**, 310.

MODERN CONCEPTS OF THE MECHANISM OF HYDROCARBON OXIDATION IN THE GAS-PHASE

N. N. SEMENOV

1. A summary of modern concepts

The mechanism of hydrocarbon oxidation was formulated in the main towards the 1960's, after prolonged and extensive investigations. It was established for certain that this is a chain mechanism involving degenerate branching and yielding both oxygen-containing products (peroxides, aldehydes, alcohols, ketones, oxides, etc.) and cracking products (hydrocarbons of a lower molecular weight, unsaturated hydrocarbons, etc.).

The basic oxidation steps are considered, as a rule, to proceed by the following radical chain mechanism:

(0) $\quad RH + O_2 \to \dot{R} + H\dot{O}_2$ $\qquad$ chain initiation

(1) $\quad \dot{R} + O_2 \to R\dot{O}_2$ $\qquad\qquad\quad$ ⎫

(1') $\quad \dot{R} + O_2 \to olefin + H\dot{O}_2$ $\qquad$ ⎪

(2a) $\quad R\dot{O}_2 + RH \to ROOH + \dot{R}$ $\qquad$ ⎬ $\quad$ chain propagation

(2b) $\qquad\quad R\dot{O}_2 \to R'CHO + R''\dot{O}$ ⎪

(2') $\quad H\dot{O}_2 + RH \to H_2O_2 + \dot{R}$ $\qquad$ ⎭

(3) $\qquad\quad ROOH \to R\dot{O} + \dot{O}H$ $\qquad$ ⎫

(3') $\quad R'CHO + O_2 \to R'\dot{C}O + H\dot{O}_2$ ⎬ $\quad$ degenerate branching

(4) $\qquad\qquad R\dot{O}_2 \to$ $\qquad\qquad\qquad$ chain termination

Reaction (0) was first suggested by Hinshelwood [1] in 1947, and then theoretically substantiated by Semenov in 1954. [2] Semenov investigated the recombination of two radicals. Recombination ($2C_2H_5 \to C_4H_{10}$) is known to occur in parallel with disproportionation ($2C_2H_5 \rightarrow C_2H_4 + C_2H_6$) and both reactions seem to involve no activation barrier ($E \simeq 0$). Disproportionation was observed also on the interaction of other alkyl and alkoxyl radicals such as $CH_3 + CH_3O \rightarrow CH_4 + CH_2O$. [3] It would be natural to suggest that interaction of two radicals would always involve disproportionation. It follows that the reverse reaction, the formation of two radicals by reaction between a saturated and an unsaturated molecule, would occur in accordance with the principle of microscopic reversibility, without an activation barrier. As a rule, all these reactions are strongly endothermic and, consequently, their activation energy would be equal to their endothermicity.

However, the pre-exponential factors for these reactions remain uncertain. Semenov [2] has calculated the steric factor for the reaction $CH_2O + O_2 \rightarrow HCO + HO_2$ from experimental data on the kinetics of methane oxidation and found it to be $1/200$. It may well be that steric factors of this kind would be found also for other reactions of this type.

Chain initiation (0) and chain branching (3′) are reactions of this type. The first is endothermic by 45–50, and the second by 32–33 kcal mole^{-1}. Branching would be of importance only when considerably faster than initiation. For instance, as $D(H\text{—}C_3H_7)$ is higher than $D(H\text{—}COCH_3)$ by 10–15 kcal mole^{-1}, the heats of these reactions and, thus, their activation energies will differ by 10–15 kcal mole^{-1}. At $T = 300$ °C this would give a difference in rates of about four orders of magnitude.

Reactions (0) and (3′) were not studied directly, as their rates are very low due to their high endothermicity.† However, other reactions of the same type are of a low endothermicity or even exothermic. These would proceed at a high rate, as their activation barrier is zero. Certain reactions of this kind were studied.

Soviet scientists [4,5] have investigated recently the kinetics of formation of atoms and radicals in bimolecular processes that are either exothermic or required only a low energy. An example would be the elementary reactions involving molecular fluorine:

$$(5) \quad F_2 + C_2H_4 \rightarrow \dot{F} + FCH_2\dot{C}H_2 + 16 \text{ kcal mole}^{-1}$$

$$(6) \quad F_2 + HI \rightarrow \dot{F} + HF + \dot{I} + 27 \text{ kcal mole}^{-1}, \text{ etc.}$$

† The formation of H_2O_2 from HO_2 at the initial steps of CH_3CHO[50] and CH_2O[51] oxidation was reported in two papers only.

and also reactions of C_2H_5Li with alkyl halides in solution

$$C_2H_5Li + (C_6H_5)_3CCl \rightarrow LiCl + (C_6H_5)_3\dot{C} + \dot{C}_2H_5 + 16\cdot5\dagger$$

$$C_2H_5Li + C_2H_5I \rightarrow LiI + \dot{C}_2H_5 + \dot{C}_2H_5 - 14\cdot5 \text{ kcal mole}^{-1}, \text{ etc.}$$

It was ascertained, making use of the ESR technique, that free radicals actually are products of reactions (5) and (6). The activation energies for these exothermic reactions do not exceed several kcal mole^{-1}. The reaction involving ethylene is so fast that no chains will form in excess ethylene, the whole amount of fluorine being consumed in the formation of radicals that recombine or disproportionate to form end products.

Reaction (1) $\dot{R} + O_2 \rightarrow R\dot{O}_2$ was studied extensively. Both for the gas- and for the liquid-phase its rate constant is high ($\simeq 10^{-13}$ cm^3 molecule^{-1} s^{-1}), [6,7] and its activation energy is close to zero.

Reaction (2) yields the main oxidation intermediates. For liquid-phase oxidation occurring usually at temperatures below 200 °C, (2a) would undoubtedly be the main reaction, and branching would occur by hydroperoxide dissociation to radicals (reaction (3)). At gas-phase oxidation temperatures ($\simeq$ 300 °C and higher) this reaction does not seem to occur. This was convincingly shown by Norrish[8] and Shtern[9] during the 1950's. It was suggested that the peroxide radical undergoes unimolecular dissociation to form aldehyde and alkoxy radicals (OH for methane oxidation), [10] subsequently converted to alcohol. Later on a suggestion was made[2,9] that decomposition follows isomerization of the peroxide radical, the activation energy being close to some 20 kcal mole^{-1}. Thus, for gas-phase oxidation the main chain-propagating reaction would be that of peroxide radical decomposition (2b) and branching would, consequently, occur by reaction (3').‡

† In hydrocarbon solutions lithium ethyl will be in an associated state. This is not allowed for in the values of heats of reaction, as the heats of association are unknown. Investigation of the reaction mechanism shows that the $\dot{C}_2H_5$ radical of the C_2H_5Li molecule will be formed in a bound state.

‡ It may be suggested that due to strong exothermicity of the isomerized peroxide radical decomposition, the activation energy for decomposition would be low and the 20 kcal mole^{-1} would be accounted for by isomerization.

$$H_3C-\dot{C}H-CH_3 + O_2 \rightarrow H_3C-CH-CH_3 \rightarrow H_3C-\dot{C}HCH_3 \rightarrow CH_3CHO + CH_3\dot{O}$$
$$\underset{O-O\cdot}{|} \qquad \underset{O-O}{| \quad |}$$

Recently Nalbandyan *et al.*[11] have studied methane oxidation photosensitized by mercury, and have determined experimentally the difference in activation energies for the two reactions involving RO_2 radicals (2a and 2b). It was found to be $\simeq$ 8 to 10 kcal mole^{-1}. Since the activation energy value for reaction (2a) seems to be $\approx$ 8–10 kcal mole^{-1}, then that for decomposition of RO_2 upon isomerization will be close to 20 kcal mole^{-1}.

Artificially initiated gas-phase oxidation occurs at low temperatures (below 200 °C) yielding alkylperoxide as primary product.[12] However, it is not certain whether hydroperoxide is formed by reaction (2a). The route of RO_2 recombination with HO_2 or with H atoms formed in great amounts by an intensive initiation is also possible.

Reaction (1′) accounts for cracking products under oxidation conditions. The ratio of cracking to oxidation products is known to increase with temperature. For instance, for propane oxidation this ratio will be 0·46 at 280 °C, and 2·45 at 420 °C.[9] The true cracking mechanism (decomposition of alkyl radicals) will be invalid here due to the considerable activation energy for this process. Reaction (1′) was proposed in 1954 by Norrish and Knox[8] and then by Satterfield.[13] There is yet no information on the rate constant of this reaction. It was only reported by various authors that the difference in activation energies for the formation of cracking and oxidation products lies between 14 and 19 kcal mole^{-1}.[14] As the competing reaction $R + O_2 \rightarrow RO_2$ (1) occurs without activation energy, that for reaction $R + O_2 \rightarrow$ olefine $+ HO_2$ (1′) should be some 14 to 19 kcal mole^{-1}.

The scheme considered accounts for many phenomena observed in hydrocarbon oxidation, namely for the composition of products, its variation with temperature, elementary reactions, the general aspect of kinetic curves, and the chemical nature of degenerate branching. The mechanism of hydrocarbon oxidation at considerably lower temperatures, in the liquid-phase, will be very near to that for the gas-phase, the only difference being that, instead of decomposing, the RO_2 radical will form a hydroperoxide (reaction 2a).

Among the saturated hydrocarbons methane has been studied most extensively with respect to the kinetics and mechanism of its oxidation. The first bi-radical chain schemes for this reaction were proposed as far back as 1934 by Norrish[15] and Semenov.[16] In 1948 Norrish[17] suggested a mono-radical scheme for methane oxidation.

Another radical chain scheme[2] was proposed later:

$$CH_4 + O_2 \rightarrow \dot{C}H_3 + H\dot{O}_2 \qquad \text{chain initiation}$$

$$\left.\begin{array}{l} \dot{C}H_3 + O_2 \rightarrow CH_2O + \dot{O}H \\ \dot{O}H + CH_4 \rightarrow H_2O + \dot{C}H_3 \\ \dot{O}H + CH_2O \rightarrow H_2O + H\dot{C}O \end{array}\right\} \quad \text{chain propagation}$$

$$CH_2O + O_2 \rightarrow H\dot{O}_2 + H\dot{C}O \qquad \text{chain branching}$$

$$H\dot{C}O + O_2 \rightarrow CO + H\dot{O}_2$$

$$\left.\begin{array}{l} H\dot{O}_2 + CH_4 \rightarrow H_2O_2 + \dot{C}H_3 \\ H\dot{O}_2 + CH_2O \rightarrow H_2O_2 + H\dot{C}O \end{array}\right\} \quad \text{chain propagation}$$

$$\left.\begin{array}{l} \dot{O}H \xrightarrow{\text{wall}} \\ CH_2O \xrightarrow{\text{wall}} \end{array}\right\} \quad \text{chain termination}$$

An expression for $[CH_2O]_{max.}$ and $w_{max.}$ was derived from this scheme. As many of the constants for the elementary reactions were known, it was possible to determine the absolute reaction rate and to compare theoretical and experimental $[CH_2O]_{max.}$ and $w_{max.}$ values, and also the activation energies. The agreement appeared to be very good, thus showing the validity of the proposed scheme.

It will be noted, however, that at reaction times exceeding that of the maximum rate, when methane consumption is still great, the theoretical curve (allowing for consumption of reactants) will go down, whereas the experimental curves for methane consumption and $[CH_2O]$ accumulation remain unchanged for a long time. Contrary to the oxidation of other alkanes, the maximum rate of methane oxidation is attained at $\simeq 10\%$ conversion, remaining constant up to $\simeq 80\%$.[18] The constancy in rates is observed for some other oxidation reactions, such as the oxidation of CH_2O, CH_3CHO, benzene, acetylene. However, this astonishing phenomenon has not yet been explained. The mechanism of methane oxidation cannot be considered as completely elucidated without proper substantiation of this phenomenon.

A number of other important facts, for which no explanation can be obtained within the scope of the scheme given in page 229, is observed for higher hydrocarbons. These are first of all the cool-flame oxidation and the negative temperature coefficient. The region of the latter is known to adjoin that of cool flames, which seems to be evidence for the same principle underlying both phenomena.

A somewhat modified oxidation mechanism was proposed by Hinshelwood and Seakins in 1963.[19] It was suggested that the main primary product was a hydroperoxide, while all other oxidation products were obtained by destruction of active centres and the hydroperoxide. Though the scheme was proposed for low temperatures, it was used in attempting to obtain an explanation for the temperature dependence of reaction orders and for the negative temperature coefficient.

The Hinshelwood and Seakins scheme is:

$$(7) \quad RH + O_2 \rightarrow \dot{R} + H\dot{O}_2$$

$$(8) \quad \dot{R} + O_2 \rightarrow R\dot{O}_2$$

$$(9) \quad R\dot{O}_2 + O_2 \rightarrow \qquad \text{chain termination}$$

$$(9') \quad R\dot{O}_2 \rightarrow \qquad \text{chain termination}$$

$$(10) \quad R\dot{O}_2 + RH \rightarrow ROOH + \dot{R}$$

$$(11) \quad ROOH \rightarrow \alpha\dot{R} \qquad \text{chain branching}$$

$$(12) \quad ROOH \rightarrow \qquad \text{chain termination}$$

$$(13) \quad \dot{R} + O_2 \rightarrow \qquad \text{chain termination}$$

$$(13') \quad \dot{R} \rightarrow \qquad \text{chain termination}$$

It will be noted that if two rather odd steps of chain termination (reactions 9 and 13) are excluded from this scheme, it becomes that given on page 229 of this paper, the only difference being the direct decomposition of hydroperoxide to end products (reaction (12)) which may occur along with RO_2 destruction at the wall.

Taking the reaction scheme on page 229 and adding reaction (12), and assuming that the products of reaction 3 eventually give two $\dot{R}$ radicals, the differential equations for variations in R,

RO_2, and hydroperoxide concentrations with time can be set up. Making use of the steady-state concentration method, i.e. assuming $d[R]/dt$ and $d[RO_2]/dt$ to be zero, we obtain

$$\frac{d[ROOH]}{dt} = \frac{n_0 a_2}{a_4} + \left[\frac{2a_3 a_2}{a_4} - (a_3 + a_{12})\right][ROOH] \qquad \textbf{(A)}$$

Here $a_2 = k_2[RH]$; $n_0 = k_0[RH][O_2]$; $a_4 = k_4$; $a_3 = k_3$; $a_{12} = k_{12}$. Integrating (A) we have

$$[ROOH] = \frac{n_0'}{f-g}[\exp(f-g)t - 1]$$

where $\qquad n_0' = \dfrac{a_2 n_0}{a_4}; \quad f = \dfrac{2a_3 a_2}{a_4} \quad$ and $\quad g = a_3 + a_{12}.$

When $f > g$, the equation obtained will be that for ordinary degenerate branching without allowance for burning up.

The case $g > f$ is also possible. Then

$$[ROOH] = \frac{n_0'}{g-f}[1 - \exp(-(f-g))t]$$

It is known from the theory for ordinary branched reactions that a limit is observed for $f = g$; from being slow, steady-state, the reaction becomes fast, autoaccelerated. A similar, though less marked, limit is observed for degenerate branching. Barring the work aimed directly at investigation of the oxidation limit, [20] all other experiments were and are made at pressures considerably in excess of this limit. Thus oxidation will be always of an auto-accelerated nature. It follows that there is no sense in discussing the case $g > f$, i.e. a reaction below the limit.

The equation obtained by Hinshelwood and Seakins for $d[ROOH]/dt$ is in complete agreement with our equation (A). However, by some misunderstanding, it is exactly the case $g > f$ that is considered. Consequently Hinshelwood and Seakins derived an equation for $[ROOH]_{max.}$ and $w_{max.}$ without allowing for the burning up. Yet with autoaccelerated reactions for which $f > g$ mathematical calculation of the decrease in the rate of acceleration and, the more so, of complete ending of acceleration (i.e. of the $w_{max.}$ value) is, in principle, impossible without allowing for the burning up. This is the case when the rate of elementary

steps is proportional to the first power of concentration of intermediates, in particular of radicals, $w \sim (f-g)n$, i.e. that to which
the scheme of Hinshelwood and Seakins refers. On the other hand,
it is known that when some elementary steps have rates proportional to the product of intermediate concentrations (for
instance, chain termination by recombination of radicals), the
concentrations of all intermediates will attain with time a maximum
stationary value, even without allowing for the burning up. The
reaction rate determined with consumption of initial products
will also reach a maximum and become constant. At first the
reaction rate will be autoaccelerated following the exp (ϕt) law,
then autoacceleration will become weaker and the reaction rate will
attain gradually a maximum value. The methane oxidation
mechanism (page 233) involving reactions of radicals with intermediates such as

$$\dot{O}H + CH_2O \rightarrow H_2O + H\dot{C}O$$

$$H\dot{O}_2 + CH_2O \rightarrow H_2O_2 + H\dot{C}O$$

is exactly one of this kind. Obviously, the true reaction route may
be established only with allowance being made for the burning up.
When $g > f$ and the rates of elementary steps are linear functions of
active centre concentrations, there will be no autoacceleration at
the start of the reaction, and its rate will obey the $[1 - \exp(-\phi t)]$
law.

The coefficient A of [ROOH] in the equation for $d[ROOH]/dt$
(see reference [19], page 326) is

$$A = (k_{11} + k_{12}) - (k_{10} \Sigma \alpha k_{11} [RH])/Q \qquad \textbf{(B)}$$

where Q is the term allowing for chain termination. For autoaccelerated reactions A will be negative, and not positive, as
suggested by Hinshelwood and Seakins. This seems to be the
misunderstanding.

This misunderstanding is due to the authors considering $A = 0$
(which corresponds to $f - g = 0$) as a condition for explosion. This
is true for non-degenerate branched-chain reactions, where the
rates of branching and chain propagation are of the same order
(oxidation of phosphorus, hydrogen, CO, etc.) and where explosion
is a result of the chain avalanche progress. For degenerate branched

chain reactions involving oxidation of hydrocarbons, the rate of branching representing a molecular reaction (for example hydroperoxide decomposition) is considerably lower than that of radical chain propagation. In this case the equalities $f - g = 0$ or $A = 0$ correspond to transition from a slow stationary reaction to a reaction with relatively slow autoacceleration. As to the explosion it will be of a thermal nature for such reactions, and will occur as a result of gradual increase in the autoaccelerated reaction rate. The limiting condition for degenerate chain branching $f = g$ corresponds to low pressures. Yet hydrocarbon oxidation, as stated above, is always conducted at considerably higher pressures and for the latter $f > g$, i.e. A will be negative.

Let us assume that A is negative, i.e. that $f > g$, in other words

$$f = k_{10} \Sigma \alpha k_{11} [\text{RH}]/Q \gg (k_{11} + k_{12}) = g$$

Then the Hinshelwood and Seakins scheme would correctly describe acceleration with time obeying the $W \sim e^{\phi t}$ law. The activation energy for the αk_{11} term in expression (**B**) may be seen to be highest, while all other terms in Q cannot involve more than $E > 40$ kcal mole^{-1}. Thus f would scarcely drop with increasing temperature, and the occurrence of a negative temperature coefficient would seem to be very doubtful.

2. Dehydrogenation scheme

It was stated above that the scheme of Hinshelwood and Seakins is a modification of the hydrocarbon oxidation scheme given on page 229 of this paper. The Knox scheme [21] represents another, in principle different, approach to the problem. It, so-to-say, revives to a certain extent the old dehydrogenization concepts of Lewis [22] suggesting that the main part is played by formation of olefines.

The appearance of olefines in hydrocarbon oxidation was detected and extensively studied before. Some ten years ago it was shown that the appearance of olefines is within the scope of the oxidation scheme mentioned above. Even a small amount of oxygen is known to decrease the cracking temperature (oxidation cracking). Knox treats the appearance of olefines as the primary step determining the further oxidation route.

Knox scheme

$$(14) \qquad C_3H_8 + O_2 \rightarrow \dot{C}_3H_7 + H\dot{O}_2$$

$$(15) \qquad \dot{C}_3H_7 + O_2 \rightarrow C_3H_6 + H\dot{O}_2$$

$$(16) \qquad C_3H_6 + H\dot{O}_2 \rightarrow \dot{C}_3H_6OOH$$

$$(17) \qquad \dot{C}_3H_6OOH + O_2 \rightarrow \dot{O}OC_3H_6OOH$$

$$(18) \quad \dot{O}OC_3H_6OOH + H\dot{O}_2 \rightarrow HOOC_3H_6OOH + O_2$$

$$(19) \qquad HOOC_3H_6OOH \rightarrow 2\dot{O}H + CH_2O + CH_3CHO$$

$$(20) \qquad \dot{O}H + C_3H_8 \qquad \rightarrow H_2O + \dot{C}_3H_7$$

etc.

It will be readily seen that the scheme suggested by Knox is not one for a branched chain reaction. Indeed, though reaction (19) yields two OH radicals at once, these are converted into two HO_2 radicals (by (20) and (15)), both taking part in the formation of one peroxide molecule by reactions (16) and (18). This is rather a scheme for an unbranched chain reaction and, as such, will not involve acceleration from low to very high rates characteristic of hydrocarbon oxidation; consequently, the reaction time will appear to be too long.

Let us try to show the validity of this general suggestion.

An asymptotic solution for the set of kinetic equations corresponding to the scheme of Knox shows that even at low degrees of conversion the reaction becomes stationary, displaying a characteristic constant rate of the consumption of initial substances

$$-d(C_3H_8)/dt \simeq 4n_0 k_{18}/k_{16} \qquad \qquad (C)$$

and the subsequent rate decrease is due solely to consumption of initial products.

An essential conclusion may be drawn from equation (C), i.e. the maximum reaction rate is proportional to that of initiation $n_0 = k_{14}(C_3H_8)(O_2)$. This contradicts all data on the kinetics of hydrocarbon oxidation.

The time of 50 % conversion in a stationary reaction, as derived from **(C)** would be $t_2 \simeq X_0 k_{16}/4 n_0 k_{18}$, where X_0 is the initial propane concentration. Now it appears from analysis of the equations that the time for the reaction to become stationary is

$$t_1 \simeq 1/k_{16} \sqrt{(k_{18}/n_0)}.$$

Thus 50% of the total reaction time '$t_{\frac{1}{2}} = t_1 + t_2$' will be at a minimum at a certain k_{16} value. By taking reasonable n_0 and k_{18}† values and assuming that $T = 600$ °K it is possible to calculate the k_{16} value and, consequently, the shortest reaction time permissible from the Knox scheme. It appeared to be $\simeq 10^4$–10^6 s. Yet according to experiment propane oxidation under these conditions is completed within several minutes (at $T = 350$ °C, $P = 282$ mm, $t_{\text{overall}} = 150$ s, [24] and at $T = 345$ °C, $P = 360$ mm, $t_{\text{overall}} = 1$ to 2 min). [25]

Full development of the reaction was followed by means of an electronic computor, the constants being varied over a wide range. The maximum reaction rate appeared to be a function of n_0, k_{16} and k_{18} in agreement with the asymptotic solution, while the total time of the reaction ($t_{\frac{1}{2}}$) was but several times that obtained asymptotically.

The Knox scheme does not allow for destruction of radicals, though in reality this is indispensable. Electronic computing making use of the Knox scheme and allowing for radical destruction both at the surface and homogeneously yielded 10^8 s, i.e. several years, for 40% of burning up. This would simply correspond to occurrence of the reaction at an initiation rate n_0.‡ And no wonder: one of the chain propagation steps in the Knox scheme is the interaction of two radicals (reaction 18). But that of

---

† The following constants were used in the calculation:
$k_{14} = f \times 10^{-10} \exp(-45,000/RT)$ cm³ molecule⁻¹ s⁻¹; at 600 °K k_{14} will be $f \times 5 \times 10^{-27}$, and, accordingly, $n_0 = f \times 1.3 \times 10^{10}$ molecules⁻¹, under the assumption that $(C_3H_8)_0 = (O_2)_0 = 100$ mm $= 1.61 \times 10^{18}$ molecule cm⁻³. The steric factor f was varied from 1 to 10^{-2} and mostly taken as unity.

The expression given by Knox[23] $k_{16} = 0.33 \times 10^{-12} \exp(-6,000/RT)$ cm³ molecule⁻¹ s⁻¹ was taken for k_{16}. At 600 °K $k_{16} \simeq 2.2 \times 10^{-15}$ cm³ molecule⁻¹ s⁻¹. k_{18} was assumed to be 10^{-11} cm³ molecule⁻¹ s⁻¹. This is the constant for recombination of alkyl radicals and seems to be the maximum possible value.

As follows from the asymptotic solution, other constants practically affect neither the maximum reaction rate, nor the duration of the reaction.

‡ The rate constants taken were those ensuring the shortest reaction time when no radicals were destroyed.

other radicals (recombination or disproportionation) would result in chain termination. For example

$$(a)\ \ H\dot{O}_2 + H\dot{O}_2 \rightarrow H_2O_2 + O_2$$

$$(b)\ \ \dot{C}_3H_7 + \dot{C}_3H_7 \nearrow^{C_3H_6 + C_3H_8}_{\searrow C_6H_{14}}$$

$$(c)\ \ \dot{C}_3H_7 + H\dot{O}_2 \nearrow^{C_3H_8 + O_2}_{\searrow C_3H_7OOH} \quad \text{etc.}$$

By making use of an electronic computor it was ascertained that in the absence of chain terminations the radical concentrations are such that the rates of reactions (a), (b), (c) become comparable with that of reaction (18). And this means that when chain terminations are involved the number of Knox cycles per one radical formed cannot be essentially higher than unity. In other words, allowing for chain termination, the total reaction rate will be, in fact, close to n_0.

As to the experimental results obtained by Knox, it seems that the only essentially new experimental fact he observed was the formation of olefines at early reaction steps, to an extent of 75 to 80 % of the hydrocarbon consumed. This very interesting fact seems to be evidence that simultaneously with homogeneous oxidation there occurs at the vessel wall some usual catalytic reaction yielding an olefine, for example

$$C_3H_8 + O_2 \rightarrow H_2O_2 + C_3H_6$$

At the initial stage, when the rate of homogeneous oxidation is very low, the catalytic reaction will be dominating, but with development of the process and as a result of usual branching occurring in accordance with the oxidation scheme, the rate of homogeneous oxidation becomes considerably higher than that of the catalytic reaction.

Heterogeneous reactions occurring in parallel with homogeneous chain branching are often observed in slow hydrocarbon oxidation. [26] They are particularly evident when determining the reaction limits for degenerate branching. [27] Enikolopyan has shown recently, [28] using the tracer technique, that heterogeneous formation of CO_2 and H_2O at a relatively constant reaction rate seems to take place along with fast autoaccelerated methane oxidation. Up to 90% of CO_2 with respect to the consumed methane appears during the induction period. In the course of violent degenerate-branching oxidation CO_2 starts forming as a result of this reaction. The relative amount of CO_2 formed heterogeneously becomes very small.

Thus, the fact discovered by Knox does not seem to have direct bearing on homogeneous oxidation. As to the other facts found by Knox, these are in essential agreement with the results obtained by other scientists, in particular by Shtern, [29] and being within the scope of the scheme on page 229.

From the above considerations the Knox scheme cannot be considered as valid for real oxidation processes.

3. Possible participation of excited species in branching

Up to now we have considered slow autoaccelerated oxidation of hydrocarbons as belonging to the class of chain reactions with degenerate branching. As these reactions are slow they cannot involve chain ignition. Chain ignition is known to occur in ordinary branched chain reactions, such as oxidation of H_2, CO, phosphorus, sulphur, etc., the branching being due to radical reactions. However, as mentioned above, a cool-flame region with marked pressure and temperature limits is observed for all hydrocarbons except methane. This fact and the occurrence of intensive luminescence are greatly in favour of the suggestion that the cool flame phenomenon is no other than chain ignition with a mechanism of branching involving radicals, and is essentially different from slow oxidation. [30] Such types of radical branching were suggested before by Lewis and von Elbe, [31] and then by Voevodskii and Vedeneev, [32] but these branchings always appeared to be connected with great energy expenditure. The fact is that for an elementary branching step to occur more bonds should be broken than formed. Usually these reactions are strongly endothermic and, as such, require a higher activation energy. The rate of such an elementary branching step would be negligibly low and chain ignition would be impossible.

At the same time many elementary chain oxidation reactions are strongly exothermic. This would suggest that the energy liberated in chain propagation reactions might be used for branching. In other words, excited particles, products of elementary reactions, capable of using the total energy or a part of it for an elementary branching act, would contribute to branching.

The general concepts of the part played by excited particles in branching have arisen long ago. Indeed, even when first considering the mechanisms of reactions between phosphorus and oxygen, [33] hydrogen and oxygen, [34] hydrogen and chlorine, [33] etc., it was suggested that the energy for chain branching was imparted by the excited molecule of an elementary act product to the molecule of the initial reagent.

However, there was then no reliable information on the nature and distribution of excitation energy in molecules formed by chemical reactions. The idea that excited particles contributed to branching was, consequently, left aside for a long time. Moreover, it was established for the hydrogen–oxygen reaction that branching as described usually

$$H + O_2 \rightarrow OH + O$$

involves no formation of excited particles.

More, though far from sufficient, information on excited particles formed by elementary acts has become available during the recent years due to a great amount of work carried out, particularly by Norrish with co-workers, [35] by Polanyi, [36] and by other scientists. [37] For instance, Norrish found that vibrationally excited oxygen molecules having a great number of vibrational quanta [35,38] were formed by reactions

$$O + O_3 \rightarrow O_2^* + O_2$$
$$O + NO_2 \rightarrow O_2^* + NO, \text{ etc.}$$

Vibrationally excited oxygen molecules may be responsible for chain propagation in photochemical ozone decomposition. For a long time this reaction was considered to be the only example of a chain reaction involving excited molecules. It is known at present that in reality the number of such reactions is great. [39]

For instance, it was shown by Soviet scientists [40-43] that gas-phase reactions between molecular fluorine and various compounds follow the chain branching mechanism, and branching may be explained only by making use of the hypothesis on contribution from excited particles.

In a simple reaction of F_2 with CH_3I [40-42] branching occurs by unimolecular decomposition of an excited CH_2FI molecule to an atom and a radical

$$\dot{F} + CH_3I \rightarrow HF + \dot{C}H_2I$$

$$\dot{C}H_2I + F_2 \rightarrow CH_2FI^* + \dot{F} + 77 \text{ kcal mole}^{-1}$$

$$CH_2FI^* \rightarrow \dot{C}H_2F + \dot{I}$$

When F_2 reacts with methylene chlorine [43] decomposition of an excited molecule may yield carbene that may be considered as a biradical. This would also result in branching

$$\dot{F} + CH_2Cl_2 \rightarrow HF + \dot{C}HCl_2$$

$$\dot{C}HCl_2 + F_2 \rightarrow CHFCl_2^* + F$$

$$CHFCl_2^* \rightarrow HCl + \dot{C}FCl \quad \text{(branching)}$$

In olefine reactions the energy may be accumulated in two elementary acts. For example

$$\dot{F} + C_2Cl_4 \rightarrow \dot{C}_2Cl_4F^*$$

$$\dot{C}_2Cl_4F^* + F_2 \rightarrow \dot{C}_2Cl_4F_2^{**} + \dot{F}$$

$$C_2Cl_4F_2^{**} \rightarrow \dot{C}l + C_2Cl_3F + \dot{F}$$

In this case immediately after addition of F to C_2Cl_4 the total energy would be concentrated in the radical formed. At least a part of this energy seems to remain unexpended until reaction with F_2, since the huge concentrations of Cl and F atoms in flames of reactions between F_2 and C_2Cl_4, as detected by the ESR technique, suggest a branching nature of chain interaction. [42]

Now, let us make use of these concepts returning to discussion of cool flames.

Long induction periods characteristic of degenerate branchings are not observed, as a rule, for ordinary branched chain reactions.

On the other hand, rather long delays are always observed for cool-flame ignition. It would be natural to suggest that in this case branching is due to a reaction of the radical with some slowly accumulating oxidation product. As the addition of acetaldehyde, or some other higher aldehyde, to the hydrocarbon–oxygen mixture is known to eliminate the cool-flame induction period, the reaction of the radical with aldehyde would seem to be that responsible for branching resulting in a cool flame. With this allowance the chain ignition mechanism will be

$$(21) \quad \dot{R} + RC\overset{\displaystyle O}{\underset{\displaystyle H}{\diagup}} \quad \rightarrow RH + R\dot{C}O$$

$$(22) \quad R\dot{C}O \quad \rightarrow \dot{R} + CO$$

$$(23) \quad R\dot{C}O + O_2 \quad \rightarrow RC\overset{\displaystyle O}{\underset{\displaystyle O\dot{O}}{\diagup}} + 25 \text{ kcal mole}^{-1}$$

$$(24) \quad RC\overset{\displaystyle O}{\underset{\displaystyle O\dot{O}}{\diagup}} + RC\overset{\displaystyle O}{\underset{\displaystyle H}{\diagup}} \rightarrow [RC\overset{\displaystyle O}{\underset{\displaystyle OOH}{\diagup}}]^* + R\dot{C}O$$

$$\downarrow$$

$$\dot{R} + CO_2 + \dot{O}H$$

Here branching is due to unimolecular decomposition of an excited molecule of peracid.

Reaction (24) involving decomposition to three radicals is endothermic by $\simeq 8$ kcal mole^{-1} and would scarcely be responsible for branching, taking into consideration that the acetylhydroperoxide radical gives away all its energy obtained in reaction (23).

Whereas, if we assume that at least a part of the energy of reaction (23) remains unexpended until the occurrence of reaction (24), i.e. RO_2 reacts in an excited state, [44] branching will become possible. (Reactions (23) and (24) are exothermic, in sum, by approximately 15 kcal mole^{-1}.)

Naturally the proposed mechanism is only hypothetical and requires experimental evidence.

It will be noted, however, that it provides a fairly good qualitative explanation for certain features of cold flames. For instance, it follows from the above scheme that addition of formaldehyde would be considerably less efficient than of acetaldehyde, as in this case reaction (24) will be endothermic by 21 kcal mole^{-1} (i.e. reactions (23) and (24) are in sum weakly endothermic, by some 4 kcal mole^{-1}). Indeed, the low efficiency of formaldehyde is consistent with experiment. [45]

It will be emphasized that for the reaction of HCO with oxygen reactions (23) and (24) will not occur, as it was shown by Norrish [46] that CO and HO$_2$ will be formed by reaction (23). In this case branching and the consequent reaction of HO$_2$ with aldehyde would be endothermic by $\simeq$ 60 kcal mole^{-1}, i.e. there will be, in fact, no reaction.

This scheme permits an approach, though yet qualitative, to the explanation of the negative temperature coefficient. As a matter of fact, the RCO radical reactions may follow two routes: the first is the unimolecular decomposition RCO → R + CO requiring an activation energy of $\simeq$ 18 kcal mole^{-1} and resulting in chain propagation only; the second would be a bimolecular reaction with oxygen

$$\dot{R}CO + O_2 \rightarrow RC \begin{array}{c} \diagup O \\ \diagdown O\dot{O} \end{array}$$

occurring with an activation energy close to zero and resulting in branching. At high temperatures decomposition of RCO will be predominant and, thus, radical branching will not take place. The reaction rate will decrease with temperature in the usual way. Both reaction routes will be observed at lower temperatures. Further lowering of temperature will make branching predominant and, consequently, the reaction rate will start increasing with decreasing temperature. At still lower temperatures, due to a decrease in the rate of step (21), both routes of RCO conversion will become unimportant and only degenerate branching will remain (as observed for high temperatures) and the reaction rate will once more start decreasing normally with temperature. It will be remembered that the main hydrocarbon oxidation chain will follow the scheme given on page 229 of this paper.

Cool flames of hydrocarbons, aldehydes, and ethers[47] are known to be capable of extending into a cooler medium without transition to a hot flame. It was suggested that the heat liberated by cool-flame ignition warms up the adjacent layer of cool gas to the temperature of chain ignition. However, investigation of cool flames of CS_2[48] at a very low content in the mixture (0·03 %), when adiabatic heating does not exceed several degrees centigrade, has shown that the flames propagate along the tube into its cool part. The difference of temperature between the ignition and flame propagation limits is $140 - 50 = 90$ °C.

This phenomenon can be given an explanation by the positive interaction of chains, i.e. by branching reactions occurring at a rate proportional to the product of two intermediate concentrations, one of the intermediates being possibly a radical. For cool-flame ignition of CS_2,[48] and for a broader limit of chain ignition in a mixture of $H_2 + O_2$,[49] occurring with increase in initial centres, this hypothesis has given quantitative agreement with experiment.

For cool-flame ignition of hydrocarbons the nature of flame propagation seems to be the same and it is consistent with the scheme proposed above.

Though oxidation has been studied for many years, strict quantitative elucidation of its mechanism is far from being achieved. This reaction is very complex and its manifestations are diverse depending upon different factors. Determination of the true mechanism of such complex processes is very difficult. The sensitivity of analytical methods available is yet insufficient for detection of free radicals and other labile intermediates and, the more so, for quantitative determination of their concentrations. Moreover, kinetic studies of oxidation are hindered by dependence of this reaction on the state of the wall, on impurities, etc.

All scientists of various countries concerned with oxidation should unite in the effort of attaining a final solution for this problem which is of exceptional importance both for everyday life and for industry.

I am happy to have the opportunity of writing a chapter for the volume dedicated to Professor Norrish's scientific activities. Professor Norrish and his school made an invaluable contribution to the solution of a number of important questions, and to the solution of general fundamental problems encountered in oxidation studies, in this most interesting field of science.

REFERENCES

1 Cullis, C. F. and Hinshelwood, C. N. *Disc. Faraday Soc.* 1947, **2**, 117.
2 Semenov, N. N. *Some Problems of Chemical Kinetics and Reactivity.* Acad. Nauk SSSR, 1954, second enlarged edition in 1958. (Russian.) English translation, Pergamon Press, 1959.
3 Walters, W. D. *Ann. Rev. Phys. Chem.* v, 1961, **12**, 411; Thynne, J. C. J. and Gray, P. *Trans. Faraday Soc.* 1963, **59**, 1149.
4 Kapralova, G. A., Rusin, L. Yu., Chaikin, A. M. and Shilov, A. E. *Dokl. Akad. Nauk SSSR*, 1963, **150**, 1282; Dyachokovskii, F. S., Bubnov, N. N. and Shilov, A. E. *Dokl. Akad. Nauk SSSR*, 1958, **123**, 870.
5 Kapralova, G. A. and Shilov, A. E. *Kinet. Katal.* 1961, **2**, 362; Dyachokovskii, F. S. and Shilov, A. E. *Zh. Obshch. Khim.* 1963, **33**, 406.
6 Hoey, G. R. and Kutschke, K. O. *Can. J. Chem.* 1955, **33**, 496.
7 Dingledy, D. P. and Calvert, J. G. *J. Am. Chem. Soc.* 1963, **85**, 856.
8 Knox, J. H. and Norrish, R. G. W. *Proc. Roy. Soc.* A, 1954, **222**, 151, Norrish, R. G. W. *Disc. Faraday Soc.* 1951, **10**, 269.
9 Chernyak, N. Ya., Antonovskii, V. L., Revzin, A. F. and Shtern, V. Ya. *Zh. fiz. Khim.* 1954, **28**, 240; Polyak, S. S. and Shtern, V. Ya. *Zh. fiz. Khim.* 1953, **27**, 351, 631; Repa, L. A. and Shtern, V. Ya. *Zh. fiz. Khim.* 1954, **28**, 414.
10 McKellar, J. F. and Norrish, R. G. W. *Proc. Roy. Soc.* A, 1961, **263**, 51.
11 Montashyan, A. A. and Nalbandyan, A. B. *Izv. Akad. Nauk armyan, SSR*, 1962, **15**, 15; Kleimenov, N. A. and Nalbandyan, A. B. *Dokl. Akad. Nauk SSSR*, 1959, **124**, 119; Montashyan, A. A., Moshkina, R. I. and Nalbandyan, A. B. *Izv. Akad. Nauk armyan, SSR*, 1961, **14**, 185.
12 Nalbandyan, A. B. and Fok, N. V. *Dokl. Akad. Nauk SSSR*, 1953, **85**, 1093; Gray, J. A. *J. chem. Soc.* 1952, p. 3150; Watson, J. S. and Darwent, B. M. B. *J. phys. Chem.* 1957, **61**, 577; Burgess, R. H. and Robb, J. C. *Trans. Faraday Soc.* 1958, **54**, 1015; Poroikova, A. I., Voevodskii, V. V. and Nalbandyan, A. B. *Dokl. Akad. Nauk SSSR*, 1959, **124**, 119; Montashyan, A. A. and Nalbandyan, A. B. *Izv. Akad. Nauk armyan. SSR*, 1961, **14**, 526; Poroikova, A. I. and Nalbandyan, A. B. *Dokl. Akad. Nauk SSSR*, 1965; Revzin,

A. F., Sergeev, G. B. and Shtern, V. Ya. *Zh. fiz. Khim.* 1954, **28**, 985; Sergeev, G. B. and Shtern, V. Ya. *Dokl. Akad. Nauk SSSR*, 1953, **91**, 1357.

13 Satterfield, C. N. and Reid, R. C. *Fifth Symposium on Combustion*, *N.Y.* 1955, p. 511.

14 Satterfield, C. N. and Reid, R. C. *J. phys. Chem.* 1955, **59**, 283.

15 Norrish, R. G. W. and Foord, S. G. *Proc. Roy. Soc.* A, 1936, **157**, 503; Norrish, R. G. W. and Wallace, J. *Proc. Roy. Soc.* A, 1934, **145**, 307.

16 Semenov, N. N. *Chain Reactions* (Russian). Moscow, ONTI, 1934. English translation, Oxford, 1935.

17 Norrish, R. G. W. *Revue Inst. fr. Pétrole*, 1948, **4**, 288.

18 Karmilova, L. V., Enikolopyan, N. S., Nalbandyan, A. B. and Il'in, V. I. *Zh. fiz. Khim.* 1961, **35**, 1435; 1957, **31**, 851.

19 Seakins, M. and Hinshelwood, C. N. *Proc. Roy. Soc.* A, 1963, **276**, 324.

20 Norrish, R. G. W. and Reagh, J. D. *Proc. Roy. Soc.* A, 1940, **176**, 429; Sadovnikov, P. Ya. *Zh. fiz. Khim.* 1937, **9**, 575; Spence, R. *J. chem. Soc.* 1932, p. 686; Neiman, M. B. and Serbinov, A. I. *Zh. fiz. Khim.* 1932, **3**, 75; 1933, **4**, 41; Koval'skii, A. A., Sadovnikov, P. Ya. and Chirkov, N. M. *Zh. fiz. Khim.* 1933, **4**, 50.

21 Knox, J. H. *Trans. Faraday Soc.* 1960, **56**, 1225; Knox, J. H. and Wells, C. H. J. *Trans. Faraday Soc.* 1963, **59**, 2786, 2801; Knox, J. H. *The Mechanism of Oxidation of Alkanes in the Gas Phase*. University of Edinburgh, 1965; Knox, J. H. and Turner, J. M. C. *Radical Selectivity*; *The Initial Stages of Alkane Oxidations*, University of Edinburgh, 1965; Knox, J. H. *Ann. Rep.* 1962, **59**, 18.

22 Lewis, B. *J. chem. Soc.* 1927, p. 1555; 1929, p. 759; 1930, p. 58.

23 Knox, J. H. *Combustion and Flame*, 1965, **9**, 297.

24 Chernyak, N. Ya., Antonovskii, V. L., Revsin, A. F. and Shtern, V. Ya. *Zh. fiz. Khim.* 1954, **28**, 240.

25 Newitt, D. M. and Thornes, L. S. *J. chem. Soc.* 1937, p. 1669.

26 Kistiakowsky, G. B. and Spence, R. *J. Am. Chem. Soc.* 1930, **52**, 4837; Koval'skii, A. A., Sadovnikov, P. Ya. and Chirkov, N. M. *Zh. fiz. Khim.* 1933, **4**, 50.

27 Spence, R. *J. Am. Chem. Soc.* 1930, **52**, 686; Neiman, M. B. and Egorov, L. *Zh. fiz. Khim.* 1932, **3**, 61.

28 Karmilova, L. V., Enikolopyan, N. S. and Nalbandyan, A. B. *Zh. fiz. Khim.* 1961, **35**, 1458.

29 Shtern, V. Ya. *Gas-Phase Oxidation of Hydrocarbons* (Russian), Akad. Nauk SSSR, Moscow, 1960. English translation, Pergamon Press, 1964.

30 Polyak, S. S., Enikolopyan, N. S. and Shtern, V. Ya. *Zh. fiz. Khim.* 1958, **32**, 2224.

31 Lewis, B. and Elbe, G. von. *J. Am. Chem. Soc.* 1937, **59**, 970.

32 Voevodskii, V. V. and Vedeneev, V. I. *Dokl. Akad. Nauk SSSR*, 1956, **106**, 679.

33 Semenoff, N. *Chem. Rev.* 1929, **6**, 347.

34 Semenoff, N. *Z. Phys.* 1927, **46**, 109; *Z. phys. Chem.* B, 1929, **2**, 161.
35 Norrish, R. G. W. *The Study of Energy Transfer in Atoms and Molecules by Photochemical Methods.* Douzième Conseil de Chimie, Bruxelles, p. 99, 1962 (transfert d'énergie dans le gaz); McGrath, W. D. and Norrish, R. G. W. *Z. phys. Chem.* 1958, **15**, 245.
36 Charters, P. E. and Polanyi, J. C. *Disc. Faraday Soc.* 1962, **33**, 107; Charters, P. E., Khare, B. R. and Polanyi, J. C. *Nature, Lond.*, 1962, **193**, 367; Polanyi, J. C. and Rosner, S. D. *J. chem. Phys.* 1963, **38**, 1028; Airey, J. R., Getty, R. R., Polanyi, J. C. and Snelling, D. R. *J. chem. Phys.* 1964, **41**, 3255; Cashion, J. K. and Polanyi, J. C. *Proc. Roy. Soc.* A, 1960, **258**, 529.
37 McKinley, J. D., Garvin, D. and Boudart, M. J. *J. chem. Phys.* 1964, **41**, 3255.
38 McGrath, W. D. and Norrish, R. G. W. *Proc. Roy. Soc.* A, 1957, **242**, 265; A, 1960, **254**, 317; Basco, N. and Norrish, R. G. W. *Can. J. Chem.* 1960, **38**, 1769.
39 Semenov, N. *On the Possible Importance of Excited States in the Kinetics of Chain Reactions.* Douzième Conseil de Chimie, Bruxelles, Université Libre de Bruxelles, 1962, p. 183.
40 Vedeneev, V. I., Chaikin, A. M. and Shilov, A. E. *Kinet. Katal.* 1963, **4**, 320.
41 Kapralova, G. A., Trofimova, E. A., Rusin, L. Yu., Chaikin, A. M. and Shilov, A. E. *Kinet. Katal.* 1963, **4**, 653.
42 Rusin, L. Yu., Chaikin, A. M. and Shilov, A. E. *Kinet. Katal.* 1964, **5**, 1121.
43 Semenov, N. N. and Shilov, A. E. *Kinet. Katal.* 1965, **6**, 3.
44 McDowell, C. A. and Thomas, J. R. *J. chem. Soc.* 1950, p. 1462; Tipper, C. F. H. *Q. Rev.* 1957, **11**, 313.
45 Shtern, V. Ya. and Polyak, S. S. *Dokl. Akad. Nauk SSSR*, 1949, **66**, 235; Antokol'skii, V. L. and Shtern, V. Ya. *Dokl. Akad. Nauk SSSR*, 1951, **78**, 303.
46 McKellar, J. F. and Norrish, R. G. W. *Proc. Roy. Soc.* A, 1960, **254**, 147.
47 White, A. G. *J. chem. Soc.* 1927, p. 498; Townend, D. T. A. and Chamberlain, E. A. C. *Proc. Roy. Soc.* A, 1937, **158**, 415; Hsieh, M. S. and Townend, D. T. A. *J. chem. Soc.* 1939, pp. 332, 337, 341; McCormac, M. and Townend, D. T. A. *J. chem. Soc.* 1940, **143**, 151; Spence, K. and Townend, D. T. A. *Nature, Lond.*, 1945, **155**, 330.
48 Voronkov, V. G. and Semenov, N. N. *Zh. fiz. Khim.* 1939, **13**, 1695.
49 Nalbandyan, A. B. *Zh. fiz. Khim.* 1946, **20**, 1959; Dubovitskii, F. I. *Acta phys.-chim. URSS*, 1935, **2**, 761; Nalbandyan, A. B. *Acta phys.-chim. URSS*, 1934, **1**, 305.
50 Sokolova, N. A., Markevich, A. M. and Nalbandyan, A. B. *Zh. fiz. Khim.* 1961, **35**, 850.
51 Thomas, J. M. and Norrish, R. G. W., *Nature, Lond.*, 1966, **210**, 728.

10

THE INTERPRETATION OF
COOL FLAME AND
LOW-TEMPERATURE COMBUSTION
PHENOMENA

JOHN H. KNOX

1. Introduction

In 1934 R. G. W. Norrish published the first of a series of important papers [1-21] on the combustion of hydrocarbons and their derivatives. This group of papers, only a part of his large scientific output, constitutes a major contribution to the field of combustion. It is a privilege to have the opportunity to review this contribution and to show how it has led to a broadening and deepening of our understanding of this group of very complex reactions.

The papers describe investigations carried out over a period of thirty years. The problems studied have been of increasing complexity and have employed successively more sophisticated experimental techniques. In the development of these techniques Norrish has often played a major part. In the earliest studies [2,4,5] the overall rate of oxidation, obtained from pressure measurements, was used as the main kinetic parameter from which the mechanism was derived, but the realization of the importance of other checks led Norrish to use progressively more advanced chemical, analytical, photochemical and spectroscopic techniques. [3,5,8,9,12] Amongst these were flash photolysis, which was soon applied to combustion studies, [14] and various sensitive and discriminatory analytical methods [13,15] culminating in gas chromatography. [21,23]

Norrish was amongst the first to develop free radical chain mechanisms based upon the principle of degenerate branching formulated by Semenov in 1930. [24] In his scheme for the oxidation of methane, formaldehyde was proposed as the branching intermediate. Since that time the view that aldehydes are the key branching intermediates in combution reactions has formed the basis of Norrish's schemes for the oxidation of alkanes, [10,11,13,15,16] olefins, [12,21] and aromatics. [17,18] While the complexities of gas-phase oxidation are now known [25,26] to be much greater than, envisaged in 1935 it seems likely that Norrish's original insight is broadly correct, even if the mode of formation of aldehydes is somewhat different from that originally proposed. [4,10]

The concept of the degenerately branching chain was applied with conspicuous success to the oxidation of methane at about 500 °C. [4] Elaborating slightly [27] on the original treatment of Semenov [24] the concept may be described as follows. Suppose that a reaction proceeds by way of straight chains of finite length ν, and that an intermediate I is formed in a fraction α of the chain steps. Suppose that after a delay θ, which may be long in comparison with the duration of the primary chain, each molecule of I can generate n new free radicals. The reaction will accelerate slowly and exponentially according to the law

$$\text{Rate} = A(e^{\phi t} - 1) \approx A e^{\phi t} \tag{A}$$

if each chain generates at least one new free radical by the branching process. The acceleration constant or net branching factor is

$$\phi = (n\alpha\nu - 1)/\theta \tag{B}$$

If only a fraction, β, of the molecules of I actually survive to produce radicals in the branching reaction, the net branching factor becomes

$$\phi = (n\alpha\beta\nu - 1)/\theta \tag{C}$$

where θ now represents the lifetime of I with respect to the branching process only.

In the oxidation of methane Norrish [2,4] showed that the reaction accelerated exponentially with time and obeyed a maximum rate law

$$\text{Rate} = k[O_2][CH_4]^2 P \tag{D}$$

where P is the total pressure. The mechanisms first proposed for the reaction [1,2,4] involved O atoms and CH_2 radicals, but it was realized later, although it had not been obvious in 1934, [1] that chains involving O atoms and CH_2 radicals were unlikely on energetic grounds, and the final form of the methane oxidation scheme,[10] incorporating Axford and Norrish's scheme for the formaldehyde oxidation,[7] involved OH and CH_3 as propagating radicals. The scheme included the kinetically important reactions (1) to (5).

$$CH_4 + OH \rightarrow CH_3 + H_2O \tag{1}$$

$$CH_3 + O_2 \rightarrow CH_2O + OH \tag{2}$$

$$CH_2O + O_2 \rightarrow 2 \text{ radicals equivalent to OH} \tag{3}$$

$$CH_2O + OH \rightarrow CO + H_2O + \text{radical equivalent to OH} \tag{4}$$

$$OH \rightarrow \text{diffusion-controlled wall reaction} \tag{5}$$

In terms of the variable defined above:

$$\left.\begin{array}{l} n = 2, \quad \alpha = 1, \quad \beta = k_3[O_2]/(k_3[O_2] + k_4[OH]) \\ \nu = k_1[CH_4]P/(k_5(S/d)), \quad \theta = 1/k_3[O_2] \end{array}\right\} \tag{E}$$

In the expression for ν the termination rate is expressed as $k_5[OH]S/Pd$, where S is the specific surface activity and d the vessel diameter. In the early stages of reaction when [OH] is low, $\beta = 1$ and the acceleration constant is

$$\phi_i = 2k_1k_3[O_2][CH_4]P/(k_5(S/d)) - k_3[O_2] \tag{F}$$

As the concentration of OH rises β and hence ϕ fall even when no allowance is made for consumption of reactants. At the maximum rate $\phi = 0$ and $n\alpha\beta\nu = 1$. Assuming long chains, this condition leads to

$$\begin{aligned} \text{maximum rate} &= k_1[OH][CH_4] \\ &= 2(k_1^2 k_3/k_4 k_5)[CH_4]^2[O_2]P/(S/d) \end{aligned} \tag{G}$$

in excellent formal agreement with experiment. Although this agreement does not prove the mechanism, subsequent work has broadly confirmed its correctness. Refinements and modifications have been made yet formaldehyde is still regarded as being the major branching intermediate below about 500 °C.[25,26]

Shortly after Norrish[2] outlined his scheme for the oxidation of methane and higher alkanes, Ubbelohde put forward a peroxidation mechanism[28] as a result of work on pentane. The key reactions were:

$$R + O_2 \rightarrow ROO \tag{6}$$

$$ROO + RH \rightarrow ROOH + R \tag{7}$$

Autoacceleration was imagined to occur from the homogeneous decomposition of the peroxide into free radicals. This scheme was developed by Hinshelwood, [29,30,31] Walsh [32] and co-workers. The key reactions in Hinshelwood's scheme [31] are (6) to (12).

$$RH + O_2 \rightarrow R + HO_2 \tag{8}$$

$$ROOH \rightarrow \text{radicals equivalent to } nR \tag{9}$$

$$ROOH \rightarrow \text{destruction} \tag{10}$$

$$ROO \rightarrow \text{termination} \tag{11}$$

$$R \rightarrow \text{termination} \tag{12}$$

It is envisaged that both termination reactions may involve oxygen so that the rate constants k_{11} and k_{12} have the general form $k = k' + k''[O_2]$.

This mechanism predicts autocatalysis if the rate of (9) exceeds that of (11)+(12) giving

$$\phi = k_9 \left\{ \frac{nk_9}{k_9 + k_{10}} \frac{k_7[RH]}{k_{11} + k_{12}k_7[RH]/k_6[O_2]} \right\} - 1 \tag{H}$$

Since the mechanism contains no reaction for the removal of branching agent which is second order in reaction products (e.g. ROOH + radical or ROOH + ROOH), β is constant throughout the reaction and no maximum rate is predicted unless the consumption of reactants is allowed for or the reaction is assumed to have ϕ negative. Hinshelwood makes the second assumption in order to derive expressions for the maximum rate. Its implications are interesting. The reaction will not be a true chain reaction unless n is very small or (10) is much faster than (9). The development of the reaction in its initial stages will not be exponential but will follow the law:

$$\text{rate} = A'(1 - e^{-\phi't}) \tag{J}$$

where ϕ' ($\equiv -\phi$) is positive. The maximum rate will be reached within a few lifetimes of the branching intermediate and will be proportional to the rate of initiation by (8).

These predicted features bear little resemblance to those found by experiment and the assumption that ϕ is negative must surely be unsound. Mulcahy has shown that if the maximum rate results

from consumption of reactants, [33] expressions similar to Hinshelwood's may be obtained for the maximum rate. However the
maximum rate is often achieved with relatively little consumption
of reactants and is almost certainly kinetically controlled in the
sense that the concentration of branching intermediate at maximum
rate results from a balance between its formation and consumption
by radical reactions. The agreement claimed by Seakins and
Hinshelwood [31] between experiment and theory is probably
fortuitous since their mechanism, under the assumed condition of
negative ϕ, fails to predict more basic properties of gas-phase
combustion.

The Norrish schemes [4,6,12] and those based upon them were
among the few which successfully predicted the important features
of gas-phase oxidations. While this argues in favour of their
validity, Norrish has always realized that kinetic tests are insufficient to verify a mechanism, and he has employed many physical,
chemical and photochemical tests to check his ideas. The proposi-
tion that formaldehyde was the key intermediate in the oxidation
of methane led Axford and Norrish to examine the kinetics of the
oxidation of formaldehyde alone. [6,7] They found that the oxidation was an unbranched chain reaction probably initiated by (3).
Norrish and Reagh [5] showed that the surface had a profound
effect upon the rates of oxidation of a number of alkanes, confirm-
ing earlier results obtained with methane. [4] The discovery of a
limiting vessel diameter below which no reaction occurred confirmed the general idea that the reactions proceeded by degenerate
branching chains and that below a certain diameter the chain
length fell so that the term $n\alpha\nu$ became less than unity. Harding and
Norrish [8,12] confirmed that formaldehyde was an important
branching intermediate in the oxidation of ethylene by showing
that the induction period of the normal slow oxidation could be
removed by adding a concentration of formaldehyde close to that
normally present at maximum rate. More important, when formaldehyde was added in excess of this concentration, the initial
reaction rate exceeded the normal maximum but then declined
towards it. This experiment proved that the maximum rate was
kinetically controlled by the balance between formation and consumption of the branching intermediate and not by the consump-

tion of reactants as supposed by Mulcahy. [33] The pressure versus time curves for this important experiment are shown in Fig. 1.

Norrish and Patnaik [9] used another technique to prove the importance of aldehydes in combustion reactions. They showed that the rates of oxidation of methane and ethylene were increased if the mixture was strongly illuminated with light absorbed by aldehydes but not peroxides. The kinetics showed that if the light-induced catalysis resulted from a reaction $I + h\nu$ = radicals, the

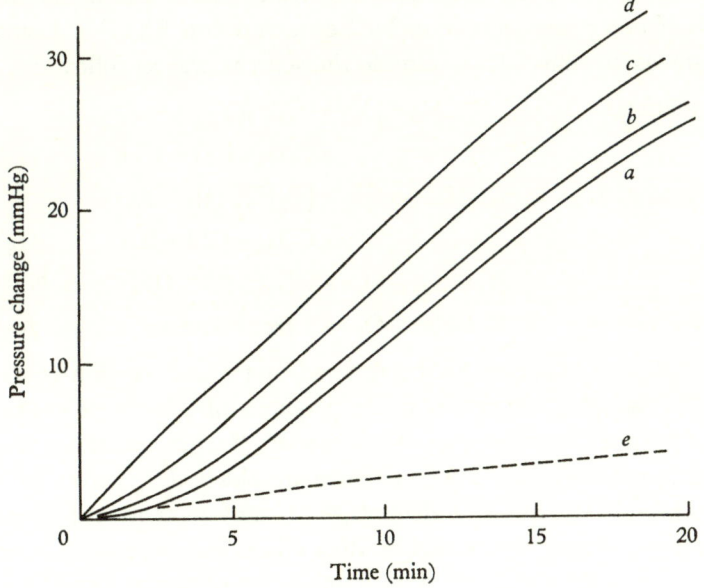

Fig. 1. Addition of formaldehyde to ethylene-oxygen mixtures. Temp. = 400 °C. Initial pressures: $p_{C_2H_4}$ = 100 mmHg; p_{O_2} = 200 mmHg. (a) No formaldehyde. (b) p_{HCHO} = 8 mmHg. (c) p_{HCHO} = 18 mmHg. (d) p_{HCHO} = 20·5 mmHg. (e) Decomposition of formaldehyde alone in reaction vessel p_{HCHO} = 20 mmHg. Quartz reaction vessel. (From [14].)

thermal branching reaction must be $I + O_2$ = free radicals. The result therefore favoured oxidation of an aldehyde rather than pyrolysis of a peroxide. More recently kinetic spectroscopy has been used to verify the presence of OH in the slow combustions of formaldehyde and acetaldehyde, [19] and in the oxidation of methyl radicals generated by photolysis of methyl iodide. [20] The key roles of formaldehyde and OH in combustion reactions were thus documented in a classic series of experiments.

The evidence for the identity of the branching intermediate in the oxidation of the higher hydrocarbons is less conclusive and much controversy has centred upon this aspect of combustion. [11,19,32]

Norrish proposed one of the first free radical schemes for the oxidation of higher alkanes in 1935. [2] Like the scheme for the oxidation of methane it involved O atoms but was later modified and reformulated in terms of OH radicals. The final form [10] included an aldehyde degradation scheme which was a modification of one proposed earlier by Lewis and von Elbe. [35] Using propane as an illustrative example the scheme was as follows:

$$\text{Oxidation:}\quad CH_3CH_2CH_2 + O_2 \rightarrow (C_3H_7OO)$$
$$\rightarrow C_2H_5CHO + OH \qquad (13)$$
$$\text{Degradation:}\quad C_2H_5CHO + OH \rightarrow (C_2H_5CO) + H_2O$$
$$\rightarrow C_2H_5 + CO + H_2O \qquad (14)$$
$$C_2H_5 + O_2 \rightarrow CH_3CHO + OH \qquad (14')$$
$$\text{Branching:}\quad R \cdot CHO + O_2 \rightarrow 2 \text{ free radicals} \qquad (15)$$

The scheme was partially substantiated by the observation of the complete range of aldehydes in the products of cool flame oxidation of hexane [13] and propane. [15] However, the formation of other products not derivable from intermediate aldehydes, particularly ring ethers, [13] olefins and alcohols, [13,15] showed that other fates for peroxy radicals were possible. Ring ethers had previously been observed by Ubbelohde [28] in the combustion of pentane and their formation in considerable yield has recently been confirmed gas-chromatographically. [36,37] They are probably formed by reactions such as (16):

$$RCH_2CH_2CH_2CH(OO)R' \rightarrow \underset{\underline{\qquad O \qquad}}{RCHCH_2CH_2CHR'} + OH \qquad (16)$$

Alcohols could be formed by reactions such as (17) and (18), and olefins by (19):

$$RCH(OO)R' \rightarrow RCHO + R'O \qquad (17)$$
$$R'O + XH \rightarrow R'OH + X \qquad (18)$$
$$R + O_2 \rightarrow AB + HO_2 \ (AB \equiv \text{olefin}) \qquad (19)$$

The original elegance of the simple scheme was to a large extent destroyed by the appearance of these side reactions, and in some ways the peroxide theory, as elaborated by Walsh, offered a simpler interpretation of the formation of complex products. [32]

In addition to explaining the reaction products any mechanism for the oxidation of alkanes (and other organic compounds) must explain the existence of a negative temperature coefficient in the rate of slow oxidation. This unusual phenomenon occurs with a wide range of oxidation systems. [16,30,38-43] Generally the overall reaction rate and the acceleration constant decrease with temperature over a range of about 50 °C in the neighbourhood of 350 °C. A typical example is shown in Fig. 2.

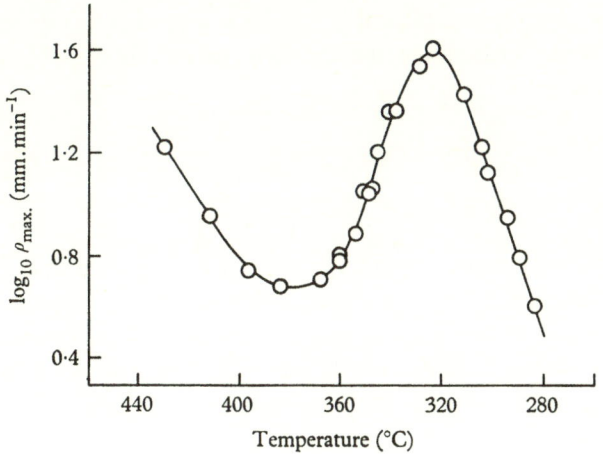

Fig. 2. The influence of temperature on the maximum rate. 60 mmHg propane and 120 mmHg oxygen. (From [31].)

Closely associated with the negative temperature coefficient are the phenomena of cool flames and two-stage ignitions. [44,45] It is generally believed that the negative temperature coefficient is symptomatic of a fundamental change of mechanism from one producing predominantly oxygenated products at low temperatures to one forming cracking products at high temperatures. [25,26] Both peroxide and aldehyde theories gave explanations of the phenomenon.

Norrish [11,16] supposed that above 350 °C the olefin-forming reaction (19) becomes more important than the aldehyde-forming reaction (13) because it has a higher activation energy. Since (19) does not directly form the branching agent the reaction rate must fall. In terms of equation (C), α declines with temperature. According to the peroxide theory, either the decomposition of ROOH to inactive products, such as aldehydes and water, increases with temperature more rapidly than the decomposition to radicals, [28,31] that is β decreases with temperature, or ROO becomes unstable and unable to abstract H before decomposing to inactive products, [32] that is α decreases with temperature.

Quite recently Benson [47] has suggested a similar explanation which derives from the idea that the ROO radical will readily decompose into $R + O_2$ at about 400 °C. He supposes that the negative temperature coefficient arises from a balance between the reactions (7) and (19) connected through the equilibrium (6).

$$R + O_2 \rightarrow AB + HO_2 \tag{19}$$

$$R + O_2 \rightleftharpoons ROO \tag{6}$$

$$ROO + RH \rightarrow ROOH + R \tag{7}$$

Chemical tests of these hypotheses should be possible and have often been attempted but with inconclusive results. In a cool flame the temperature of the reacting gas first rises through the region of the negative temperature coefficient and then falls to near its original value, the complete cycle taking a fraction of a second. If conditions are correctly chosen several successive cool flames may be observed. According to the two theories either the aldehyde or peroxide concentrations should drop sharply during the passage of the flame. Newitt and Thornes [48] observed a fall in the higher aldehyde concentration during the cool flame of propane but the fall was hardly greater than their experimental error. Bardwell and Hinshelwood [30] observed a much greater fall in the peroxide concentration during the cool flame of butanone, but they failed to identify their peroxide, which was most likely to have been hydrogen peroxide. [15]

Recently Bonner and Tipper [49] have demonstrated conclusively with heptane and cyclohexane that the concentrations of hydroperoxides fall during the passage of a cool flame, but with propane only minute traces of hydroperoxide can be detected at any stage in the reaction. Undoubtedly the mono- and polyhydroperoxides found by Tipper and co-workers[49,50] in the slow combustion products of C_4 and higher hydrocarbons will cause autocatalysis as was clearly demonstrated by the pre-war work of Neiman and co-workers,[51] yet their presence in measurable quantities does not seem to be an essential for cool flame formation. Since the most recent work[52] has failed to confirm Newitt and Thornes's observations, the explanation of the negative temperature coefficient and of the origin of cool flames is still uncertain. Later in this paper a new explanation of the phenomenon is considered. However, it is first necessary to describe the main experimental features of cool flames and describe some recent work on slow oxidation.

2. Experimental features of cool-flame combustion

Cool flames were discovered by Humphrey Davy[53] in 1812 and were observed by several workers in the nineteenth century. Perkin[54] carried out a systematic study and established that cool flames could be obtained from a wide variety of hydrocarbons, alcohols, aldehydes, acids, oils and waxes. The phenomenon was most clearly demonstrated with diethyl ether and acetaldehyde. The flames could be initiated by holding a metal ball heated to 100–300 °C in the vapour of the substance. They were quite cool and did not char paper or ignite carbon disulphide. They emitted a pale blue light visible only in a darkened room and produced quantities of partial oxidation products including aldehydes and acids. Little further work was carried out until the 1930's when the phenomenon was rediscovered by Prettre[55] and Townend.[56] Cool flames occur at lower pressures and temperatures than hot ignitions, within a lobe-shaped region as shown in Fig. 3. Towards the lower temperature limit an ignition peninsula in the hot-flame limit is usually observed; the upper limit is normally associated with the negative temperature coefficient in the rate of slow oxidation. Multiple cool flames have also been observed by numerous workers including Townend,[13,15,16,41,42,44,45,56–58] and as many as eight successive light and pressure pulses have been observed in static systems[57] although the usual maximum is five. The light emitter, where it has been identified[18,28,59–61] is electronically excited formaldehyde, the minimum excitation energy being about 77 kcal mole^{-1}. Since the same emission has been observed by Gaydon and Wolfhard [62] in a low-pressure flame of methanol, it is likely that the excited molecules arise from a reaction of CH_3O with another radical. That only about one formaldehyde molecule in 10^6 actually emits radiation [63] is in accord with this view, since radical–radical reactions between reactive species such as CH_3O will be rare. The same type of light seems to be emitted during slow combustion and is probably a general feature of low temperature oxidation reactions.

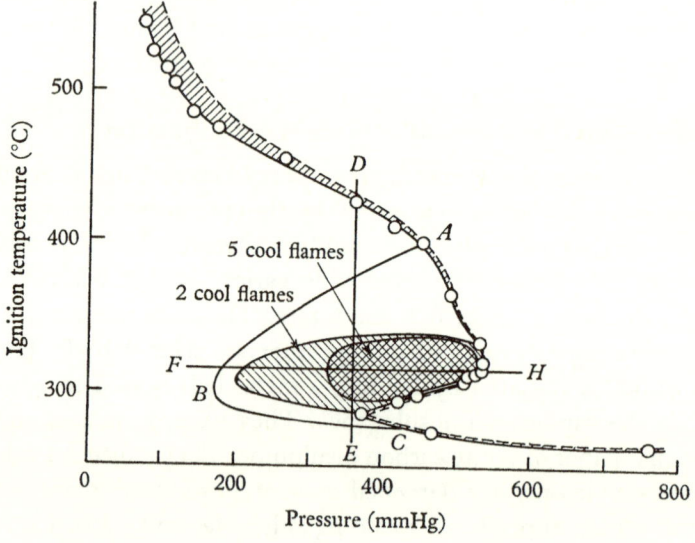

Fig. 3. Cool-flame and ignition limits for equimolar propane + oxygen mixtures in a silica vessel. (From [48].)

Recently the 'morphology' of slow oxidation reactions has been studied in detail by Lucquin and co-workers. [64] They have shown that the phenomena are more complex than previously thought, and in particular they have observed a *pic d'arrêt* towards the end of many slow oxidations in oxygen-deficient conditions. The *pic d'arrêt* is a sharp pressure rise accompanied by a light pulse just before the oxygen is exhausted. The phenomenon is probably quite different from a cool flame and may be connected with the exothermic recombination of radicals which are normally oxidized.

During a cool flame the temperature of the gas rises to about 450 °C. [15,52] The maximum temperature is only slightly affected by the starting temperature of the gas in which the flame develops although it rises slightly with wall temperature. [15] The pressure pulse is associated almost entirely with the temperature rise and hardly at all with an increase in the number of molecules in the

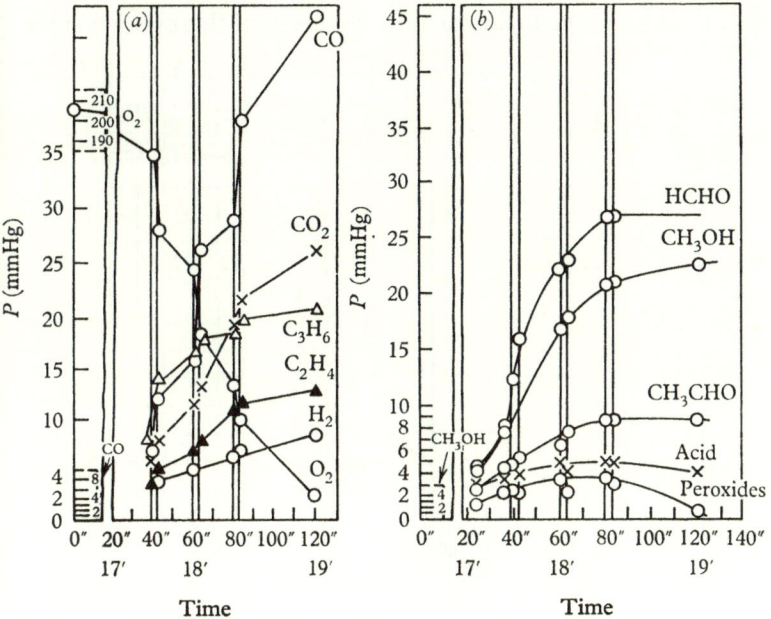

Fig. 4. The products of cool-flame oxidation of propane. The occurrence of a cool flame is shown by double vertical lines. Mixture: $C_3H_8 + O_2$; $P_{init.} = 420$ mmHg; $T = 280$ °C. (a) Products insoluble in water; (b) products soluble in water. (From [26].)

system. Since the flame moves at only about 15 cm s^{-1} [25,52,65] its duration in a moderately sized reaction vessel is about a second. The actual flame thickness is, however, only about 0·2 cm and therefore the time for a given element of gas to pass through the flame is about 15 ms. During this period there is only a slight increase in the overall reaction rate and the total reaction occurring in a cool flame is small. This is clearly shown by the data of Agnew and Agnew for ether [52] and of Shtern [66] for propane (see Fig. 4). If the flame is assumed to be nearly adiabatic it is readily shown that only a very small part of the total exothermicity of the reaction is liberated in the flame. On the basis of a 200 °C temperature rise the heat release in the propane cool flame is about 6 kcal mole^{-1}. Since the major overall process is conversion to CO and H_2O for which ΔH per mole of propane is -286 kcal mole^{-1}, only about 5 % of the total heat release occurs within the flame. Conceivably the flame could result from a minor exothermic process which plays little part in the main reaction. Shtern is of this opinion since

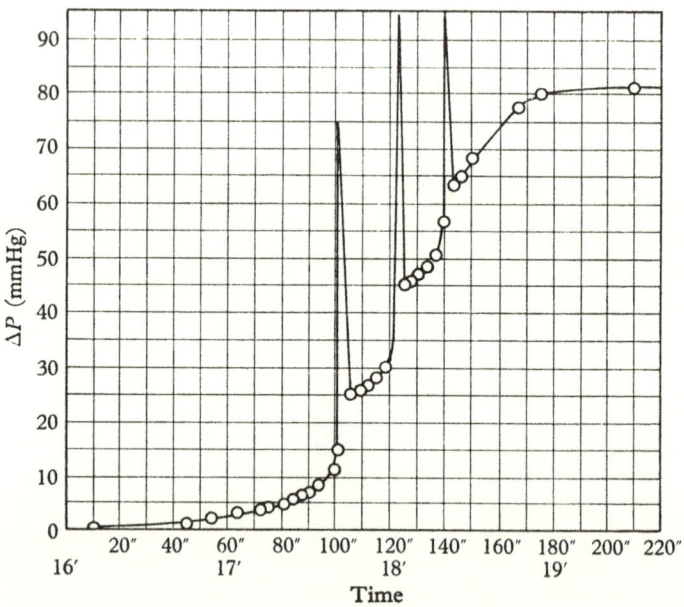

Fig. 5. The kinetic curve of cool-flame oxidation of a mixture of $C_3H_8 + O_2$; $P_{init.} = 420$ mmHg; $T = 280$ °C. (From [26].)

he states that the pressure–time curve with the cool-flame pulses 'cut off' is very close to the expected curve for the slow reaction (Fig. 5). A view similar to this is developed below.

There is good evidence that cool flames arise out of the normal slow oxidation process. Malherbe and Walsh [67] showed that the induction period (measured to the maximum rate of slow oxidation in diethyl ether) decreased as the pressure increased, and was continuous with the induction period for cool flames. Thus at the cool-flame boundary the cool flame appears to develop precisely when the slow reaction reaches its maximum rate. This result is most readily interpreted in terms of a thermal theory of cool flames.

It is worth examining to what extent such a theory can explain cool-flame phenomena. According to Semenov[24] thermal instability in a chemical reaction whose rate depends only upon temperature, will occur when the change in heat output of the reaction with

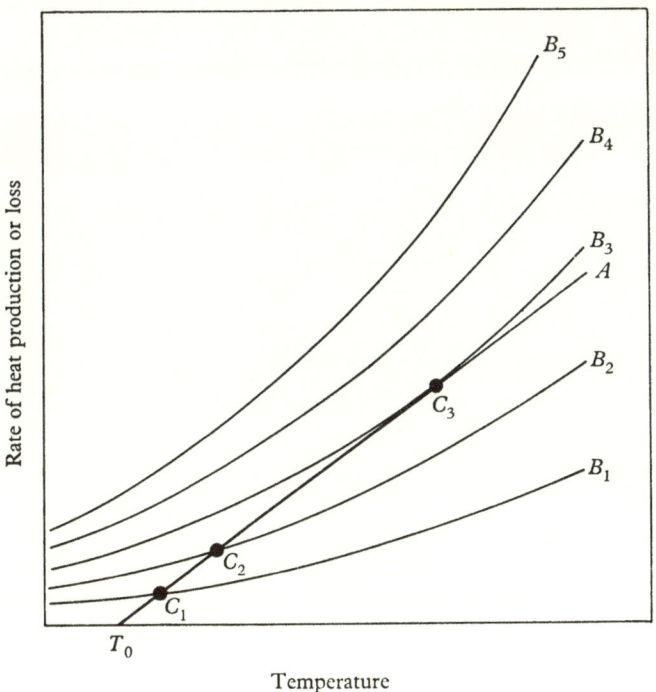

Fig. 6. Heat production and removal as a function of temperature for a normal chemical reaction.

temperature exceeds the capacity of the walls of the vessel to remove the heat. The idea is illustrated in Fig. 6. Each of the lines, B_1–B_5, represents the change of heat output of the reaction with temperature for a particular reaction mixture. The straight line A represents the rate at which the reaction vessel can remove heat if the wall temperature is T_0 and the gas temperature anything above T_0. When the line A cuts a heat production line such as B_1, the reaction is thermally stable after a short warm-up period; the reacting gas assumes a temperature T_1 and a rate R_1 represented by point C_1. If the line A fails to cut the heat production line, as with B_5, the walls can never remove heat sufficiently fast and thermal explosion results. The limiting case is represented by B_3 which is tangential to A; the point C_3 represents the limit of thermal stability.

In a reaction which accelerates with time the curves B_1–B_5 may be taken to represent the variation in the rate of heat production with temperature for different stages in the reaction. Thus during development the rate and temperature will follow points such as

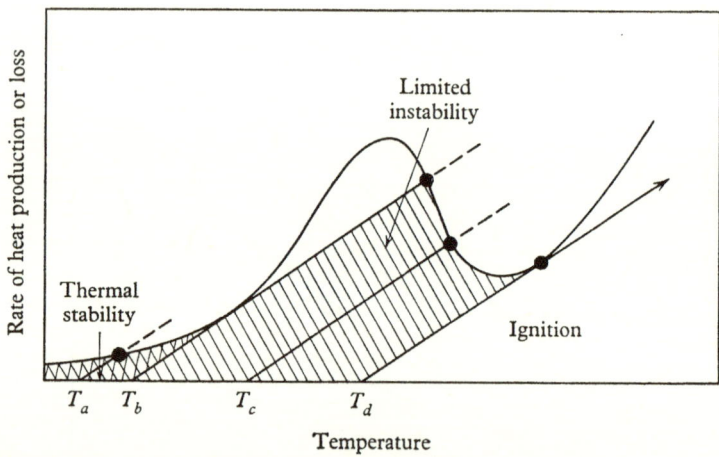

Fig. 7. Heat production and removal as a function of temperature for a reaction with a negative temperature coefficient.

C_1, C_2, etc. At C_3 the system becomes thermally unstable and explodes. This may occur some time after the initiation of the reaction. In degenerately branching chain reactions which do not explode the heat production curve for the maximum reaction rate never becomes tangential to A, that is it falls below B_3.

Cool flames may be partially explained if the heat-production curve shows a negative temperature coefficient. Suppose first that the reaction rate depends only upon temperature (Fig. 7) and consider various wall temperatures.

At low temperatures such as T_a, the reaction self-heats slightly but settles down to a steady state. At higher wall temperatures, T_b, the heat-removal curve becomes tangential to the heat-production curve and but for the operation of the negative temperature coefficient the reaction would explode. There is in fact a rather larger temperature jump and the temperature is stabilized at a much higher value than before. Only when the wall temperature is such that the heat-removal curve is tangential to the right part of the heat-production curve will true explosion occur (T_d). We might therefore identify the first zone of temperature stability up

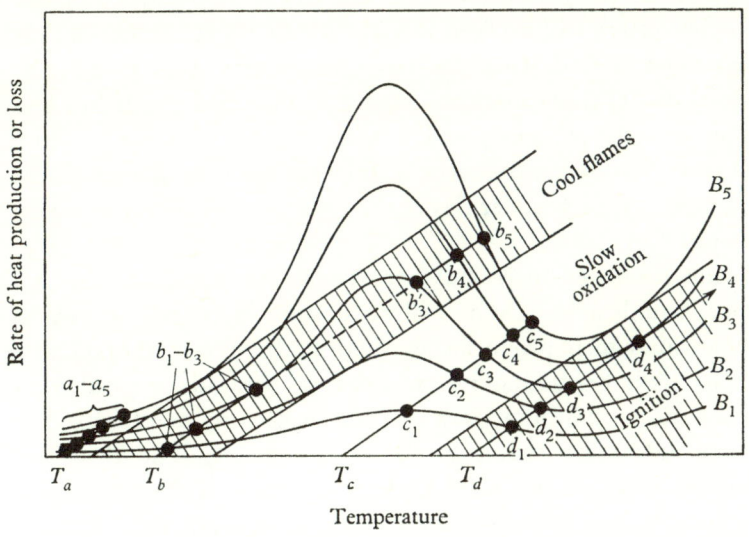

Fig. 8. Heat production and removal as a function of temperature and extent of reaction for an accelerating reaction with a negative temperature coefficient.

to T_b as the slow reaction zone, from T_b to T_d as the cool-flame zone of limited instability, and above T_d as the ignition region. According to this model there is no slow oxidation region between the cool-flame zone and ignitions.

Suppose now that the reaction accelerates and that for each stage in the reaction there is an appropriate heat-production curve with a negative temperature coefficient as in Fig. 8. The variation of the maximum rate with temperature is represented by curve B_5. Consider a wall temperature T_a. The reaction will remain thermally stable with a small temperature difference up to maximum rate: the temperature during reaction follows the points a_1 to a_5. This is a typical slow oxidation. At a wall temperature T_b the reaction will be thermally stable initially and the gas temperature will increase gradually from T_b till the point b_3 is reached. Here the system becomes thermally unstable and the temperature jumps suddenly to the point represented by b_3'. It rises somewhat further to b_5 but then declines and eventually drops back to T_b when all reaction ceases. The sudden temperature jump at a particular stage in the reaction is very similar to a cool flame. If the initial temperature is T_c there is no temperature jump in the course of the reaction, but at T_d the system will proceed to a hot ignition after an initial slow acceleration period when the gas temperature rises to d_4. The thermal model thus predicts a region of slow reaction below and above a cool-flame region, but it does not predict a low temperature ignition peninsula nor the occurrence of multiple cool flames. These must arise out of peculiarities of the chemistry of the combustion reaction.

Frank-Kamenetskii[69] and later Walsh[32] supposed that cool flames could arise isothermally through a purely chemical mechanism which produced alternating changes in the concentrations of a branching intermediate (a peroxide according to Walsh) and an inhibitor (formaldehyde according to Walsh). Any thermal instability was regarded as secondary. The theory is formally attractive but since there is no evidence for the cycling of concentrations of intermediates which can be identified as catalyst and inhibitor, few have accepted this view of cool-flame phenomena.

It is more likely that cool flames originate from thermal instability which is rapidly quenched because of the failure of some vital reaction when the temperature rises, coupled with a time lag in processes involving some vital intermediate. The idea may be illustrated in more detail by reference to Fig. 9. We suppose that the reaction starting isothermally at 320 °C produces a branching intermediate (or other critical substance), Z, in a fraction $\alpha = 0.2$ of the chain steps and that α decreases to a very low value above about 400 °C.

The reaction is thermally stable in the initial stages, and the temperature follows AB in Fig. 9a. The concentration of Z builds up exponentially during this period (GH, Fig. 9b): the concentration at any time results from a balance between the nearly equal rates of formation and destruction of Z (PQ, Fig. 9c). Since the overall rate of reaction is roughly proportional to the overall rate of destruction of Z, PQ may also be taken to represent the overall reaction rate. The progress of the reaction as a whole may be represented by the curve VW for the consumption of reactant in Fig. 9d.

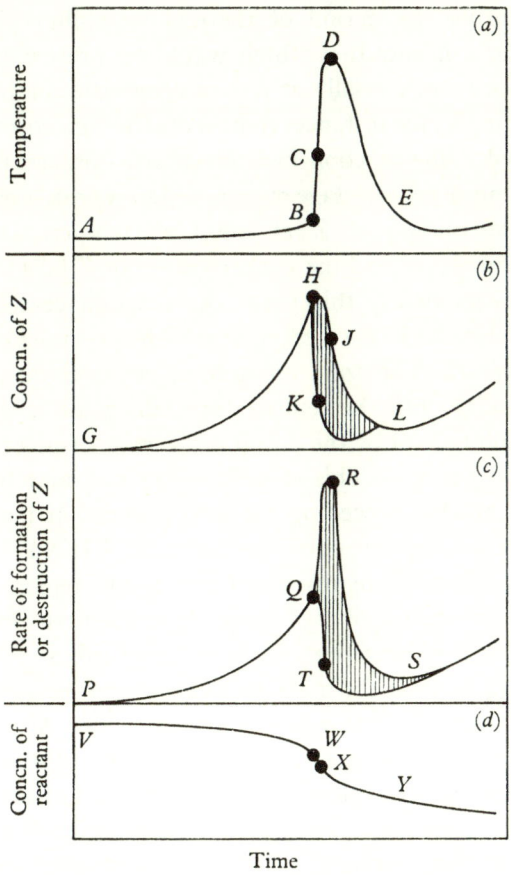

Fig. 9. Changes in various parameters during a cool flame (see text).

At a particular stage in the reaction (point B in Fig. 9a) the system becomes thermally unstable for reasons explained above. The temperature rises rapidly in say 10 ms to about 450 °C (BCD in Fig. 9a). During this short time the concentration of Z will hardly change (HJ in Fig. 9b), although its rate of decomposition will increase rapidly (QR in Fig. 9c). Simultaneously with the increases in temperature and rate of decomposition of Z, the rate of formation of Z falls drastically (QT in Fig. 9c): there is then a serious imbalance between the rates of formation and destruction of Z as shown by the shaded area in Fig. 9c. The imbalance may also be represented by the difference between the actual concentration of Z during the period of thermal instability (HJL in Fig. 9b) and the concentration which would be present had the reaction developed isothermally at any intermediate temperature (HKL). The point K, for instance, represents the concentration of Z to be expected if the reaction had developed isothermally at a temperature C and reached a stage of completion represented by X.

The rapid destruction of Z at the high temperature soon eliminates the imbalance and the concentration of Z falls to a low value L. Somewhat before this point the reaction ceases to be thermally unstable and the temperature drops quickly after reaching its peak D. The total amount of reaction during the period of instability until the temperature falls again is approximately that brought about by the decomposition of a concentration of Z given by the point H. This amount of reaction may be quite small compared to that preceding the development of instability and a reactant consumption curve of the form $VWXY$ may be obtained which shows little discontinuity in the region of the thermally unstable reaction or cool flame. After the first cool flame the whole process may be repeated a number of times so long as sufficient reactant remains for thermal instability to reassert itself.

Norrish[11,13] believes that cool flames arise out of the fast decomposition of accumulated peroxide, but he distinguishes between this rather special role of peroxides and that of aldehydes as branching intermediates in the slow oxidation. The distinction between the branching process in the slow oxidation and that leading to cool flames seems to be rather a fine one and it is more likely that the same processes are responsible for both effects, the balance between aldehyde and peroxide branching depending upon the substance being oxidized and the temperature.

The precise reactions leading to cool flames are still a matter for discussion and little progress is likely to be made until we have a clearer understanding of the elementary reactions which occur in slow isothermal combustions. Only then can the conditions for thermal or chain instability [70] be clearly formulated and a computation of the development of the reaction after the onset of instability carried out.

3. Recent work on the slow oxidation mechanism of lower hydrocarbons†

The view that the primary product of oxidation of alkanes changes between 300 and 450 °C was a direct result of the careful analysis of reaction products. Analyses taken after moderate consumptions of reactants (usually over 10 %) showed that the high temperature reaction produced predominantly cracking products (olefins, lower alkanes, hydrogen), while the low-temperature reaction produced predominantly oxygen-containing products (oxides of carbon, aldehydes, alcohols, epoxides, acids, peroxides, etc.). Thus Shtern [71] claims that the ratio of cracking to oxidation in the combustion of propane rises from about 0·3 at 300 °C to about 2·5 at 450 °C, while Satterfield and Reid [72] in 1955 taking all available data for the oxidation of propane showed that the ratio 'cracking/oxidation' rose with temperature with an activation energy of about 19 kcal mole^{-1}.

The author believes that this widely held view is basically unsound and has arisen because analysis has been carried out at too advanced a stage in the reaction (see ref. [73]), when primary processes are no longer dominant.

Some time after the investigations of Shtern, [26] Falconer and Knox [74] examined the oxidation of propane between 435 and 475 °C using gas chromatography for analysis. Propene and water were the major initial products with ethylene next. Since the initial ethylene/propene ratio decreased with increase in oxygen pressure, it was argued that propene must be formed by oxidation of propyl radicals while ethylene was formed by pyrolysis, i.e. by reactions (21) and (22). The propane itself was considered to be removed by (20)

$$C_3H_8 + HO_2 \text{ (or OH)} \rightarrow C_3H_7 + H_2O_2 \text{ (or } H_2O) \qquad (20)$$

$$C_3H_7 + O_2 \rightarrow C_3H_6 + HO_2 \qquad (21)$$

$$C_3H_7 \rightarrow C_2H_4 + CH_3 \qquad (22)$$

† The term 'lower alkane' used here refers specifically to ethane, propane and butane.

The methyl radicals were thought to react with oxygen to give formaldehyde and OH as proposed by Norrish [10] or to abstract H to give methane.

This mechanism differs from that of Shtern [75] who supposes that all 'pyrolysis products', including propene, are formed without the intermediacy of oxygen. This is unlikely since the reactions of alkyl radicals with oxygen [76,77] must be very much faster than pyrolysis to olefin and H [78] at 450 °C.

Falconer and Knox [74] observed that the yield of propene reached a nearly stationary level as the reaction proceeded, and from experiments in which part of the propane was replaced by propene it was concluded that this quasi-stationary state resulted from a balance between formation and consumption of propene. Propene also had an interesting effect on the kinetic curves. For a simple degenerately branching chain reaction plots of rate of reaction, $d(\Delta p)/dt$, against the extent of reaction, Δp, should be straight lines of gradient ϕ, the net branching factor. The experimental plots showed two straight portions. The change in gradient occurred roughly when the propene concentration reached its maximum. On replacing propane by the concentration of propene normally present at maximum rate, curves were obtained without any change of gradient (see Fig. 10). This showed that propene played an important rôle in the later development of the reaction, and that its own oxidation led to further catalysis of the propane oxidation.

In view of the controversy regarding the low-temperature oxidation mechanism, the rôle of propene was next examined [27] at 318 °C. Unexpectedly, below 1 % consumption of propane, 75–80 % of the alkane consumed appeared in the products as propene: on the simple aldehyde and peroxide theories the initial products should have contained virtually no olefin. However, at quite an early stage in the reaction the percentage conversion to propene fell and carbon monoxide appeared in the products. After about 10 % consumption the propene concentration again reached a kinetically controlled stationary value. Very similar results have now been obtained from studies of the low-temperature oxidations of ethane [79] and isobutane. [80,81] In the early stages of oxidations there is therefore no fundamental change in the nature of the reaction: the primary oxidation process for alkyl radicals is the same at about 300 °C as at 450 °C.

Confirmation that the stationary concentration of olefin resulted from a kinetic balance was obtained in the study of the oxidation of isobutane by Turner.[83] Before describing this work it is necessary to consider briefly the main features of olefin oxidation. Harding and Norrish[12] showed that formaldehyde was the

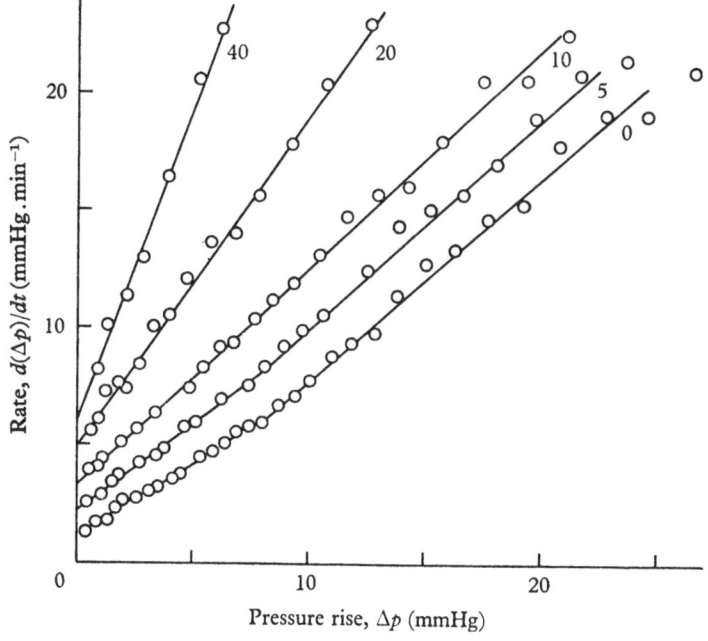

Fig. 10. Effect of replacement of propane by propylene at 435 °C. $P_{C_3H_8} + P_{C_3H_6} = 100$ mmHg; $P_{O_2} = 50$ mmHg. Pressures of C_3H_6 (mmHg) marked on lines. (From [74].)

branching intermediate in the oxidation of ethylene at 400 °C and proposed a mechanism in which it was the sole product formed directly from ethylene. Knox and Wells[79] confirmed this by showing that ethylene is initially converted 85 % to formaldehyde and 10 % to ethylene oxide above 320 °C. Several other studies of olefin oxidation have shown that this type of reaction is general. Ethylene, propene,[84,85] isobutene,[81,82] but-2-ene,[20,86] n-hexene[87] and methyl pent-2-ene[88] all give initially high yields of carbonyl compounds and lesser yields of epoxides according to the reactions (23) and (24):

$$A{=}B + O_2 \rightarrow A{=}O + B{=}O \qquad (\sim 80\%) \qquad (23)$$

$$A{=}B + \tfrac{1}{2}O_2 \rightarrow \underline{A{-}O{-}B} \qquad (\sim 20\%) \qquad (24)$$

In the oxidation of isobutene,[81] which is not autocatalytic at 300 °C, acetone is the major product at all stages as shown in Fig. 11. Initially acetone is formed with an equimolar yield of (formaldehyde + carbon monoxide). In the oxidation of isobutane it is therefore possible to follow any consumption of isobutene by observing the formation of acetone which is a minor initial product from isobutane. The results of Turner[83] (Fig. 12) show that a marked increase in the acetone yield in the later stages of the isobutane oxidation coincides with the levelling out in the isobutene yield in accord with the overall process.

$$C_4H_{10} \rightarrow 80\% \; C_4H_8 \rightarrow 80\% \, CH_3COCH_3$$
$$\searrow \qquad\qquad \searrow \qquad\qquad\qquad (25)$$
$$20\% \text{ Minor} \quad 20\% \text{ Minor}$$
$$\text{products} \qquad \text{products}$$

Olefins must therefore play an important part in the intermediate and later stages of alkane oxidations contrary to the view expressed by Shtern.[89]

The major source of aldehydes in alkane combustion reactions is now seen to be the oxidation of the primary olefin and not reaction (12) as originally proposed by Norrish.[10] It also becomes clear why acetaldehyde and formaldehyde are the predominant aldehydes in the oxidation of propane, and why lower aldehydes are more plentiful in combustion products than higher aldehydes. In the oxidation of propane, as suggested by Shtern,[90] acetaldehyde is the most likely branching agent in the absence of appreciable yields of propyl hydroperoxides.[49] In the oxidation of ethane and isobutane the answer is less simple. These two hydrocarbons are less easy to oxidize than propane and the olefin oxidations, unlike that of propene,[84] are not noticeably autocatalytic at low temperatures. This is probably because the only aldehyde formed from ethylene and isobutene is formaldehyde which according to Walsh is an inhibitor of low-temperature oxidations. [32] One is therefore tempted to identify the olefin as the branching intermediate in the oxidations of ethane and isobutane, although some

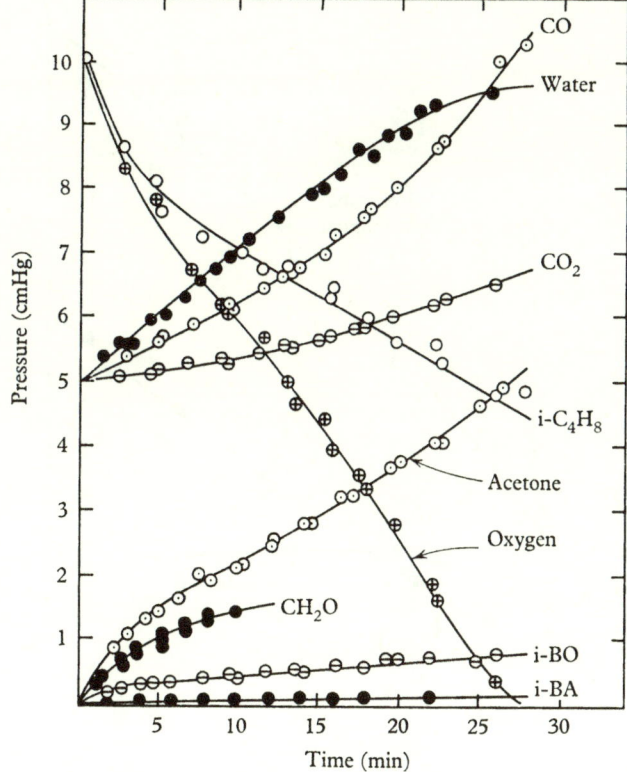

Fig. 11. Development of products for oxidation of 200 mmHg of a 1/1 iso-butene/oxygen mixture at 293 °C in clean Pyrex. The complete reaction is shown. i-BO=isobutene oxide, i-BA=isobutyraldehyde. (From [81].)

assistance in branching may be given by formaldehyde and the traces of higher aldehydes and hydroperoxides which may be formed. [91]

While 80% of the initial product of alkane oxidation is the 'conjugate' olefin, the remaining 20% comprises a complex mixture which has only been analysed fully with the aid of gas chromatography.[80,81] Hay, Knox and Turner[81] showed that while the ratio of major to minor products is little affected by reaction conditions the composition of the minor products varies widely and unpredictably. The key to this variability seems to lie in the surface of the reaction vessel since wide variations in the

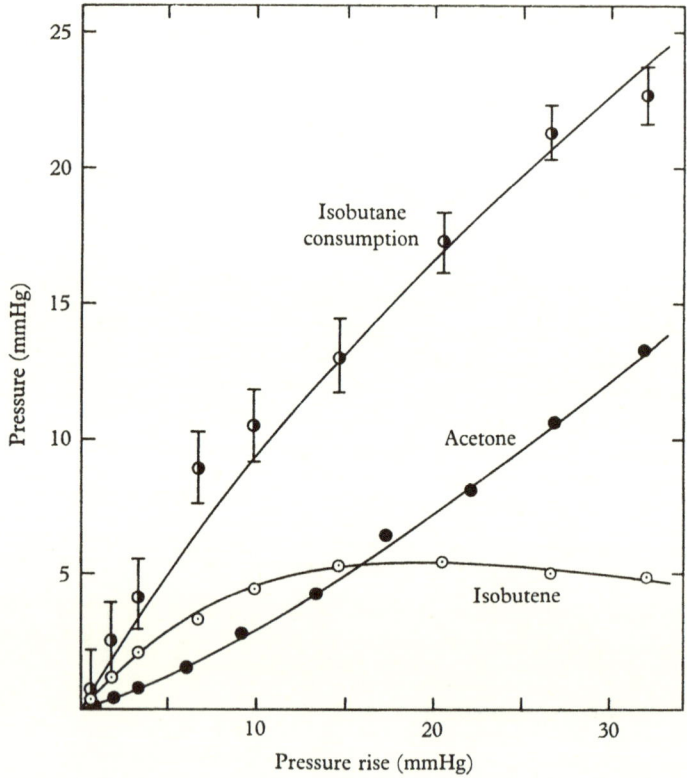

Fig. 12. Comparison of yields of acetone and isobutene with consumption of isobutane in the oxidation of 133 mmHg oxygen + 67 mmHg isobutane at 300 °C. Curve for isobutane consumption calculated from mechanism, assuming 85 % conversion of isobutane to isobutene and 85 % conversion of isobutene to acetone.

fractional initial yields of minor products occur when the surface is changed. Data for isobutane are shown in Fig. 13. It appears that two main homogeneous processes occur, one forming olefin and the other an unknown species Z which may be derived from ROO. If Z is sufficiently stable to diffuse to the walls of the reaction vessel it may decompose into a mixture whose composition depends critically upon the nature of the surface. The suggested reactions are (19) and (6) (see p. 258), (26) and (27):

$$ROO \rightarrow Z \qquad\qquad (26)$$

$$Z \rightarrow \text{diffusion to wall and decomposition} \qquad (27)$$

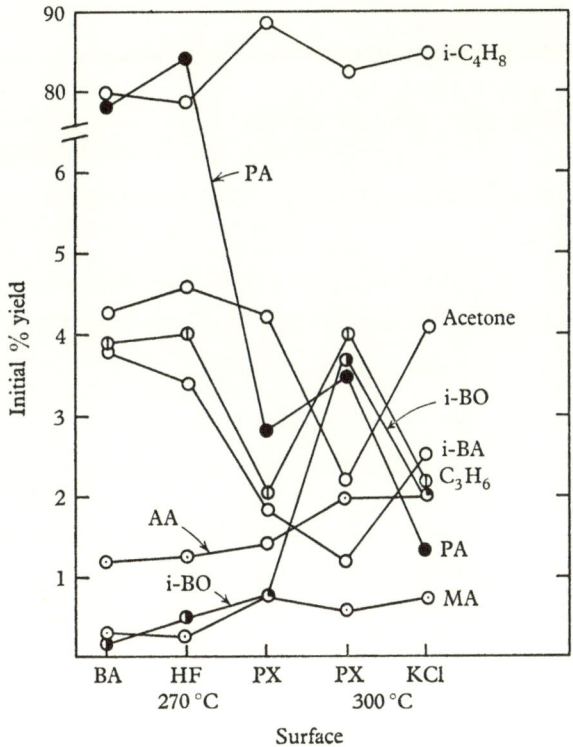

Fig. 13. Effect of surface on the initial percentage yields of products from the oxidation of 150 mmHg isobutane + 75 mmHg oxygen at 270 and 300 °C. PX = clean Pyrex; HF = washed with hydrofluoric acid; BA = coated with boric acid; KCl = coated with potassium chloride. AA, PA, and i-BA = acet-, propion-, and butyl-aldehydes, MA = methacrolein, i-BO = isobutene oxide. (From [81].)

Since one of the wall reaction products may well be the olefin formed in (19) the relative rates of (6) and (19) cannot be accurately determined until a surface is found upon which Z reacts to give a single or small number of distinctive products.

4. The mechanism of the slow oxidation of lower hydrocarbons

In discussing the overall mechanism of the slow oxidation, three types of reaction must be distinguished:

(1) The main product-producing or chain-propagating reactions, for which the possibilities are severely limited once the initial products are known.

(2) The chain-propagating reactions which form the minor products, and which in alkane oxidations are strongly influenced by the nature of the reaction vessel surface.

(3) Reactions which produce little identifiable product but which are of great kinetic importance, chain-branching and chain-terminating reactions.

4.1. *Chain-propagating reactions.* The main overall reactions in the oxidations of lower alkanes above 300 °C are (28) and (29). [92]

$$RH + O_2 \rightarrow AB + H_2O_2 \tag{28}$$

$$RH + \tfrac{1}{2}O_2 \rightarrow AB + H_2O \tag{29}$$

Where RH is an alcohol, AB is a ketone or aldehyde. [93]

Since the formation of alkyl radicals by hydrogen abstraction from RH can hardly be doubted and, since these radicals do not pyrolyse until well above 300 °C, the initial reaction in the absence of reaction products must occur by the elementary steps (30) and (19).

$$RH + HO_2 \rightarrow R + H_2O_2 \tag{30}$$

$$R + O_2 \rightarrow AB + HO_2 \tag{19}$$

In the oxidation of olefins the major process (23) suggests no obvious pair of elementary reactions, and three different chain mechanisms have been proposed: Skirrow and Williams [82] and Norrish and Porter [21] propose mechanisms based upon the initial addition of OH to AB, while Knox and Wells [79] propose the initial addition of HO$_2$ to AB. In addition to these a molecular mechanism has been proposed by Cullis et al. [88]

Independent experimental tests can be brought to bear on reaction (30). One of the simplest ways of obtaining information about chain-propagating reactions in complex systems is to carry out competitive experiments in which the chosen radical is made to react with a pair of similar reactants, for instance alkanes. From the pattern of selectivity over a range of alkanes the basic reactivity of the radical may be deduced. Highly selective radicals will be those whose reactions with alkanes are endothermic, since activation energy differences from one alkane to another will be close to the bond strength differences, whereas unselective radicals will be those whose reactions are exothermic.

Thermochemistry leads one to expect that HO_2 will be selective and similar to Br, while OH will be unselective and similar to Cl (the bond strengths of the hydrides are respectively 89, [94] 87, 119, 102 [95]). Experiments carried out by Knox, Smith and Trotman-Dickenson [96] and by Falconer, Knox and Trotman-Dickenson [97] showed that with a mean alkane consumption of about 20 % the radical (or mixture of radicals) responsible for removing alkanes in mixed oxidations was unselective and rather like Cl. It was therefore more likely to be OH or RO than HO_2. This result was in direct conflict with the mechanism derived from the analytical data and led us to suspect that HO_2 might be converted into a more reactive radical in the intermediate stages of the reaction. However, in the early stages there is no product with which HO_2 could react and reaction (30) seems unavoidable. The selectivity in the early stages should therefore be much greater than later. The idea was tested by Knox and Turner [98] who measured selectivities in the earliest stages of alkane oxidations using the relative rates of formation of olefins. Two propane + isobutane mixtures were examined, a 1:1 mixture and a 122:1 mixture. Within experimental error the two mixtures gave the same initial selectivity or rate constant ratio for removal of the two alkanes $k_{C_4H_{10}}/k_{C_3H_8} = 2\cdot8 \pm 0\cdot2$ at 300 °C compared with $1\cdot3 \pm 0\cdot1$ obtained previously for an average consumption of 20 %. Similar experiments with propane + ethane mixtures [99] gave an initial ratio $k_{C_3H_8}/k_{C_2H_6} = 5\cdot0 \pm 0\cdot4$ at 360 °C compared to $2\cdot3 \pm 0\cdot1$ obtained previously. The marked decline in selectivity between the early and later stages of the reaction supports the hypothesis that the reaction is at first propagated by the unreactive radical HO_2 which soon gives way to a more reactive species such as OH.

Since only the olefin is present in sufficient quantity to effect a conversion, and since it is consumed in the later stages of the reaction, the conversion process must start with the reaction

$$HO_2 + AB \rightarrow ABOOH \qquad (31)$$

(an abstraction from AB is ruled out by the nature of the major oxidation products of olefins).

Only one mechanism for converting HO_2 to OH has been firmly established: [100] the kinetics of the slow oxidation of hydrogen in boric acid-coated vessels demand that HO_2 is predominantly removed by the reaction (32).

$$2HO_2 \rightarrow H_2O_2 + O_2 \qquad (32)$$

At about 500 °C most of the peroxide decomposes to OH

$$H_2O_2 + M \rightarrow 2OH + M \qquad (33)$$

so that the conversion $2HO_2 \rightarrow 2OH + O_2$ is achieved for nearly all HO_2 radicals. The same conversion process is used by Sampson [101] to explain the kinetics of oxidation of ethane in a fast flow system at about 600 °C. Mechanisms involving this conversion have the property that the reaction rate is equal to the rate of pyrolysis of the peroxide present. The validity of the mechanism can therefore be verified independently knowing the rate constant for peroxide decomposition [102] and the concentration in the mixture.

Below 450 °C this conversion cannot operate since hydrogen peroxide decomposes heterogeneously. [102] In search for a low temperature conversion mechanism the author [103] has suggested a series of reactions in which the intermediate is a di-hydro-peroxide, since these should decompose at reasonable rates above about 270 °C. [104]

$$HO_2 + AB \rightarrow HOOAB\cdot \qquad (31)$$

$$HOOAB\cdot + O_2 \rightarrow HOOABOO\cdot \qquad (34)$$

$$HOOABOO\cdot + HO_2 \rightarrow HOOABOOH + O_2 \qquad (35)$$

$$HOOABOOH \rightarrow HO + A{=}O + B{=}O + OH \qquad (36)$$

Overall reaction

$$2HO_2 + AB \rightarrow 2OH + AO + BO \qquad (37)$$

This mechanism is analogous to Baldwin's [100] for the hydrogen + oxygen reaction. Although it appears to involve a chain termination step (35) this reaction is inevitably followed by what would normally be called a chain-branching step (36). Thus the net effect is that the number of radicals is unaltered and the reaction proceeds by way of a 'delayed chain' in which molecular and free radical intermediates alternate. The proper equivalents of chain-termination steps are the removal of HOOABOOH by radical attack or diffusion to the walls of the reaction vessel, or the disproportionation of HO_2 by (32). The equivalent of branching reactions are the generation of new radicals from AO or BO, or from the decompositions of hydroperoxides which are *not* formed by disproportionation of HO_2 and RO_2.

The above reactions cannot be the only ones involving the olefin since each RH produces one HO_2 and one AB while each AB removes $2HO_2$. To redress the balance it is necessary to add a set of reactions involving addition of OH to AB, namely (38) to (42).

$$OH + AB \rightarrow ABOH \qquad (38)$$

$$ABOH + O_2 \rightarrow HOABOO \cdot \qquad (39)$$

$$HOABOO \cdot + HO_2 \rightarrow HOABOOH + O_2 \qquad (40)$$

$$HOABOOH \rightarrow HOA + BO + OH \qquad (41)$$

$$HOA + O_2 \rightarrow AO + HO_2 \qquad (42)$$

Overall reaction

$$HO_2 + OH + AB + O_2 \rightarrow HO_2 + OH + AO + BO \qquad (43)$$

Reactions (38) to (42) produce the same products as (31) and (34) to (37) but effect no conversion of HO_2 to OH. A balance between HO_2 and OH concentrations can thus be set up whereby the rates of consumption of alkane and olefin are balanced and the concentrations of HO_2 and OH are stationary.

Apart from the necessity for conversion of HO_2 to OH on grounds of changing selectivity throughout the reaction there are kinetic reasons why HO_2 cannot be the main radical attacking the alkane in the later stages of reaction. The rate constant for disproportionation of HO_2 is about 0·1 times the collision number [100] and its abstraction reactions from alkanes are quite endothermic. If the A factor for H-abstraction by HO_2 is only 10 times lower than

that for disproportionation (giving an upper limit for A), disproportionation is favoured relative to abstraction from most alkanes
if the rate of reaction exceeds about 0·1 mmHg s^{-1}. The maximum
rate of typical slow oxidations often exceeds this and the rate
during the development of a cool flame is certainly much greater.
Chain oxidation by (19) and (30) cannot therefore occur at rates
often met with in slow oxidations. If HO_2 is continually being
generated by (19) there must be an efficient conversion mechanism
if chain oxidation is to proceed at reasonable rates.

4.2. *Reactions forming minor products.* There are broadly two
theories for the formation of minor products. Zeelenberg, who
first determined the complex products from the oxidations of
isobutane [80] and neopentane, [105] considers that the majority of
them can arise from unimolecular decompositions of peroxy
radicals. This type of mechanism has been elaborated by Fish [106]
who shows how they could be formed by a series of internal re-
arrangements involving internal abstraction of H atoms or CH_3
groups followed by O—O bond split. This theory, in the author's
opinion, is open to the serious objection that it is improbable that
a single species, such as a peroxy radical, can decompose or
isomerize via a large number of very different transition states at
rates which are comparable over a wide range of temperatures.

The results of Hay, Knox and Turner [81] suggest the much
simpler, although formally less attractive, explanation, that the
minor products arise on the walls of the reaction vessel. The
variety of products then reflects the variety of sites on the surface
rather than a variety of transition states available to a single
radical in the gas phase.

In this connection neopentane is of particular interest since, as
pointed out by Zeelenberg, [107] there is no reaction equivalent to
(19), and the yields of major as well as minor products should be
sensitive to the condition of the walls. Turner [108] in Edinburgh
and Drysdale [103] in Cambridge have carried out independent
investigations of the oxidation of neopentane with the effect of
surface particularly in mind. It is now clear that the yields of major
products isobutyraldehyde, isobutene and acetone are indeed
strongly sensitive to the nature of the surface. That the variation is

not due to experimental error is shown by the good agreement between the results of Turner and of Drysdale when the same surface is used.

With olefins the surface appears to have much less influence on the relative yields of minor products [81] and the evidence is that they arise mainly if not entirely homogeneously. The formation of epoxides may result from a direct addition of HO_2 across the double bond via a complex.

$$HOO\cdot + \begin{matrix} A \\ \| \\ B \end{matrix} \rightarrow H-O-O \begin{matrix} \diagup A \\ \| \\ \diagdown B \end{matrix} \rightarrow HO\cdot + O \begin{matrix} \diagup A \\ | \\ \diagdown B \end{matrix}$$

4.3. *Branching and termination reactions.* The nature of these reactions is still a matter for speculation although some progress has been made in defining the temperature régimes within which hydroperoxide branching may be important.

In the oxidation of propane Shtern [90] has shown convincingly that acetaldehyde is the likely branching agent. This seems very reasonable in view of its formation as a secondary product from propene. In oxidations which do not form a higher aldehyde the olefin itself may be the branching agent.

Peroxides when formed by reactions other than disproportionation of ROO and HO_2 will certainly act as branching agents. Their formation in the low-temperature oxidation of hydrocarbons containing four or more carbon atoms has been clearly demonstrated by Tipper and co-workers. [49, 50] In their formation internal hydrogen abstraction via 6-membered ring transition states seems to play an important part.

Chain-termination reactions almost certainly involve disproportionation of HO_2 by (32), since, being unreactive, it will be the radical present in highest concentration. Since hydrogen peroxide decomposes homogeneously above about 470 °C, and must be regarded as equivalent to 2OH radicals, the appropriate termination reaction is then heterogeneous destruction of H_2O_2.

5. An interpretation of the negative temperature coefficient

Using thermodynamic arguments Benson [47] and Knox [103] have shown independently that the bond dissociation energies in

R—OO and HOO—AB are about 29 and 14 kcal mole^{-1}. Since the entropy changes in reactions (6) and (31)

$$R + O_2 \rightarrow ROO \qquad (6)$$

$$HOO + AB \rightarrow HOOAB \qquad (31)$$

are likely to be similar [79,103] and around -32 cal deg^{-1} mole^{-1}, the equilibrium constants for the two association reactions can be found. While the ROO radical is stable up to nearly 500 °C with an oxygen concentration of 0·1 atm, the ABOOH radical is dissociated to more than 99 % even at 250 °C with [AB] = 0·1 atm, and it could be argued that mechanisms based upon ABOOH are unlikely. This is not so since there may be no other fate for HO$_2$ except reaction (31). The conversion of HO$_2$ to OH is limited not by the position of the equilibrium (31) but by the relative rates of disproportionation of HO$_2$ (32) and oxidation of ABOOH (34). Using reasonable values for the various rate constants the author has shown [103] that below 370 °C oxidation of ABOOH will predominate over disproportionation of HO$_2$ but above this temperature disproportionation takes over. Since reaction (32) removes radicals permanently from the system its increased efficiency above 370 °C will tend to reduce the maximum possible reaction rate. The negative temperature coefficient is therefore seen as a direct consequence of the instability of the ABOOH radical, and its generality from one oxidation system to another is explained.

Multiple cool flames can now be explained upon the basis of the mechanism using the ideas previously discussed, if the dihydroperoxide is identified as the intermediate whose concentration lags during the self heating. During the isothermal induction period before the cool flame a gradually increasing concentration of dihydroperoxide builds up.

Around 300 °C the lifetime of the dihydroperoxide may be of the order of 1 s. [104] If the reaction becomes thermally unstable this accumulated dihydroperoxide will survive largely intact during the rapid self heating which may take 10 ms. Nevertheless, at the end of this period it will be decomposing much faster than at the low temperature, although it will not be reformed because of the instability of the ABOOH radical. The dihydroperoxide is therefore soon exhausted and the reaction must slow down and the mixture cool. During the flame reaction the total amount of reaction will be hardly greater than the quantity of dihydroperoxide destroyed. The complete cycle may therefore occur a number of times with the major part of the oxidation occurring in the intervals between cool flames. The mechanism is a development of that described above when it was assumed that α declined rapidly with temperature.

The mechanism proposed thus gives an interpretation of many of the main features of the slow combustion of alkanes. While it differs in some important respects from the original mechanism proposed by Norrish it still maintains many of the features which he emphasized, the important part played by OH, the idea that the initial reaction product was itself an important reactant when the maximum rate was reached, and the important part played by aldehydes in the branching process. It remains a remarkable feature of the work of Norrish that so much of his original insight into the nature of chemical reactions has proved to be correct in principle if not always in detail.

REFERENCES

1   Norrish, R. G. W. and Wallace, J. *Proc. Roy. Soc.* A, 1934, **145**, 307.
2   Norrish, R. G. W. *Proc. Roy. Soc.* A, 1935, **150**, 36.
3   Carruthers, J. E. and Norrish, R. G. W. *J. chem. Soc.* 1936, p. 1036.
4   Foord, S. G. and Norrish, R. G. W. *Proc. Roy. Soc.* A, 1936, **157**, 503.
5   Norrish, R. G. W. and Reagh, J. D. *Proc. Roy. Soc.* A, 1940, **176**, 429.
6   Axford, D. W. E. and Norrish, R. G. W. *Nature, Lond.*, 1947, **160**, 437.
7   Axford, D. W. E. and Norrish, R. G. W. *Proc. Roy. Soc.* A, 1948, **192**, 518.
8   Harding, A. J. and Norrish, R. G. W. *Nature, Lond.*, 1949, **163**, 797.
9   Patnaik, D. and Norrish, R. G. W. *Nature, Lond.*, 1949, **163**, 883.
10 Norrish, R. G. W. *Revue Inst. fr. Pétrole*, 1949, **4**, 288.
11 Norrish, R. G. W. *Disc. Faraday Soc.* 1951, **10**, 269.
12 Harding, A. J. and Norrish, R. G. W. *Proc. Roy. Soc.* A, 1952, **212**, 291.
13 Bailey, H. C. and Norrish, R. G. W. *Proc. Roy. Soc.* A, 1952, **212**, 311.

14 Norrish, R. G. W. *Disc. Faraday Soc.* 1935, **14**, 16.
15 Knox, J. H. and Norrish, R. G. W. *Proc. Roy. Soc.* A, 1954, **221**, 151.
16 Knox, J. H. and Norrish, R. G. W. *Trans. Faraday Soc.* 1954, **50**, 928.
17 Norrish, R. G. W. and Taylor, G. W. *Proc. Roy. Soc.* A, 1956, **234**, 160.
18 Norrish, R. G. W. and Taylor, G. W. *Proc. Roy. Soc.* A, 1957, **238**, 143.
19 McKellar, J. F. and Norrish, R. G. W. *Proc. Roy. Soc.* A, 1960, **254**, 147.
20 McKellar, J. F. and Norrish, R. G. W. *Proc. Roy. Soc.* A, 1961, **263**, 51.
21 Norrish, R. G. W. and Porter, K. *Proc. Roy. Soc.* A, 1963, **272**, 164.
22 Macrae, K. and Norrish, R. G. W. Unpublished work.
23 Drysdale, D. and Norrish, R. G. W. Unpublished work.
24 Semenov, N. N.: (*a*) *Z. phys. Chem.* 11B, 1930, **464**. (*b*) *Chemical Kinetics and Chain Reactions*, p. 68. Oxford University Press, 1935.
25 Minkoff, G. J. and Tipper, C. F. H. *Chemistry of Combustion Reactions*. Butterworths, London, 1962.
26 Shtern, V. Ya. *The Gas Phase Oxidation of Hydrocarbons*, trans. Mullens, ed. Mullens. Pergamon, London, 1964.
27 Knox, J. H. *Trans. Faraday Soc.* 1959, **55**, 1362; 1960, **56**, 1225.
28 Ubbelohde, A. R. *Proc. Roy. Soc.* A, 1935, **152**, 354, 378.
29 Cullis, C. F. and Hinshelwood, Sir C. N. *Disc. Faraday Soc.* 1947, **2**, 117.
30 Bardwell, J. and Hinshelwood, Sir C. N. *Proc. Roy. Soc.* A, 1950, **201**, 26; 1951, **205**, 375.
31 Seakins, M. and Hinshelwood, Sir C. N. *Proc. Roy. Soc.* A, 1963, **276**, 324.
32 Walsh, A. D. *Trans. Faraday Soc.* 1946, **42**, 269; 1947, **43**, 279, 305.
33 Mulcahy, M. F. R. *Disc. Faraday Soc.* 1951, **10**, 259.
34 Karmilova, L. V., Enikolopyan, N. S., Nalbandyan, A. B. and Semenov, N. N. *Zh. fiz. Khim.* 1960, **34**, 1176 (Engl. transl. p. 562).
35 Lewis, B. and Von Elbe, G. *J. Am. Chem. Soc.* 1937, **59**, 970.
36 Sandler, S. and Beech, J. A. *Can. J. Chem.* 1960, **38**, 1455.
37 Kyriacos, G., Menapace, H. R. and Boord, C. E. *Analyt. Chem.* 1959, **31**, 222.
38 Pease, R. N. and Munro, W. R. *J. Am. Chem. Soc.* 1934, **56**, 2034.
39 Aivazov, B. V. and Neiman, M. B. *Nature, Lond.*, 1935, **135**, 655; *Zh. fiz. Khim.* 1936, **8**, 88.
40 Chernyak, N. Ya. and Shtern, V. Ya. *Dokl. Akad. Nauk SSSR*, 1951, **78**, 91.
41 Shu, N. W. and Bardwell, J. *Can. J. Chem.* 1955, **33**, 1415.
42 Burgoyne, J. H. *Proc. Roy. Soc.* A, 1940, **174**, 394.
43 McGowan, I. R. and Tipper, C. F. H. *Proc. Roy. Soc.* A, 1958, **246**, 52, 64.
44 Townend, D. T. A. *Chem. Rev.* 1937, **21**, 259.

45 Kane, C. P., Chamberlain, E. A. C. and Townend, D. T. A. *J. chem. Soc.* 1937, p. 436.
46 Ref. [26], p. 480.
47 Benson, S. W. *J. Am. Chem. Soc.* 1965, **87**, 972.
48 Newitt, D. M. and Thornes, L. S. *J. chem. Soc.* 1937, p. 1656.
49 Bonner, B. H. and Tipper, C. F. H. *Tenth Symposium (Int.) on Combustion*, p. 145. Combustion Institute, Pittsburgh, 1965.
50 Cartlidge, J. and Tipper, C. F. H. *Combustion and Flame*, 1961, **5**, 87.
51 Blat, E. I., Gerber, M. I. and Neiman, M. B. *Acta Phys. Chem. URSS*, 1939, **10**, 273.
52 Agnew, W. G. and Agnew, J. T. *Tenth Symposium (Int.) on Combustion*, p. 123. Combustion Institute, Pittsburgh, 1965.
53 Davy, H. *Gmelin's Handbook of Organic Chemistry*, 1812, **8**, 179.
54 Perkin, W. H. *J. chem. Soc.* 1882, **41**, 363.
55 Prettre, M., Dumanois, P. and Laffitte, P. *C. r. hebd. Séanc. Acad. Sci., Paris*, 1930, **191**, 329, 414.
56 Townend, D. T. A. and Mandlekar, M. R. *Proc. Roy. Soc.* A, 1933, **141**, 484; **143**, 168.
57 Day, R. A. and Pease, R. N. *J. Am. Chem. Soc.* 1940, **62**, 2234.
58 Ref. [26], p. 316.
59 Emeleus, H. J. *J. chem. Soc.* 1926, p. 2948; 1929, p. 1733.
60 Kondrat'ev, V. N. *Acta phys. Chem. URSS*, 1936, **4**, 556.
61 Agnew, W. G. and Agnew, J. T. *Ind. Engng Chem.* 1956, **48**, 2224.
62 Gaydon, A. G. and Wolfhard, H. G. *Third Symposium (Int.) on Combustion*, p. 504. Williams and Wilkins, Baltimore, 1949, *Proc. Roy. Soc.* A, 1952, **213**, 366.
63 Topps, J. E. C. and Townend, D. T. A. *Trans. Faraday Soc.* 1946, **42**, 345.
64 Ben-aim, R. and Lucquin, M. *Oxidation and Combustion Reviews*, vol. 1, p. 1. Elsevier, Amsterdam, 1965; Lucquin, M. *J. chem. Phys.* 1958, p. 827.
65 Bradley, J. N., Jones, G. A., Skirrow, G. and Tipper, C. F. H. *Tenth Symposium (Int.) on Combustion*, p. 139. Combustion Institute, Pittsburgh, 1965.
66 Ref. [26], p. 321.
67 Malherbe, F. E. and Walsh, A. D. *Trans. Faraday Soc.* 1950, **46**, 824, 835.
68 Ref. [23], p. 80.
69 Frank-Kamenetskii, D. A. *Zh. fiz. Khim.* 1940, **14**, 80.
70 Semenov, N. N. *Some Problems of Chemical Kinetics and Reactivity*, trans. Bradley. Pergamon, London, 1959.
71 Ref. [26], p. 324.
72 Satterfield, C. N. and Reid, R. C. *J. phys. Chem.* 1955, **59**, 283.
73 Ref. [25], p. 158.
74 Falconer, J. W. and Knox, J. H. *Proc. Roy. Soc.* A, 1959, **250**, 493.
75 Ref. [26], p. 332.
76 Dingledy, D. P. and Calvert, J. G. *J. Am. Chem. Soc.* 1963, **85**, 856.

286 J. H. KNOX

77 Goldfinger, P., Huybrechts, G., Martens, G., Meyers, L. and Olbregts, J. *Trans. Faraday Soc.* 1965, **61**, 1933.
78 Kerr, A. and Trotman-Dickenson, A. F. *Progress in Reaction Kinetics*, ed. Porter, vol. 1, p. 105. Pergamon, London, 1961.
79 Knox, J. H. and Wells, C. H. J. *Trans. Faraday Soc.* 1963, **59**, 2786, 2801.
80 Zeelenberg, A. P. and Bickel, A. F. *J. chem. Soc.* 1961, p. 4014.
81 Hay, J., Knox, J. H. and Turner, J. M. C. *Tenth Symposium (Int.) on Combustion*, p. 331. Combustion Institute, Pittsburgh, 1965.
82 Skirrow, G. and Williams, A. *Proc. Roy. Soc.* A, 1962, **268**, 537.
83 Turner, J. M. C. Ph.D. Thesis, Edinburgh, 1964.
84 Mullen, J. D. and Skirrow, G. *Proc. Roy. Soc.* A, 1958, **244**, 312.
85 Ref. [26], p. 527.
86 Blundell, A. and Skirrow, G. *Proc. Roy. Soc.* A, 1958, **244**, 331.
87 Skirrow, G. *Proc. Roy. Soc.* A, 1958, **244**, 345.
88 Cullis, C. F., Fish, A. and Turner, D. W. *Proc. Roy. Soc.* A, 1961, **262**, 318.
89 Ref. [26], p. 373.
90 Ref. [26], p. 325.
91 Taylor, G. W. *Can. J. Chem.* 1958, **36**, 1213.
92 Knox, J. H. *Ann. Rep. Prog. Chem.* 1962, **59**, 18.
93 Cullis, C. F. and Newitt, E. J. *Proc. Roy. Soc.* A, 1961, **262**, 392.
94 Foner, S. N. and Hudson, R. L. *J. chem. Phys.* 1962, **36**, 2681.
95 Cottrell, T. L. *Strengths of Chemical Bonds*, 2nd edn. Butterworths, London, 1958.
96 Knox, J. H., Smith, R. F. and Trotman-Dickenson, A. F. *Trans. Faraday Soc.* 1958, **54**, 1509.
97 Falconer, W. E., Knox, J. H. and Trotman-Dickenson, A. F. *J. chem. Soc.* 1961, p. 782.
98 Knox, J. H. and Turner, J. M. C. *J. chem. Soc.* 1965, p. 3491.
99 Irving, G. W. and Knox, J. H. Unpublished work.
100 Baldwin, R. R. and Mayor, L. *Trans. Faraday Soc.* 1960, **65**, 80, 93, 103.
101 Sampson, R. J. *J. chem. Soc.* 1963, p. 5095.
102 Baldwin, R. R. and Bratten, D. *Eighth Symposium (Int.) on Combustion*, p. 110. Williams and Wilkins, Baltimore, 1962.
103 Knox, J. H. *Combustion and Flame*, 1965, **9**, 297.
104 Kirk, A. D. and Knox, J. H. *Trans. Faraday Soc.* 1960, **56**, 1296.
105 Zeelenberg, A. P. *Rec. Trav. Chem. Pays-Bas*, 1962, **81**, 720.
106 Fish, A. *Q. rev. Chem. Soc., Lond.*, 1964, **18**, 243.
107 Zeelenberg, A. P. *Tenth Symposium (Int.) on Combustion*, p. 340. Combustion Institute, Pittsburgh, 1965.
108 Turner, J. M. C. Unpublished data.

11

THE SENSITIZATION AND INHIBITION OF IGNITIONS

P. G. ASHMORE

1. Introduction

Norrish's early work in photochemistry, catalysis and combustion brought to his attention many fast reactions and gave him a lasting interest in explosions and ignitions. Generations of students have shared his pleasure in his demonstrations of ignition phenomena. At one extreme, the catalytic effect of water on the combustion of carbon monoxide was shown by the dramatic snuffing of the flame in dry air; at the other, a trace of ammonia prevented the explosive combination of hydrogen and chlorine when irradiated by an arc lamp—or rather postponed it, tantalizingly, until the inhibitor was destroyed. Somewhere between, the glow limits of phosphorus were contrasted with the screaming detonation in a mixture of carbon disulphide and nitric oxide; and, perhaps best remembered of all, a simple coal-gas flame sitting on a tin departed, not with a whimper, but with a bang. These demonstrations were often left to speak for themselves, to arouse curiosity rather than to give opportunity for ready explanation. To anyone sufficiently interested to read more deeply, it became evident that an explanation of the ignition phenomena depended upon the unravelling of some very complex reaction kinetics. Here lay the fascination for Norrish. He was very well aware of the practical importance of combustion and ignition, and indeed many of his research topics were prompted by thoughts about practical requirements like the elimination of knock in petrol engines or the reduction of fire hazards in combustible mixtures. But the recurring theme in all his work is a search for the chemical basis of the phenomena.

Of the three major fields of work on ignitions—flame movement, initiation by sparks or similar sources, and the transition from slow thermal or photochemical reaction to ignition—only the last really appealed to Norrish. He recognized that any study of flame movement involved as much knowledge of fluid mechanics as of chemistry, and, apart from a brief excursion into the measurement of flame speeds in the late 1930's, he realistically left alone such mathematical topics. Nor had he much interest in inflammability limits with spark initiation, although at the beginning of World War II he examined, with the author and Dainton, the means of reducing the hazards of inflammation of hydrogen in barrage balloons. These were not lasting interests, however. He preferred to combine his studies of combustion and other slow reactions with the kinetic aspects of theories of ignition, aiming at the elucidation of the reactions that led to ignition. Acutely aware of the difficulties of establishing reproducible kinetics of chain reactions in highly purified gases, especially when the vessel surface is active, he chose deliberately to add agents that were likely to introduce homogeneous initiating, branching, or terminating reactions, in the hope of simplifying the kinetics of the main reaction. This approach perhaps underestimated the complexity of the reactions introduced by some of the additives, but in his hands, and through later work along the same lines, it has provided a good deal of useful kinetic information and the means of testing our theories of the nature and cause of many ignition phenomena. The purpose of this chapter is to outline these developments from Norrish's early work, dealing with reactions in which he was actively interested; the discussion is therefore selective and is not intended to be a comprehensive survey.

2. Spontaneous ignition boundaries

Spontaneous ignitions, characterized by the rapid consumption of reactants, sharp rises in temperature and pressure, and much chemiluminescence, are found when suitable mixtures of gaseous fuels and oxidants are exposed to certain conditions of temperature and pressure. Similar spontaneous explosive reactions occur in the exothermic decompositions of ozone, acetylene, hydrazine, nitrous oxide, nitrogen trichloride, azomethane, and some alkyl nitrates. Experiment has shown that for many reactions the ignitions occur when the temperature and pressure lie above a simple boundary like A in Fig. 1; the actual position of the boundary depends upon the compositions chosen for the reacting mixture, and the size and shape of the reaction vessel. For other reactions, however, the boundaries have more complicated shapes like the low-pressure, low-temperature peninsulae (*B*) found in the oxidation of hydrogen, phosphorus, carbon monoxide, and carbon disulphide, or the 'lobes' enclosing regions of cool flames (*C*) found in the oxidation of higher hydrocarbons. An example of these lobes is shown in Fig. 3 of

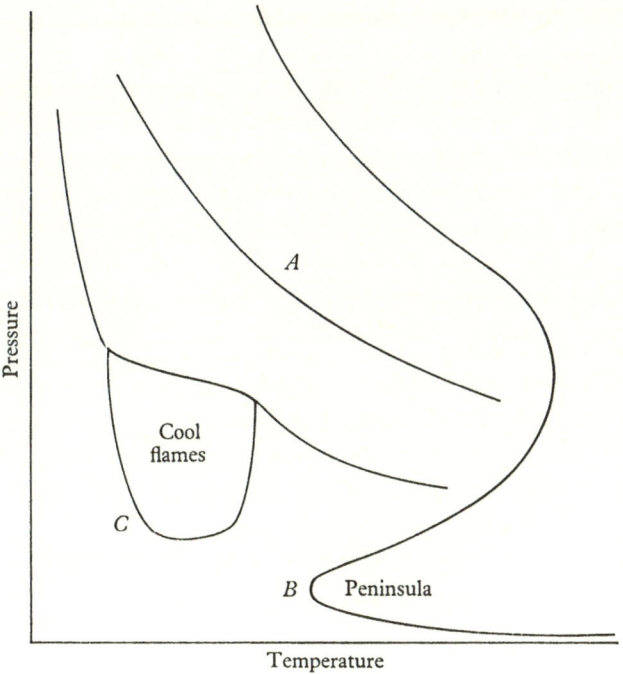

Fig. 1. Typical ignition boundaries.

Chapter 10 (page 260). The variety of boundaries is caused in part by the variety of mechanisms of these reactions. It is probably true that all the reactions of interest are chain reactions, and, speaking broadly, simple monotonic boundaries will be found in simple, non-branching chain reactions, while the complex boundaries are associated with branching or degenerate-branching chain reactions.

With either branched or non-branching chain reactions the positions and shapes of the ignition boundaries can be changed markedly by the addition to the reactants of traces of sensitizers or inhibitors. In interpreting these effects, attention must be paid to the detailed reactions introduced by the additives and to the precise nature of the ignitions.

There are two fundamentally different ways in which the very high reaction rates necessary for ignition can be attained. In the first, open to non-branching and branching chain reactions alike, the reacting mixture is self-heating and the rate accelerates, because the temperature rises, until a *thermal ignition* occurs. In the second, available only to branching chain reactions, the centres multiply isothermally because the net rate of branching is positive, so that the rate builds up to an *isothermal ignition*.

At the time when Norrish began his researches, the theories of thermal and isothermal ignitions were being developed rapidly, particularly by Semenov. It is important to recall the main features of these theories, because they not only influenced Norrish's early work but have proved the basis for much of the later work on ignitions.

2.1. *Thermal ignitions.* Semenov developed [1] in detail the earlier suggestions of Van't Hoff about the nature of thermal ignitions, relating the rate of production of heat by an exothermic reaction to the rate of loss of heat to the walls of the vessel in which the reaction occurs.

For simple gas reactions the rate law can be written

$$w_T = A \exp{(-E/RT)}p^n \tag{A}$$

where A, E are the Arrhenius parameters of the rate constant, n is the order of reaction, and p is the total pressure of mixtures of chosen proportions of reactants.

The rate of loss of heat might be controlled by conduction through the gas, with the gas temperature falling from the centre of the vessel to the walls, or by the rate of heat transfer to the walls with the same average temperature of the gas, $\bar{T}$, throughout the vessel. With the second condition, the rate of rise of $\bar{T}$ is given by

$$Vc_v\rho(d\bar{T}/dt) = VQw_T - hS(\bar{T} - T_0) \tag{B}$$

In this equation, V is the volume of the vessel, S its surface area, and T_0 the temperature of the walls; ρ is the density and c_v the heat capacity of the mixture, Q the heat evolved in unit reaction and h the coefficient of heat transfer to the walls.

It was found that the solution of equations (A) and (B) led either to a maximum temperature rise $(\bar{T} - T_0)$ of at most a few tens of degrees or, with comparatively small changes in the initial value of p or the temperature T_0, to a very rapid increase in the temperature and so to ignition after a short induction period. (See Fig. 9a for the results of comparable later computations.) For the first-order decompositions of azomethane and methyl nitrate, numerical integration of equations (A) and (B) gave product-time curves from which trends in the 'calculated' induction periods could be determined, and these were found to be in reasonable agreement with trends in the 'experimental' induction periods—i.e. the observed times between admission of the gases to the reaction vessel and the onset of ignition.

Semenov took as the condition of ignition that the curves for heat production and heat removal should just touch each other, as shown at C_3 in Fig. 6 of chapter 10 (page 263). This led to expressions connecting the total pressure of reactants and the temperature along an ignition boundary of the form

$$\ln (p^n/T_0^2) = E/RT_0 + \text{constant}$$

where the 'constant' included Q, A, V, S, h, etc., but is independent of p and T_0. This expression was tested against experimental results for a number of reactions by plotting $\ln (p^n/T_0^2)$ against $1/T_0$. This produced satisfactory linear plots, but the values of the 'constants' could not be checked because the heat transfer coefficient h could not be determined independently. In addition, a similar expression can be derived from the theory of isothermal chain ignitions, and in the absence of any means of calculating the 'constants' independently, these plots could not characterize uniquely thermal ignitions.

An important development of the thermal theory of ignitions was made by Frank-Kamenetskii. [2] He took account of the actual variation of temperature across the reaction vessel, assuming that heat was lost by conduction only, and found the conditions under which stationary states are possible for vessels of different shapes. He showed that ignitions occurred when a dimensionless parameter

δ exceeded a critical value δ_c which depends upon the shape of the vessel and the order of the reaction. For cylindrical or spherical vessels, of radius r,

$$\delta = QEr^2 w_T/\lambda RT_0^2 \tag{C}$$

and for reactions of zero order $\delta_c(0)$ takes the values 3·32 for spherical and 2·00 for cylindrical vessels. For reactions of higher order, (n), $\delta_c(n)$ takes higher values depending upon the heat capacity C_v as well as on Q and E.

The great advantage of Frank-Kamenetskii's treatment is that it eliminates the indeterminate heat transfer coefficient and yields expressions which contain measurable properties of the reacting mixture and the vessel. Hence, if the rate law is known, the ignition boundary can be predicted.

Comparisons of predicted and experimental ignition temperatures, for selected pressures of reactants, have been made for reactions such as the decompositions of azomethane, methyl nitrate, and nitrous oxide, and for the oxidation of hydrogen sulphide. [2] The comparison of ignition temperatures is not a very sensitive test of the theory, however, and only recently have more stringent comparisons been made between observed and calculated ignition boundaries, using sensitized reactions where the kinetic law and the factors in (C) are well established. These tests, carried out in Norrish's laboratory, are described in section 6.

2.2. *Isothermal ignitions.* Semenov also showed [1] that a branched chain reaction could lead to an exponential growth of chain centres, leading to a very rapid rate of reaction at the end of an isothermal induction period. Writing the *net* rate of branching by reactions which are first order in $[X]$, the concentration of chain centres, as $\phi[X]$, and the rate of initiation of centres as θ, the equation for $d[X]/dt$ is

$$d[X]/dt = \theta + \phi[X]$$

If it is assumed that θ and ϕ are constant, then, when $\phi > 0$ and $\phi t > 1$, the equation integrates to give

$$[X] = \frac{\theta}{\phi} e^{\phi t} \tag{D}$$

The rate of reaction is proportional to $[X]$, and it is easy to show that

$$\ln [\text{product}] = \phi t + \text{constant} \qquad (E)$$

provided again that θ and ϕ are constant. If the end of the induction period is defined by the rate of reaction, or the amount of the product, reaching specified values, equations (D) and (E) show that ϕ and τ, the length of the induction period, should obey the equation

$$\phi\tau = \text{constant} \qquad (F)$$

The reaction between hydrogen and oxygen provided several tests of these relationships. Ignitions occur between a first and a second pressure limit of ignition (B in Fig. 1). In 1933 Kowalskii [3] measured the rates of reaction of mixtures near the first limit with a rapidly responding membrane manometer. Measuring the concentration of products by $-(\Delta p)$, the decrease in pressure, he showed that the results fitted equation (E), and that $\phi\tau$ was approximately constant as the reactant pressure was varied at constant temperature, in accordance with (F).

The isothermal branched-chain theory of ignitions was particularly successful [4] in explaining the occurrence and the properties of the upper boundary of the low-pressure, low-temperature ignition peninsula (B, Fig. 1). The principal reactions occurring in this system, to which frequent reference will be made, are (1) to (6).

$$OH + H_2 \rightarrow H_2O + H \quad \text{(propagation)} \qquad (1)$$

$$H + O_2 \rightarrow OH + O \quad \text{(branching)} \qquad (2)$$

$$O + H_2 \rightarrow OH + H \quad \text{(branching)} \qquad (3)$$

$$H + O_2 + M \rightarrow HO_2 + M \quad \text{(termination)} \qquad (4)$$

$$HO_2 \rightarrow \text{removal at wall} \qquad (5)$$

$$HO_2 + HO_2 \rightarrow H_2O_2 + O_2 \qquad (6)$$

Provided that no other reactions of HO_2 or H_2O_2 occur, e.g. with reactant, the effective terminating step is (4). The upper boundary occurs because of competition for H atoms between the positive branching reaction (2) and the homogeneous terminating reaction (4); as (2) increases only linearly with pressure, whereas (4) increases as the square, reaction (4) predominates at higher reactant pressures and ignitions are suppressed.

Similar explanations can be found for the upper boundary of low-temperature ignition peninsulae in other reactions. The thermal theory of ignition cannot account for such upper pressure limits.

The relationship (F) played an important part in studies of ignitions in branched-chain reactions, because it provides an attractively simple connection between the observed quantity τ and the unknown value of ϕ. It is only valid, however, when ϕ remains constant during the induction period; as will be indicated later this is often not true.

3. The branched-chain thermal theory of ignitions

Norrish soon saw the necessity of combining the two main theories of ignition which were evolving at the time he was first investigating sensitized ignitions. He was impressed by the elegance and economy of the isothermal theory of ignition, but he was also aware that if the branched-chain reactions were exothermic, heating effects would accompany the growth of chain centre concentration. He was also very conscious, from his work on photochemical reactions, of the importance of mutual (or quadratic) termination at high centre concentrations. In order to interpret the results of his work upon sensitized ignitions, he found it necessary to incorporate all these points into a branched-chain thermal theory of ignitions. [5,6,7]

3.1. *Norrish's early work.* Norrish's research on ignitions began with studies of the photosensitized explosions of hydrogen and oxygen mixtures in the presence of chlorine, [8] a phenomenon first observed during work on the effects of oxygen on the photochemical reaction between hydrogen and chlorine. He also investigated the combination of hydrogen and oxygen photosensitized by nitrogen dioxide, [9] and later the thermal ignition of methane and oxygen in the presence of nitrogen dioxide [10] and the thermal ignition of hydrogen and oxygen sensitized by traces of nitrogen dioxide, [5,11,12] nitrosyl chloride, [13] or chloropicrin. [14] He was particularly interested in the occurrence with the last three reactions of lower and upper sensitizer limits of ignition. Outside these limits, slow reactions forming water were observed: both slow reactions and ignitions were found at temperatures far below the lowest temperatures of ignition of mixtures of hydrogen and oxygen alone, and were preceded by induction periods of many seconds during which no pressure changes occurred. These results were recognized as characteristic of branched-chain ignitions, but some observations pointed to thermal effects. In particular, with many mixture compositions, brief pressure *increases* were observed before the pressure fall attributed to the main slow reaction, indicating the occurrence of self-heating; and the order of effectiveness of inert gases like A, He, N_2 and CO_2 in quenching the ignitions suggested that a thermal conductivity effect was superposed upon the action of the gases in deactivating chain centres or catalysing the recombination of centres. The changes in the values of the sensitizer limits and the induction periods when systematic changes were made in the temperature, the partial pressure of the reactants or of added gases, the diameter of the vessel, and the intensity of irradiation, were established in a classical series of papers. [5,6,9,12,13] One of the most striking results was the demonstration that, with NO_2 as sensitizer in a narrow vessel, ignitions occurred within a closed boundary (Fig. 2). [12]

It was assumed that during the induction period there was little change in the concentration of nitrogen dioxide (or nitrosyl chloride) and that the value of ϕ remained constant and characteristic of the initial mixture, the vessel and the temperature. The isothermal theory was modified by including quadratic termination† at a rate $\delta[X]^2$, giving the equation

$$d[X]/dt = \theta + \phi[X] - \delta[X]^2$$

If both ϕ and δ are positive, $[X]$ reaches a limiting value, $[X]_e$, and when θ is small compared with the other two terms on the right of equation (J), $[X]_e = \phi/\delta$. Under these conditions, Norrish argued, ignition would occur if some critical volume element became self-heating, that is when $[X]_e$ exceeds some critical value determined by $\phi \geqslant \phi_c$. As the early growth of chain centres will depend on ϕ, and not on δ, the induction periods will be controlled by ϕ, and the relation $\qquad \phi\tau = $ a constant $\qquad\qquad$ **(F)**

† Quadratic branching was not excluded, but was thought to be rare [5,6] (see page 307). The δ used here (in conformity with common practice) has no connection, of course, with the δ of page 292.

will hold. Thus investigations of the induction periods near the ignition limits could still be used to make inferences about ϕ_c and the effect on it of changing the dimensions and surface of the vessel or the thermal conductivity and capacity of the mixtures.

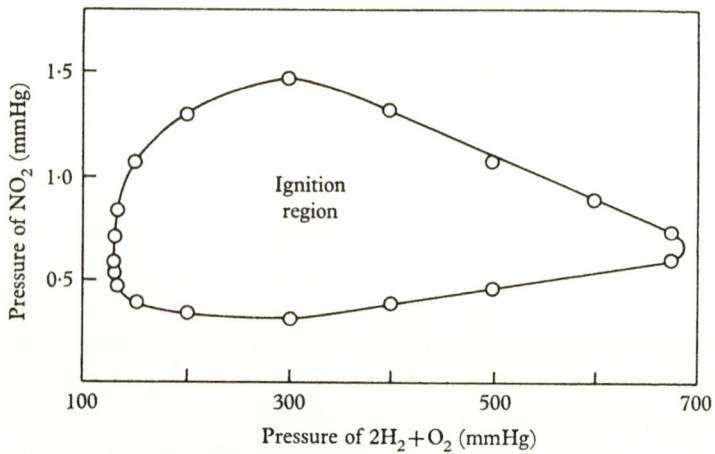

Fig. 2. Ignition boundary in $H_2/O_2/NO_2$ mixtures. Pyrex vessel 7 mm i.d., 364 °C.

A combination of this theory with a kinetic scheme involving nitrogen dioxide in the initiating, branching and terminating steps suggested that the chain-thermal theory was superior to a chain-isothermal theory of the sensitized ignitions. [6]

Some experimental results remained obstinately at variance with the theory, however. For some mixtures, τ increased on passing from slow reaction to ignition; from the relation $\phi \propto 1/\tau$, this meant that ϕ *decreased* in passing from slow reaction to ignition, and this seemed impossible with either a thermal or isothermal theory. Again, at the upper and lower sensitizer limits of ignition ϕ_c should take the same value, because the thermal properties are only trivially changed when only the sensitizer concentration is changed; hence τ should be the same at both limits, whereas it was invariably (and often very substantially) shorter at the lower limit. Lastly, photolysis markedly shortens τ without altering the position of the limits, so that τ and ϕ_c seem independent.

A complete change of outlook [15] was foreshadowed by experiments which showed that the length of the induction period was shortened by allowing the sensitizers chloropicrin or nitrosyl chloride to decompose to various extents in the vessel before adding the main reactants. It was found [16] that the sensitizer limits are virtually identical in the systems $H_2 + O_2 + NO_2$, $H_2 + O_2 + NOCl$, and $H_2 + O_2 + CCl_3NO_3$. Somewhat later, photometric studies showed that nitrogen dioxide, when added as a sensitizer, is removed during the induction period by a chain reaction with hydrogen. [17] This fast reaction [18] proceeds by the propagating steps

$$OH + H_2 \rightarrow H_2O + H \qquad (1)$$

$$H + NO_2 \rightarrow NO + OH \qquad (7)$$

and gives an overall reaction without pressure change

$$H_2 + NO_2 \rightarrow H_2O + NO \qquad (8)$$

It is easy to see that in the earlier studies, where induction periods and limits were investigated through pressure changes, reaction (8) was not suspected. At low concentrations of NO_2, reaction (8) is effectively opposed by reactions which reform NO_2, so that at the end of the induction period a mixture of NO and NO_2 is left. An obvious reforming reaction is the termolecular reaction (9), although other reactions are important as will be explained.

$$2NO + O_2 \rightleftharpoons 2NO_2 \qquad (9), (-9)$$

It became clear, therefore, that in all the sensitized ignitions investigated by Norrish the induction period was the time taken for an inhibitor to be removed, yielding a common final sensitizing mixture of NO and NO_2. This situation had been anticipated by Norrish and Griffiths. [9] 'It is therefore possible that nitric oxide is the true catalyst in the thermal reaction, whilst nitrogen peroxide (*sic*) is an inhibitor....' The role of nitric oxide was revealed by later work described in the following section. The work so far described, besides indicating the true cause of the induction periods, removed the paradoxes arising from application of the relationship **(F)**

$$\phi\tau = \text{constant} \qquad \textbf{(F)}$$

The relative amounts of inhibitor and sensitizer change during the induction periods, and hence ϕ can no longer be regarded as constant and characteristic of the initial mixture. Consequently the integrations which led to equations (D), (E) and (F) are not valid, and equation (F) cannot be applied to sensitized ignitions.

3.2. *Recent work on the system* $H_2 + O_2 + NO_2 + NO$. The first photometric studies [17-19] of this system showed that the rate of removal of the nitrogen dioxide increases sharply at the end of the induction period, when p_{NO_2} reaches 'p_e', and if the initial pressure of nitrogen dioxide (p_0) lies inside the sensitizer ignition limits, ignition follows when p_{NO_2} reaches 'p_i'. If, however, p_0 lies above the upper sensitizer limit, the accelerated removal of nitrogen dioxide declines and p_{NO_2} reaches a stationary value 'p_s'.

Further studies with a more sensitive photometer [20] revealed that below the lower limit p_{NO_2} also reaches a stationary value p_s very close to the value of the upper sensitizer limit for the same re-actant pressures. By examining mixtures with reactants at pressures just too low for ignition it was shown that p_s is indeed constant over the range of p_0 values of interest in the ignitions.

The crucial requirement for ignition appears to be that p_i lies above p_s. This simple condition allows the prediction of ignition limits by examination of the ways in which p_i and p_s vary with p_0 and the total reactant pressure $p_{2H_2+O_2}$. Experimental studies of the variation of p_i with p_0 (at fixed total reactant pressure $P_{2H_2+O_2}$) and with $P_{2H_2+O_2}$ (at fixed p_0) have given the curves in Fig. 3 a and b respectively. Figure 3 b also shows the variation of p_s with $P_{2H_2+O_2}$, p_s being independent of p_0 as already explained. When the simple condition is applied to these results, the closed curve in Fig. 3 c is obtained. [21] It is clear that in general shape this curve is in remarkably good agreement with the experimental curve, shown in Fig. 2, found in earlier work.

More detailed study [22] of p_s has revealed the reactions that control p_s. At high values of p_0, when high pressures of NO are formed, the principal reactions are (8) and (9), giving a

$$H_2 + NO_2 \rightarrow H_2O + NO \tag{8}$$

$$2NO + O_2 \rightarrow 2NO_2 \tag{9}$$

catalysed formation of water. Nearer the limit, and also as p_0 is decreased in passing through the ignition region to below the

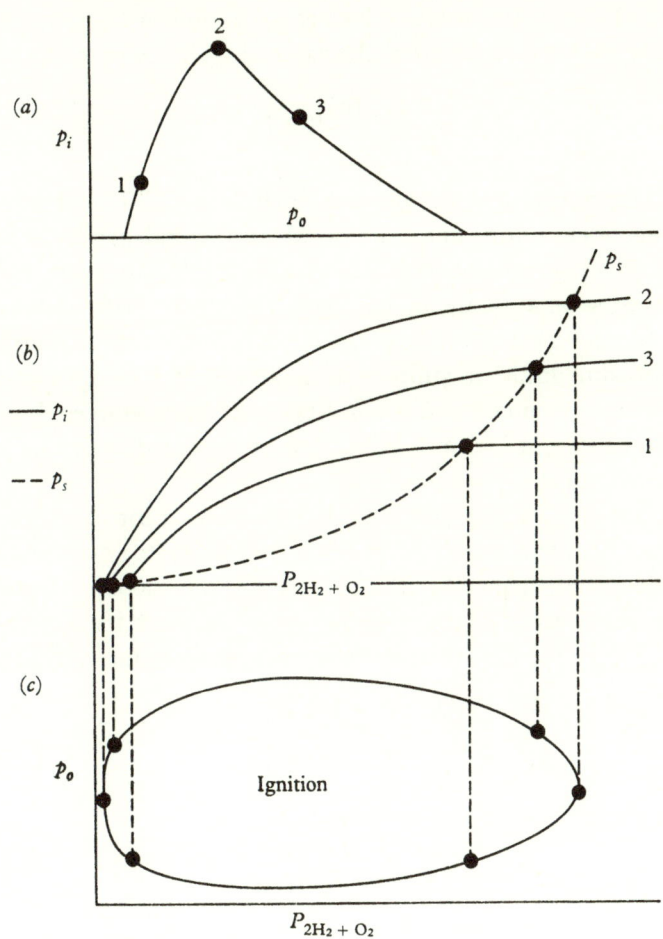

Fig. 3. Illustration of the factors controlling ignition limits in $H_2/O_2/NO_2$ mixtures. (a) p_i against p_0 derived as a vertical section of (b). (b) p_i against $P_{2H_2+O_2}$ for the values of p_0 shown as 1, 2, 3 in (a); also p_s against $P_{2H_2+O_2}$. (c) p_0 against $P_{2H_2+O_2}$ if limits are given by $p_i = p_s$ in (b).

lower limit, p_s is controlled chiefly by one of the propagating steps of (8), namely

$$H + NO_2 \rightarrow NO + OH \tag{7}$$

and the reforming reactions, more important than (9) when p_{NO} is low,

$$H + O_2 + M \rightarrow HO_2 + M \tag{4}$$

$$HO_2 + NO \rightarrow NO_2 + OH \tag{10}$$

A detailed study of p_s as low values of p_0 has shown [22b] that the reactions (7), (4) and (10) control the value of p_s. They predict that

$$p_s = \frac{\Sigma k_{4,\mathrm{M}}[\mathrm{M}]}{k_7}[\mathrm{O_2}]$$

where M represents any gas present, including $\mathrm{O_2}$. The experimental dependence of p_s on $\mathrm{O_2}$ and on M, and its independence of p_0, are in good agreement with this equation. [22b] Moreover, the relative values of $k_{4,\mathrm{M}}$ are in excellent agreement with revised values determined from studies of the effect of various gases on the second pressure limit of ignition of mixtures of hydrogen and oxygen. This agreement, and further evidence from studies [20] of the reaction between NO and $\mathrm{H_2O_2}$, can be regarded as evidence for the reaction (10).

It now appears [20] that reaction (10) can account for the sensitizing effect of nitric oxide and it is not necessary to invoke any positive branching steps other than (2) and (3). A reaction scheme including reactions (1)–(10), together with the terminating reactions proposed for the reaction between hydrogen and nitrogen dioxide,

$$\mathrm{OH + NO + M \rightarrow HNO_2 + M} \tag{11}$$

$$\mathrm{OH + NO_2 + M \rightarrow HNO_3 + M} \tag{12}$$

and the well-established fast reaction

$$\mathrm{O + NO_2 \rightarrow NO + O_2} \tag{13}$$

can account for the trends of p_e and p_i with p_0 and $P_{2\mathrm{H_2+O_2}}$ (shown in Fig. 3). The scheme has been analysed on a computer, with the assumption that at $p_{\mathrm{NO_2}} = p_e$ the net branching factor ϕ passes from negative values through zero to cause the accelerated removal of nitrogen dioxide, through (7), by increase in [H]. By adopting the values that are known for many of the rate constants, and assuming reasonable values of the others, the changes of p_e can be predicted quantitatively. [20] As p_i follows p_e very closely, it seems very likely that p_i could be predicted similarly by setting ϕ equal to some positive critical value.

Some detailed points [21] that bear on earlier observations may be mentioned. Experiments with inert gases of different thermal conductivity show that, near the lower limit, the limit position is dependent upon the thermal conductivity of the mixture, that the maximum rate outside the lower limit is high and is greater in mixtures of greater thermal conductivity, and that the maximum rate is greater in smaller reaction vessels. All these results point to a lower sensitizer ignition limit which is *thermal* in nature. Thus the pressure pulses previously observed at this limit were correctly attributed to thermal effects. In contrast, at the upper limit only very small pulses are found, the maximum rates just above the limit are far smaller than near the lower, and the effects of thermal conductivity changes and changes of vessel diameter are small. There is thus a transition from near-thermal ignitions near the lower limit to near-isothermal ignitions near the upper limit, and this transition is more marked in larger vessels or at moderate (rather than very high or very low) reactant pressures. The kinetic reasons for the transition can tentatively be attributed to the following changes. At the lower limit, where p_{NO} at the end of the induction period is small, termination is mainly quadratic, by reaction (6). Hence isothermal ignitions are not possible, and chain-thermal ignitions as proposed by Norrish are found at a critical rate corresponding to the lower limit rate. As p_0, and hence the value of p_{NO} at the end of the induction period, is increased, reaction (10) becomes more and more effective, converting HO_2 to HO and therefore necessitating linear termination by reactions (11) and (12). In the absence of quadratic termination, isothermal ignitions take place and thermal effects become negligible.

During the induction periods the branching reactions (2) and (3) are held in check by (13). Their lengths can be accounted for very satisfactorily by the time taken to reduce p_{NO_2} to p_e by reaction (8); the effects of adding inert gases, changing the relative partial pressures of reactants, and irradiation all receive natural explanations which are quite independent of any effects of these variables on the limit positions.

Thus recent work has brought about a complete change in our interpretation of these sensitized ignitions. They are not the outcome of simple branched-chain reactions to which equation (F) can be applied. The length of the induction period depends upon the rate of removal of the inhibitor, NO_2, by reaction with hydrogen;

there is no single parameter of branching, ϕ, which can be related to τ. As the initial concentration of NO_2 is increased, so the small residual amount of NO_2, p_s, increases and eventually causes the upper sensitizer limit. Between the lower and the upper limits there is a subtle change from thermal to near-isothermal control of ignitions, caused by a change from quadratic to linear terminating reactions, a change in which NO plays a large part.

It is worth emphasizing two aspects of these new and more successful interpretations of the sensitized ignitions. They were only possible after the application of a more discriminating experimental technique—the photometric studies—to the problem, in keeping with Norrish's insistence on the prime importance of experimental information. Secondly, the detailed chemistry of the reactions can markedly affect and perhaps control the physical conditions which determine the onset of ignitions; this phenomenon will be met again in the next section.

3.3. *Hydrocarbons and ignitions in $H_2 + O_2$.* While investigating the effects of inert gases in quenching ignitions in mixtures of hydrogen, oxygen and nitrogen dioxide, Foord and Norrish discovered [5] that methane exerted a much more powerful effect than other gases, inhibiting the ignitions in a most striking manner. They decided that the effect must be chemical and speculated upon the possibility that it was caused by formaldehyde, an oxidation product of methane, rather than methane itself. Some related effects of hydrocarbons were investigated by Norrish during World War II, when it was found that certain hydrocarbons considerably reduced the flash from gun-muzzles, but Norrish did not follow up the kinetic implications of these studies. It is therefore interesting to note that some similar wartime studies of the effect of hydrocarbons on exhaust gases from aircraft led Baldwin to initiate a series of careful investigations upon the effect of hydrocarbons on ignitions in hydrogen and oxygen. [23] These studies have led to the discovery of some extremely interesting ignition phenomena and, through the interpretation of these phenomena, to the evaluation of the rate constants of some abstraction reactions of the type

$$H + RH \rightarrow H_2 + R \tag{14}$$

There appear to be two distinct patterns of inhibition, associated with different mechanisms. [24] The simplest behaviour is shown by formaldehyde, ethane, propane, n- and i-butane. As the concentration of any one of these inhibitors is increased the ignition limit is reduced in an almost linear manner (Fig. 4). The amount of

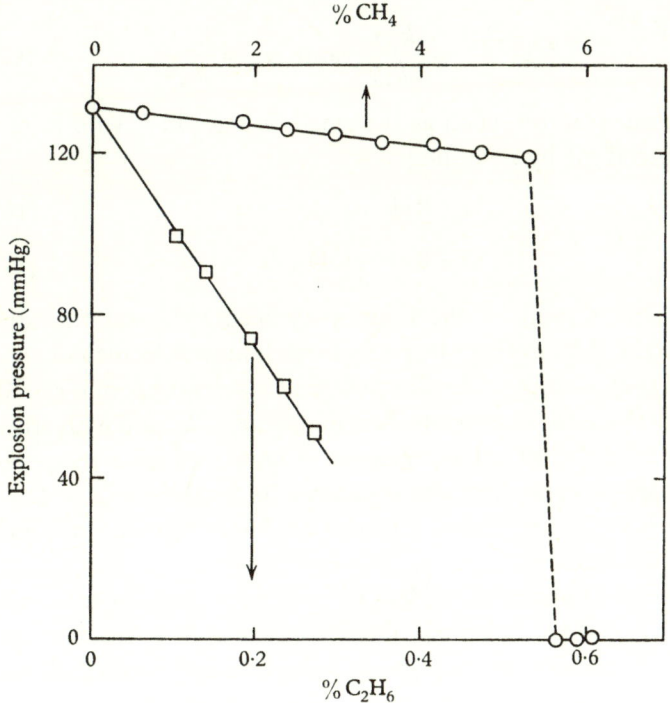

Fig. 4. Comparison of inhibiting action of CH_4 and C_2H_6; 35 mm. i.d. KCl-coated vessel, 540 °C, $H_2 = 0.28$, $O_2 = 0.14$.

inhibitor ($i_{\frac{1}{2}}$) required to reduce the limit to one-half of its uninhibited value is small, around 0.1–1.0 mole %. It is almost proportional to the mole fraction (y) of oxygen, it is much less dependent on the mole fraction (x) of hydrogen, and it is effectively independent of the vessel diameter. These results, with others that have been detailed, suggest competition between the hydrocarbon and oxygen for a chain centre. The simplest successful scheme adds reaction (14) to those describing the second limit in KCl-coated vessels (reactions (1)–(5)), assuming that the radicals R do not give rise to chain centres but are removed by reactions like

$$C_2H_5 + O_2 \rightarrow C_2H_4 + HO_2 \tag{15}$$

$$CHO + O_2 \rightarrow CO + HO_2 \tag{15a}$$

Assuming the limit can be accounted for by setting $\phi = 0$, this scheme gives

$$\frac{P_{2,0} - P_{2,i}}{P_{2,0}} = \frac{k_{14}i}{2k_2y} \quad \text{and} \quad i_{\frac{1}{2}} = \frac{k_2y}{k_{14}} \tag{G}$$

Some minor features, such as the small changes in $i_{\frac{1}{2}}$ with x, can be accounted for by including

$$OH + RH \rightarrow H_2O + R \tag{16}$$

$$O + RH \rightarrow OH + R \tag{17}$$

The equations describe the limits very accurately and the ratio k_{14}/k_2 can be determined for the hydrocarbons and formaldehyde. Baldwin et al. have described [25] how their results at 540 and 520°C can be combined with separate determinations of k_2, and with other studies of these abstraction reactions, at lower or higher temperatures, to give accurate Arrhenius parameters for the reactions (14). Some rather more restricted information about reactions (16) and (17) can be obtained.

The uninhibited limit is undoubtedly isothermal; it is well known that inert gases alter it through their influence as M in reaction (4), and the effectiveness of each gas is related directly to its collision diameter. When the ignitions are inhibited by the hydrocarbons that depress the limit linearly, it is found that the same mole fraction of any of the gases CO_2, N_2, A, and He has exactly the same effect on the limit. Equation (G) is independent of inert gas, and the experimental findings are therefore in agreement with the assumption that $\phi = 0$ at the limit, i.e. that the limits of the inhibited reaction are also isothermal.

When methane is used instead of the higher hydrocarbons, completely different effects are observed. Increasing the amount of methane added has very little effect until a critical concentration i_c is reached at which ignitions are abruptly quenched (Fig. 4). The value of i_c is about 2–10 mole % (it is much lower with neopentane which behaves like methane), and it behaves quite differently from $i_{\frac{1}{2}}$ when various changes are made in the reaction conditions.

(1) Inert gases, which have no effect on $i_{\frac{1}{2}}$, have a marked effect on i_c; for fixed mole fractions of hydrogen and oxygen, i_c is found to be *lowest* with CO_2, next with He, then with N_2 and then with A. If the effect of inert gases operates through collisions (either as M in some third-order reaction or by hindering diffusion) the order would be CO_2, N_2, A, He. The position of He with the methane inhibition clearly shows that *thermal* effects are operating.

(2) The value of i_c increases as the diameter of the vessel increases, whereas $i_{\frac{1}{2}}$ is independent of diameter. Moreover, i_c is nearly the same in clean or in KCl-coated vessels. The diameter effects thus point to a thermal condition for ignition with methane present.

(3) The values of i_c, like those of $i_{\frac{1}{2}}$, are nearly proportional to the mole fraction of oxygen present. Unlike $i_{\frac{1}{2}}$, however, i_c clearly passes through a maximum value as the mole fraction of hydrogen is increased. Thus it appears that competition for hydrogen atoms again causes inhibition, but that additional thermal effects due to the high conductivity of hydrogen are operating.

(4) There is a small kick in the pressure when methane concentrations just above the critical limit are used, very reminiscent of the kicks found for $H_2/O_2/NO_2$ mixtures near the lower sensitizer limit of ignition, and similarly suggesting thermal effects.

Baldwin concludes that methane behaves quite differently from the higher hydrocarbons and formaldehyde. He accounts for the results by assuming that formaldehyde, formed by oxidation of the methane, is the true chain terminator. Methane itself is unlikely to terminate chains because the methyl radical, formed in (14), cannot give stable radicals and molecules by reaction with oxygen.

$$H + CH_4 \rightarrow H_2 + CH_3 \tag{14a}$$

$$CH_3 + O_2 \rightarrow CH_2O + HO \tag{15b}$$

Thus the chain continues until the formaldehyde builds up to act as an inhibitor.

In support of this view, Baldwin later found that neopentane, which like methane gives a radical R that cannot yield stable products by reaction with oxygen, behaves just like methane.

Suppose that a reactant R forms an intermediate Z by reaction with a chain centre. This intermediate Z is destroyed by reaction with chain centres and some fraction at least of these reactions do not produce further chain centres. Suppose the reaction mixture is subjected to conditions where the net branching factor ϕ is initially positive. The differential equations for the centre concentration [X] and the intermediate concentration [Z] then become:

$$d[X]/dt = \theta + \phi[X] - A[Z][X] \tag{H}$$

$$d[Z]/dt = B[R][X] - C[Z][X] \tag{J}$$

If all reactions destroying the intermediate also remove a chain centre, $A = C$, and from the second equation the stationary concentration of intermediate $= B[R]/C$. If $\phi > AB[R]/C$, $d[X]/dt$ is always positive and an isothermal explosion occurs. If

$$\phi < AB[R]/C,$$

$d[X]/dt$ is negative when the equilibrium concentration of intermediate is reached, and isothermal explosion is impossible. Since $d[X]/dt$ is initially positive, however, the $[X] - t$ curve must pass through a maximum, as shown diagrammatically in Fig. 5. The resultant maximum reaction rate may be sufficient to cause a thermal explosion.

The striking feature of the mechanism is the transition from isothermal ignitions at the uninhibited limit to thermal ignitions in the presence of quite small proportions of methane. As with mixtures of hydrogen, oxygen and nitrogen dioxide, the reactions introduced by the inhibitor bring about surprising changes in the conditions regulating the onset of ignition.

3.4. *Self-inhibition in* $H_2 + O_2$ *ignitions.* Baldwin has also examined [24] the self-inhibition of the hydrogen/oxygen ignitions by the product water. This self-inhibition is evidenced by the reduction of the second limit of ignition in certain vessels when mixtures of pressure well above the limit are withdrawn at different rates. There is a rather critical balance between the increase in ϕ, because the pressure is being decreased, and a decrease in ϕ because water, which is more effective than the reactants as M in reaction (4), is being formed during the withdrawal. At slow rates of withdrawal the effect of water predominates, and the system remains just outside a (changing) ignition boundary throughout the withdrawal. This behaviour is not found in KCl-coated vessels, but in Pyrex or aged boric-acid vessels. At first sight this suggests differences in surface termination, but current understanding of

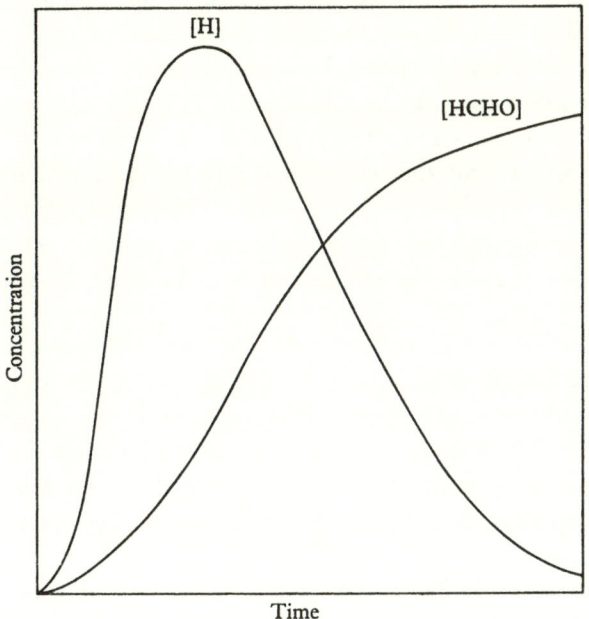

Fig. 5. Sketch of concentration–time relationship given by equations (H) and (J), with X = H and Z = HCHO.

the second limits argues in favour of a rather more subtle effect of the surfaces. The boric-acid surfaces preserve HO_2 and H_2O_2, and so encourage quadratic branching by reactions such as

$$H + HO_2 \rightarrow 2OH \tag{18}$$

$$HO_2 + HO_2 \rightarrow H_2O_2 + O_2 \tag{19}$$

$$H_2O_2 + M \rightarrow 2OH \tag{20}$$

If the quadratic branching predominates over quadratic termination, the growth of centres is controlled by the equation

$$d[X]/dt = \theta + \phi[X] + \delta'[X]^2$$

When δ' is positive, ignitions can occur isothermally even when

ϕ is negative, provided $4\theta\delta' > \phi^2$. For a steady rate, ϕ must be negative and $[X]_s = 2|\phi|/\delta'$. If $[X]$ is high enough, the rate of water formation can depress ϕ at a rate which at least compensates for the positive increase in ϕ caused by the steady fall in pressure upon withdrawal.

There are probably no very simple tests of this hypothesis. It is now possible, however, to compute the growth of chain centres during the withdrawal period, because the rate constants involved are either known or can be estimated with reasonable accuracy. These computations allow the prediction of the effects of varying the composition and the withdrawal rate on the limit, for comparison with experiment. The results of these computations and experiments support the hypothesis, and so indicate yet another complication of chain branching that must be taken into account.

3.5. *Halogenated compounds and ignitions.* Halogenated compounds are usually effective fire extinguishers and it is known that bromine compounds are more effective than chlorine compounds. Part of their action is through chemical inhibition, and it is therefore of interest to study the effect of halogens, hydrogen halides and halogenated hydrocarbons on some simple combustion reactions.

The problems are not at all simple in the oxidation of hydrocarbons. Whereas the flame speeds of many mixtures of oxygen and hydrocarbons, or their derivatives, are reduced by small molar proportions of many halogen derivatives, and the inflammability limits are correspondingly narrowed and even closed up, the effect of some derivatives on many slow oxidations is a pronounced acceleration and a corresponding expansion of the region of thermal ignitions. Thus hydrogen bromide slows down the flame speed of many oxidations, but it catalyses the low-temperature oxidation of many hydrocarbons and particularly the formation of peroxides and hydroperoxides. On the other hand, it inhibits the ignitions of hydrogen and oxygen. We see again that quite delicate balances of competing reactions can tilt the balance in branched or degenerately branched-chain reactions, and the balance may be quite different from reaction to reaction and at flame temperatures or in low-temperature oxidations.

In suppressing ignitions in hydrogen and oxygen, hydrogen bromide behaves in some ways like the hydrocarbons as inhibitors. [26] In barium bromide-coated vessels it depresses the second limit almost linearly, and it behaves like ethane, etc., in that (1) the inhibited limit increases almost linearly with the mole fraction of oxygen, (2) the exchange of inert gases between helium, argon, nitrogen, and carbon dioxide has very little effect on the inhibited limits, (3) change of vessel diameter has little effect on the inhibited limits. However, the HBr-inhibited limits increase much more pronouncedly with the mole fraction of hydrogen than do the hydrocarbon-inhibited limits.

Now the studies of the uninhibited limit in barium bromide-coated vessels show that quadratic branching probably occurs via (18).

$$H + HO_2 \rightarrow 2OH \tag{18}$$

The inhibition by hydrogen bromide has been accounted for by adding the inhibiting reaction which corresponds to (14).

$$H + HBr \rightarrow H_2 + Br \tag{21}$$

However, the Br atom can regenerate H by the reverse of (21).

$$Br + H_2 \rightarrow HBr + H \tag{-21}$$

(The reactions (21) and (−21) thus behave like reactions (28) and (30) in the inhibition of reactions between hydrogen and chlorine, see page 321.) Thus only a fraction of the bromine atoms terminate the chains by the competing reaction

$$Br + HO_2 \rightarrow HBr + O_2 \tag{22}$$

Thus the inhibition results from increased quadratic termination if (22) controls the rate of termination, or linear termination if (21) controls it. As might be imagined, solution of the equations for the growth of chain centres is very complicated, but computer studies have again proved extremely useful. Parallel studies have been made on the inhibitory action of hydrogen chloride in barium chloride-coated vessels, but it is a weaker inhibitor than hydrogen bromide because the reaction corresponding to (−21) is relatively much faster.

3.6. *Other sensitizers and inhibitors in* $H_2 + O_2$ *ignitions.* Several other compounds containing nitrogen act as sensitizers for these ignitions, but few modern studies have been made of their action. Sensitization by nitrous oxide has been attributed to the formation of nitric oxide. Ammonia also sensitizes the ignitions, and nitric oxide is known to be found in its thermal oxidation. Evidence from the flash photolysis studies [27] of the explosive oxidation of ammonia show nitric oxide as the major product, perhaps formed from breakdown or reaction of HNO which appears in a propagating step

$$NH_2 + O_2 \rightarrow HNO + OH$$

$$OH + NH_3 \rightarrow H_2O + NH_2$$

The flash photolysis studies also showed that in the oxidation of hydrazine very little nitric oxide was formed, nitrogen being produced according to the overall equation

$$N_2H_4 + O_2 = N_2 + 2H_2O$$

It is therefore interesting to note that equimolar mixtures of hydrazine and oxygen give a simple ignition boundary like A in Fig. 1, but mixtures richer in oxygen give an additional peninsula, like B in Fig. 1, in which delayed ignitions occur. [28] These delayed ignitions are attributed to the reaction between oxygen and hydrogen formed by breakdown or partial oxidation of the hydrazine, the peninsula being the familiar hydrogen/oxygen peninsula, presumably modified in detail by the presence of nitrogen, water, and ammonia. The delays are due to the effective inhibition of the hydrogen/oxygen reaction by the hydrazine. It would be interesting to explore this inhibiting effect in more detail, on the lines used by Baldwin.

4. Sensitized ignitions of methane and oxygen

In some early studies on the oxidation of methane, Norrish and Foord [29] showed that formaldehyde is the intermediate responsible for autocatalysis or degenerate branching at temperatures around 500–600 °C. In particular, they showed that formaldehyde reaches a maximum concentration when the rate of methane oxidation is fastest, and that addition of formaldehyde shortens the induction period without altering the maximum rate of oxidation. Later, Norrish and Patnaik [30] demonstrated that photolysing the reacting mixtures with light of wavelengths absorbed by formaldehyde (but not by methyl hydroperoxide) led to a shortening of the induction period and to an increase in the maximum rate. The rôle of formaldehyde as the branching agent has been widely accepted, and Enikolopyan [31] has recently shown that the rate of oxidation fits an exponential rate law which can be accounted for quantitatively. There are, however, several reaction schemes differing in detail about the precise reactions involved in the oxidation.

At higher temperatures, above 600 °C, and at moderate pressures the reactants ignite. In mixtures with less than about 60% of oxygen, there is a simple P, T boundary of ignitions. These ignitions are thought to be thermal from experimental examination of the boundary, and also because the form of the degenerate branching law, for the reaction below the limit, precludes isothermal ignitions

$$d[CH_2O]/dt = a_1 + a_2[CH_2O] - a_3[CH_2O]^2$$

With leaner mixtures, additional explosions are found within a low-temperature peninsula. These additional explosions are attributed to ignitions of mixtures of oxygen with hydrogen and carbon monoxide, formed by partial oxidation of the methane, taking place when the methane concentration falls below a critical value (cf. section 3.4).

Bone and Allum [32a] reported that the addition of traces of nitrogen dioxide can accelerate the oxidation. Interest in this sensitization has been revived because it gives a useful method for the preparation of formaldehyde from methane. A limiting yield, independent of vessel surface conditions, is attained with about 0·2% of nitrogen dioxide. Enikolopyan has shown that the pressure–time curves show two maxima in the rate, corresponding to the two temperature maxima observed in an investigation by differential calorimetry of the nitric oxide-sensitized oxidation. Karmilova, Enikolopyan and Nalbandyan [32b] investigated the rate of formation of formaldehyde and methyl alcohol (which showed a yield about half that of the formaldehyde) in a flow system at 600–700 °C., with between 0·02 and 2·0% of nitric oxide as catalyst. The formation of formaldehyde fits an equation for a branching reaction, such that

$$d[CH_2O]/dt = a_0 + \phi[CH_2O]$$

where

$$a_0 \propto [O_2][NO]/(1 + b[NO]) \quad \text{and} \quad \phi \propto [O_2]/(1 + b[NO])$$

Correspondingly, in the reaction scheme, they assume that NO takes part in a surface initiating step

$$CH_4 + NO + O_2 \xrightarrow{\text{surface}} CH_3 + OH + NO_2$$

$$CH_3 + O_2 \rightarrow CH_3OO \rightarrow CH_2O + OH$$

and in a terminating step

$$CH_3OO + NO \rightarrow CH_3NO_3$$

the propagating and branching steps being independent of the oxides of nitrogen.

The effect of nitrogen dioxide on the ignitions was investigated by Norrish and Wallace, [10] who found a substantial lowering of the spontaneous ignition temperatures of equimolar mixtures of methane and oxygen, in quartz vessels, by partial pressures of nitrogen dioxide up to 20% of the total pressure. It was later discovered [33] that chloropicrin, nitrosyl chloride, or chlorine also substantially lower the ignition temperatures, giving simple ignition boundaries with very short induction periods. With nitric oxide (placed in the reaction vessel, prior to adding the reactants) the induction periods are longer, and at each temperature the ignition pressure passes through a minimum as the sensitizer concentration is increased (Fig. 6). There is a lower and an upper sensitizer limit for certain temperatures and pressures of combustible mixture, just as for hydrogen and oxygen. With nitrogen dioxide, the same minimum in ignition pressure is reached, with rather shorter induction periods, but at higher pressures of nitrogen dioxide the curve falls away from the nitric oxide curve, so that over a certain range of pressures three ignition limits are observed (Fig. 6). The third limit may be a different ignition régime superimposed on the U-shaped curve which the two lower limits generate, but it does not correspond to the boundary of ignition of methane and nitrogen dioxide in the absence of oxygen.

A more detailed study of the sensitized ignitions with NO_2 has been made recently. [34] It combined measurements of pressure changes, analytical study of the extents of product formation and reactant consumption, and photometric determination of the

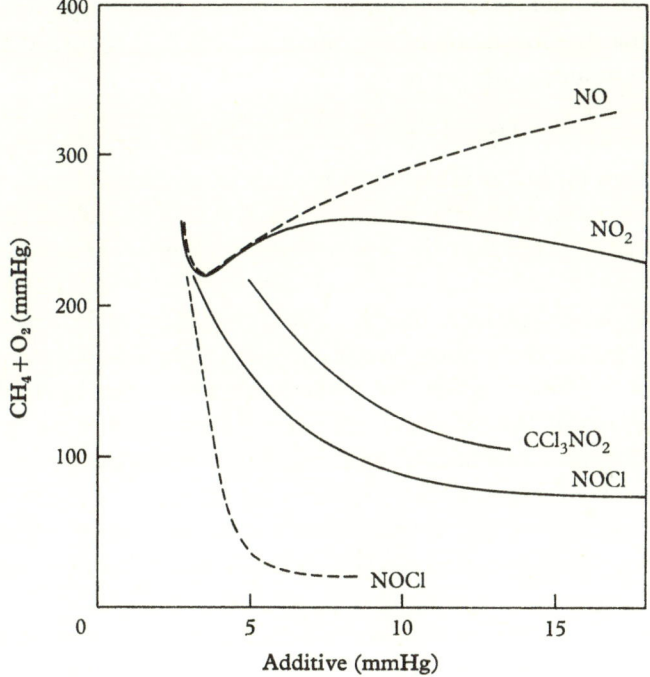

Fig. 6. Effect of additives on ignition of $CH_4 + O_2$ at 490 °C. ———, Additives premixed; – – – –, placed in vessel before admitting reactants.

nitrogen dioxide. The general pattern of the limits has been confirmed. In addition, the photometric studies reveal that where virtually identical ignition boundaries are found with NO and NO_2, the same ratio of nitrogen dioxide and nitric oxide is reached by the end of the induction period, regardless of whether nitric oxide or nitrogen dioxide was the additive. With nitrogen dioxide as additive the NO_2 concentration falls rapidly, probably by reaction with CH_4, to a minimum and then rises to a shallow maximum,

$p_{NO_2}^{max.}$; with nitric oxide, NO_2 is formed, reaching the same maximum value $p_{NO_2}^{max.}$. Thus the situation is in some respects remarkably like that in the oxidation of hydrogen sensitized by nitrogen dioxide, the induction periods again being the time taken for the additive to react to give the final sensitizing mixture. Also, there is evidence of reactions like (10) taking place, for the rate of formation of nitrogen dioxide, when starting from nitric oxide, is many times greater than could be achieved by the termolecular reaction (9). It is possible that the formation of the nitrogen dioxide is catalysed by peroxy compounds, similar to the CH_3O_2 envisaged

$$CH_3O_2 + NO \rightleftharpoons CH_3O + NO_2$$

by Enikolopyan, but it is also possible that fast reactions such as (10) are occurring, with HO_2 formed by oxidation of formaldehyde.

The rates of reaction at the first and second limits, obtained by extrapolation of values of reaction rates measured outside the limits, are consistent with the ignitions being thermal in origin. Displacement of the ignition boundary caused by the additions of inert gases of different conductivity also indicates thermal ignitions at first and second limits. Two maximum rates of pressure rise are attained during a reaction outside the first limit, starting with either NO or NO_2, just as the Russian workers found. Ignition is believed to occur when the first maximum rate exceeds a critical value. With first maximum rates just below this critical value, cool flames or degenerate explosions arise when the second maximum rate is fast enough. Ignition at the third limit arises as a result of a fast initial reaction which is observed with nitrogen dioxide, but not with nitric oxide, as additive. The ignition takes place toward the end of an initial period of rapid removal of nitrogen dioxide, when several mm of nitrogen dioxide are still present in the reacting mixture. The effect of changing the thermal conductivity on the position of the third limit suggests that ignition there, as at the first and second limits, is thermal.

The reasons for the appearance of a second limit of ignition do not appear as clear cut as in the sensitized hydrogen–oxygen reaction. If the ignitions are indeed thermal, the maximum rates of reaction must pass through a maximum with increase of initial sensitizer concentration between the first and second limits. This, in fact, is the case for total pressures of methane and oxygen (P_T) below the minimum of the U-shaped ignition boundary. Furthermore, kinetic expressions of the type

$$\text{rate of reaction} = k_1 p^n x^a / (1 + k_2 x^b)$$

(where $a < b$ and $x = [NO_2]$) can account satisfactorily for the displacement of the ignition boundary with change of temperature.

A reaction mechanism that gives rise to the above rate expression and embraces all of the experimental facts has not yet been devised. It is possible that degenerate branching reactions of formaldehyde, such as

$$CH_2O + NO_2 \rightarrow CHO + HNO_2 \tag{23}$$

may be responsible for acceleration of the reaction rate to the first maximum value coinciding with maximum observed concentrations of both formaldehyde and nitrogen dioxide ($p_{NO_2}^{max}$). Subsequent acceleration of the rate to the second maximum rate, at the time of appearance of a maximum concentration of methyl alcohol, is tentatively ascribed [34] to degenerate branching reactions of the type

$$CH_3OH + X \rightarrow CH_2OH + HX$$

where X is either NO_2 or O_2. It is conceivable that reaction (23) is responsible for the onset of ignition at both first and third limits, critical conditions at the third limit arising in the presence of a much larger concentration of nitrogen dioxide, although at a considerably smaller concentration of formaldehyde, than at the first limit.

It is clear that the large number of possible intermediates in the oxidation of methane and the variety of their reactions with NO_2, NO, and O_2, makes interpretation of these sensitized ignitions much more complicated than those of hydrogen and oxygen. An unexpected feature of the ignitions in $CH_4/O_2/NO_2/NO$ mixtures is that elementary nitrogen is formed in surprisingly large yields, although none was detected during the slow reactions outside the limits. Calculations showed that an unlikely duration (compared with the time of cooling due to conduction) of the maximum temperature rise after adiabatic ignitions would have to occur for the bimolecular decomposition of nitric oxide to account for the nitrogen, so that it seems likely that it is formed by the decomposition of poly-nitroso derivatives, or reactions between NO and fragments such as CN or NH. The radicals CN and NH have been detected by flash photolysis [35] during the combustion of hydro-carbon/oxygen/NO_2 mixtures. Christie [36] has suggested reaction sequences of the type

$$R + NO \rightarrow RNO \xrightarrow{2NO} \underset{\underset{N=O}{|}}{RN}—O—N=O \rightarrow R + N_2 + NO_3$$

to account for the formation of nitrogen during the photolysis of alkyl iodides in the presence of nitric oxide. However, the forma-tion of nitrogen during the ignitions, and not during the fast reaction above the second limit, suggests that the interaction of radicals is the more likely source in these sensitized ignitions.

5. Sensitized ignitions of carbon monoxide and oxygen

Norrish was interested in the combustion of carbon monoxide principally because carbon monoxide is a common product of the oxidation of many hydrocarbons or their derivatives. It is, however, much more resistant to oxidation, and even when oxidation occurs it appears to be inhibited by some intermediate or product and is rarely complete. Norrish did not publish any work on the reaction between the pure gases, in spite of the kinetic and spectroscopic interests raised by the partial completion of the reaction. He was, however, very intrigued by the effects of traces of water and hydrogen on the reaction, and many audiences have seen his effective lecture demonstration of the difficulty of burning dry carbon monoxide in dry air. Following reports of the reduction in the intensity of the radiation from carbon monoxide ignitions upon the addition of hydrogen, Norrish decided to investigate the effect of hydrogen on the ignition limits. His work with Buckler [37] provides a telling demonstration of the unexpected effects that addition of another fuel can have on a combustion reaction.

It was known that the ignition of carbon monoxide occurs within a boundary shaped like the ignition peninsula of hydrogen and oxygen, but lying at much higher temperatures.

Norrish and Buckler found that the extent of formation of carbon monoxide in ignitions was greatly increased by the addition of very small proportions of hydrogen ($\sim$ 0·1 %). At temperatures above the tip of the carbon monoxide/oxygen peninsula, replacement of carbon monoxide by hydrogen raises the second limit of ignition rapidly. With quite low proportions of hydrogen, ignitions occur at much lower temperatures, characteristic of the hydrogen/oxygen peninsula. The effect of vessel diameter is small. Later work [38] has confirmed that the 'mixed' limit at around 540 °C passes through a maximum at about 10% of hydrogen; at the other extreme of fuel composition, the replacement of hydrogen by carbon monoxide slowly raises the second limit of the hydrogen/oxygen ignitions, probably because CO and H_2 have different efficiencies as an inert gas M in reaction (4). The general pattern of sensitization requires that the reactions in carbon monoxide/oxygen mixtures are replaced by those found in hydrogen/oxygen mixtures.

In spite of many spectroscopic and kinetic studies of the ignitions, the limits, and the slow reaction outside the limits for CO/O_2 mixtures, complete agreement has not been reached on the precise propagating and branching steps in the oxidation. It seems certain, however, that a relatively slow reaction between CO and O gives an excited CO_2 molecule which then gives rise to branching. The kinetic effect of the hydrogen is to circumvent this slow reaction of oxygen atoms by introducing the sequence

$$O + H_2 \rightarrow OH + H \tag{3}$$

$$H + O_2 \rightarrow OH + O \tag{2}$$

$$OH + CO \rightarrow CO_2 + H \tag{25}$$

$$OH + H_2 \rightarrow H_2O + H \tag{1}$$

Together with the terminating reactions

$$H + O_2 + M \rightarrow HO_2 + M \tag{4}$$

$$O + CO + M \rightarrow CO_2 + M \tag{24}$$

these reactions can describe Norrish and Buckler's results. If the condition for the second limit is that ϕ is zero, the reaction scheme leads to an equation describing the limits:

$$2k_2 = k_4[M]\left\{1 + \frac{k_{24}[CO][M]}{k_3[H_2]}\right\}$$

When the partial pressure of CO is low, the equation reduces to the condition for ignition at the second limit of hydrogen/oxygen mixtures and the effect of CO is simply as an inert gas.

$$2k_2 = k_4[M]$$

At the other extreme, the relation between the low partial pressure of hydrogen, p_{H_2}, and the total pressure P_T of a given mixture is

$$p_{H_2} = \frac{k'P_T^3}{k_2 - k''P_T}$$

Norrish and Buckler found this equation fitted their results well.

Later work has suggested that reactions of CO with H atoms and with HO_2 radicals should be taken into account. The most interesting of this work is concerned with the rates of formation of CO_2 and of H_2O in slowly reacting mixtures of hydrogen and oxygen with about 1 % of carbon monoxide, in boric acid-coated vessels. [25] In these vessels the mechanism of the hydrogen/oxygen reaction is well established so that some definite information about the reactions of CO with O, OH, and HO_2 can be obtained. The evidence strongly suggests that reactions of O atoms with CO can be neglected in the presence of hydrogen, as assumed by Norrish and Buckler. The rate constant of the important reaction

$$OH + CO \rightarrow CO_2 + H \qquad (25)$$

can be compared with the rate constant for reaction (1), and at 500 °C k_{25}/k_1 is 0.30. The same ratio has been determined at higher temperatures in the course of flame studies, and the combined results suggest that $E_{OH+H_2} - E_{OH+CO}$ is about 5.2 kcal mole^{-1}. As the best present values of E_{OH+H_2} are about 5.5 ± 0.5 kcal mole^{-1}, it appears that E_{OH+CO_2} is close to zero. This is in conflict with earlier studies which put the value at 7–8 kcal. It is rather important to resolve this discrepancy, as reaction (25) has frequently been used to estimate the concentration of OH radicals in flames by adding CO and measuring the rate of formation of CO_2.

Another fuel with marked influence on the oxidation of carbon monoxide is methane. Additions of small proportions of methane expand the glow and ignition regions, and increase the rate of slow reaction. [39] With higher proportions, however, the rate of slow reaction passes through a maximum and, correspondingly, the ignition region is contracted. It is likely that initially the addition of methane introduces reactions similar to those introduced by adding hydrogen; for example

$$O + CH_4 \rightarrow CH_3 + OH \qquad (26)$$

and so the slow step in the CO/O_2 reaction is by-passed. It is not clear, however, whether the inhibition found on adding more methane is due to reactions of centres with methane itself, or with the formaldehyde which is formed by its oxidation. As formaldehyde is known to contract the ignition region (and also in view of its function [24] in the inhibition of hydrogen/oxygen ignitions by methane) it is tempting to ascribe the inhibitory effect to the formaldehyde.

Buckler and Norrish also examined the effect of adding nitrogen dioxide to CO/O_2 mixtures, following reports by Crist and Roehling [40] that at about 500 °C the initial rates of oxidation rise with increase in the initial amount of nitrogen dioxide added, fall again to a broad minimum when more NO_2 is used, and later rise again. It seems fairly certain that, at higher pressures of NO_2, the reaction is a direct molecular reaction between NO_2 and CO. Norrish and Buckler observed the maximum initial rate at lower pressures of NO_2, established that it was not due to the presence of H_2, but thought it might require the presence of traces of water. They also found that ignitions are produced in mixtures of CO/O_2 with small additions of H_2 and NO_2 which show all the characteristics of the ignitions in H_2/O_2 mixtures with traces of NO_2, as might be expected.

There are frequent references to 'transient' or 'pulsed' glows in descriptions of work on the oxidation of carbon monoxide. Norrish and Ashmore [41] found a remarkable phenomenon while investigating the effect of chloropicrin on ignitions of carbon monoxide and oxygen. The principal effects of small amounts (< 1%) of chloropicrin was to raise the ignition boundary to higher temperatures. After these inhibition experiments, mixtures of nominally pure carbon monoxide and oxygen were admitted to pressures and temperatures well inside the normal ignition region, and a series of separate and distinct flashes were observed. Each flash was accompanied by a sharp pressure drop. This intriguing 'lighthouse effect' was observed many times in 1939, and again in a quite different apparatus, after using chloropicrin, in 1948 and 1949. Unfortunately it proved impossible to control the frequency and intensity of the flashes and so to identify their cause.

6. Sensitized ignitions of hydrogen and chlorine

In a series of classical papers Norrish investigated the effects of water, [42] of oxygen [43] and of nitrogen-containing compounds [44] upon the rate of the photochemical reaction between hydrogen and chlorine. In particular, he elucidated the role of nitrogen trichloride as the true inhibitor formed from nitrogenous compounds present adventitiously or added deliberately, and showed that during the induction periods the inhibitor was decomposed by a photo-sensitized reaction with chlorine. [45] It is interesting to note that many years later a similar explanation was arrived at for the cause of the induction periods, and the fate of the inhibitor, in ignitions of hydrogen–oxygen mixtures with nitrogen dioxide or similar additives.

Investigations of the action of chloropicrin (CCl_3NO_2) on ignitions of hydrogen and oxygen led Norrish and Ashmore to suggest that chlorine atoms were formed during the decomposition of the additive and to test the effect on ignitions of hydrogen and chlorine. [46,47] The normal ignition region was extended to very low pressures (region X in Fig. 7) by chloropicrin, giving a region shaped like the cool-flame region of hydrocarbon oxidations (B in Fig. 1). Further investigations [33] showed that two decomposition products of chloropicrin, nitric oxide and nitrosyl chloride, had very different effects on the hydrogen/chlorine ignitions. Nitric oxide sensitized the ignitions (curve V, Fig. 7) while nitrosyl chloride inhibited them at low temperatures (curve W, Fig. 7).

The upper temperature boundary with chloropicrin depends upon a delicate balance of initiating reactions such as

$$CCl_3NO_2 \rightarrow Cl + NO + COCl_2 \tag{27}$$

$$NO + Cl_2 \rightarrow NOCl + Cl \tag{28}$$

$$NOCl + M \rightarrow NO + Cl + M \tag{29}$$

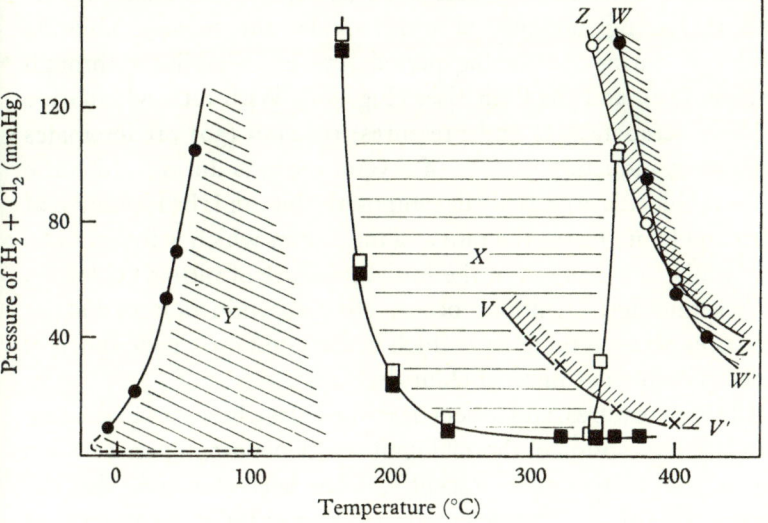

Fig. 7. Regions of ignition of $H_2 + Cl_2$. (a) Alone, above ZZ'; (b) with 0·5 mm NOCl, above WW'; (c) with 0·5 mm NO, above VV'; (d) with 0·5 mm CCl_3NO_2, within X (short entry, ■; long entry, □); (e) with 0·5 % NCl_3, within Y.

and terminating reactions such as

$$NOCl + Cl \rightarrow NO + Cl_2 \tag{30}$$

$$NO + Cl + M \rightarrow NOCl + M \tag{31}$$

The relative balance of these depends upon the amount of additive present initially and its reactions in the premixed gases during the mixing periods, the admission, and the induction periods. A curious feature of the upper temperature boundary of the ignitions with chloropicrin, demonstrating this balance, was its different position when the gases were admitted to the vessel by long or by short paths through narrow, heated inlet tubes.

The main features of the thermal reaction between hydrogen and chlorine in the presence of nitric oxide and nitrosyl chloride were established by Ashmore and Chanmugam, [48] who showed that there is an intimate connection between the sensitized ignition boundaries and the kinetics of the slow reaction outside them. With nitric oxide, the initial rate of reaction is enhanced because reaction (28) increases the rate of initiation; correspondingly, the ignition boundary is lowered and there is no detectable induction period; with increasing amounts of nitric oxide, the nitrosyl chloride produced during the mixing period acts as an inhibitor through reaction (30), and the limit rises (Fig. 8a). With nitrosyl chloride at lower temperatures and pressures, reaction (30) predominates and causes induction periods of several seconds during which the nitrosyl chloride is partially decomposed; this, and the formation of nitric oxide, increases the rate to a maximum which is found to be practically independent of the initial pressure of nitrosyl chloride. Correspondingly, ignitions occur, after an induction period, at pressures above a boundary which is the same for a wide range of initial pressures of nitrosyl chloride.

All these features, and others described elsewhere, are in keeping with the assumption of thermal ignitions depending upon a critical rate at each temperature. Recently, it has been shown [49] that the application of the Semenov–Frank-Kamenetskii treatment of thermal ignitions (outlined earlier in this chapter) to a complex reaction scheme for these mixtures, is remarkably successful. When conditions are chosen to lead to non-autocatalytic reactions, for example when nitric oxide is added and [NOCl] is low, the simple Frank-Kamenetskii condition (C) predicts pressure limits of ignition accurately. Figs. 8a and 8b show the excellent agreement between the experimental and calculated limits of ignition. It must be emphasized that the calculations are absolute and do not depend on 'anchoring' the curve by any assumption of equality of experimental and calculated limits at any point.

When nitrosyl chloride is added initially and the reaction is autocatalytic, the basic equations for the rate of heat production (B) and the rates of formation of intermediate free atoms and of products can be solved on a computer. [48] The values of the rate constants required, and the thermal properties of the system, are

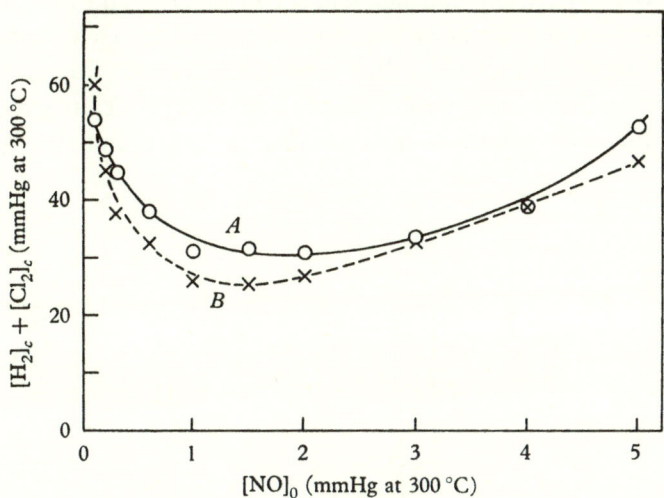

Fig. 8*a*. Comparison of calculated (×) and experimental (○) limits for $H_2/Cl_2/NO$ mixtures with $[NO]_0$ placed in mixing vessel. 15 min. mixing time. Reaction vessel at 300 °C.

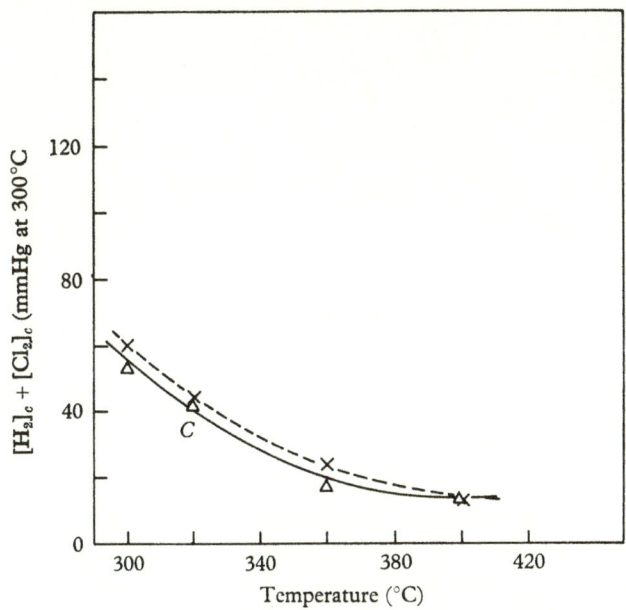

Fig. 8*b*. Comparison of calculated (×) and experimental (△) limits of $H_2/Cl_2/NO$ mixtures, 0·1 mm. NO, Pyrex vessel 28 mm i.d.

known from completely independent experiments. The rate of temperature rise clearly shows a transition from stationary to non-stationary conditions as the initial pressure is raised (Fig. 9a). The ignition limits determined from these curves are in good agreement with the experimental limits (Fig. 9b), for a wide variety

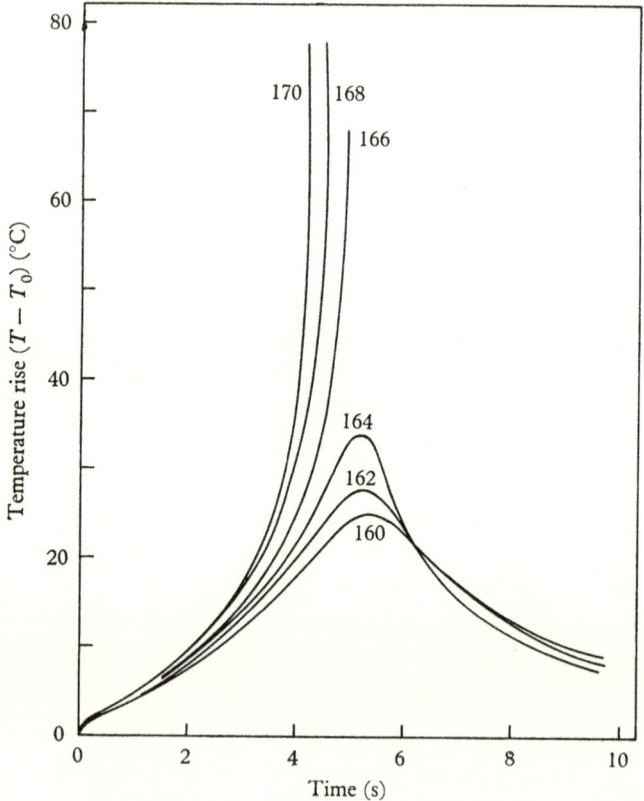

Fig. 9a. Calculated temperature rise in $H_2/Cl_2/NOCl$ mixtures, 1 mmHg NOCl, 360 °C. The figures on the curves give $[H_2]+[Cl_2]$ in mmHg.

of conditions of temperature or additive concentration. The induction periods are of the correct order of magnitude.

This treatment provides an accurate test of the validity of previous assumptions that in applying thermal theories of ignition to autocatalytic reactions the maximum isothermal rate should be used rather than the initial rate.

In addition, it can be said that it appears unnecessary to invoke chain branching to account for the sensitized ignitions in hydrogen and chlorine. There have been suggestions of energy chains in which some specific flow of energy takes place between the products M* of the exothermic reaction step

$$H + Cl_2 \rightarrow HCl^* + Cl^*$$

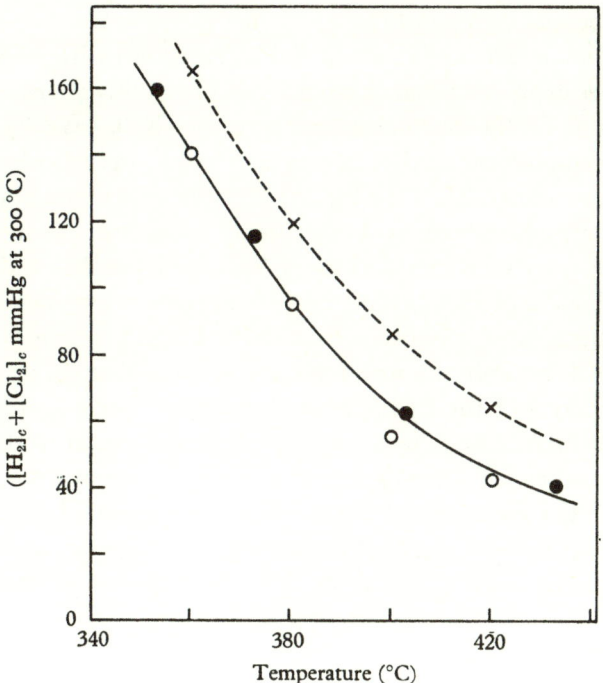

Fig. 9b. Calculated ($\times$) and experimental (O and ●) limits of H$_2$/Cl$_2$/NOCl mixtures, 1 mmHg NOCl.

and the decomposition

$$Cl_2 + M^* \rightarrow 2Cl + M$$

However, it would not appear that the exchange of vibrational energy between HCl* and Cl$_2$ would be particularly favourable, in view of the very different vibration frequencies. Recent work by Polanyi suggests that (a) only about 7 % of the heat of reaction is converted into vibration in the HCl bond, [50a] (b) Cl* loses its excess energy of translation very rapidly; [50b] this may be chiefly by collision with Cl_2, rather than with H_2, but in this case the distinction between specific effects and 'thermal' effects seems too fine to be useful. Taking into account the comparisons of the experimental ignition limits with those computed assuming a thermal theory of the ignitions, the conclusion is that under the conditions of the sensitized ignitions the main hydrogen–chlorine reaction proceeds by a non-branched chain.

One surprising effect [51] in sensitized ignitions of hydrogen and chlorine does depend on chain branching of the additive, however. The inhibitor of the photochemical reaction, NCl_3, can by its thermal decomposition 'ignite' mixtures of hydrogen and chlorine at room temperature (cf. Y in Fig. 7). Ignitions occur within an ignition peninsula which is undoubtedly associated with the peninsula of spontaneous explosions of NCl_3 itself, as observed by Apin. [52] Such a peninsula can only be explained by branched chain reactions, and it appears that the branching provides Cl atoms or NCl_x radicals at a rate sufficient to cause thermal or iso-thermal ignitions of the hydrogen and chlorine. These ignitions would repay further study in the light of Norrish's recent work [53] on the photochemical decomposition of NCl_3 in the presence of chlorine. As the time-scale of the thermal decomposition is quite long, it would be well worth investigating the temperature–time relationships near the upper pressure limit of the ignition peninsula.

7. Conclusions

There are some clear trends in the studies of sensitized ignitions over the past thirty-five years. Improved experimental methods have shown that the reactions introduced by the additives are much more subtle than was at first thought, involving in many cases their removal by chemical reaction to give other active sensitizers or inhibitors, with consequent changes in the net rates of branching. Theories of the transition from slow reaction to ignition are still based on the thermal and isothermal theories originally developed by Semenov and by Norrish, but there is surprisingly facile transition from one criterion to the other with small changes of conditions within a reacting system. This subtle interplay has meant the abandonment of simple approaches such as that summarized in the equation $\phi\tau$ = constant. Fortunately the complete equations for the rates of reaction and the rates of heat production and loss can now be solved, using computer methods, for those reactions where there is sufficient information about the kinetic mechanism, the rate constants, and the thermal conductivities of the reactant mixtures. The combination of improved experimental methods and the powerful new computing techniques has resulted in a much fuller understanding of the sensitized ignitions discussed in this chapter, and quantitative predictions of the ignition limits have been made successfully for reactions with quite complex autocatalytic mechanisms.

REFERENCES

1 Semenov, N. *Chemical Kinetics and Chain Reactions.* Oxford University Press, 1935.
2 Frank-Kamenetskii, D. A. *Diffusion and Heat Exchange in Chemical Kinetics* (translated by N. Thon). Princeton University Press, 1955.
3 Kowalskii, A. A. *Phys. Z. Sov.* 1933, **4**, 723.
4 Lewis, B. and Von Elbe, G. *Combustion, Flames and Explosions of Gases.* Academic Press, New York, 1951.
5 Foord, S. G. and Norrish, R. G. W. *Proc. Roy. Soc.* A, 1935, **152**, 196.
6 Norrish, R. G. W. and Dainton, F. S. *Proc. Roy. Soc.* A, 1941, **177**, 421.
7 Dainton, F. S.: (*a*) *Trans. Faraday Soc.* 1942, **38**, 227; (*b*) *Chain Reactions.* Methuen, London, 1956.
8 Norrish, R. G. W.: (*a*) *Nature, Lond.*, 1931, **127**, 853; (*b*) *Proc. Roy. Soc.* A, 1932, **135**, 334.
9 Griffiths, J. G. A. and Norrish, R. G. W. *Proc. Roy. Soc.* A, 1933, **139**, 147; *see also* Norrish, R. G. W. and Porter, G. *Proc. Roy. Soc.* A, 1952, **210**, 439.
10 Norrish, R. G. W. and Wallace, J. *Proc. Roy. Soc.* A, 1934, **145**, 307.
11 Norrish, R. G. W. *J. Am. Chem. Soc.* 1938, **60**, 1513.
12 Norrish, R. G. W. and Dainton, F. S.: (*a*) *Nature, Lond.*, 1939, **144**, 30; (*b*) *Proc. Roy. Soc.* A, 1941, **177**, 393.
13 Norrish, R. G. W. and Dainton, F. S. *Proc. Roy. Soc.* A, 1941, **177**, 411.
14 Norrish, R. G. W. and Ashmore, P. G. *Proc. Roy. Soc.* A, 1950, **203**, 454.
15 Ashmore, P. G., Dainton, F. S. and Norrish, R. G. W. *Nature, Lond.*, 1955, **177**, 546.

16 Ashmore, P. G. *Trans. Faraday Soc.* 1955, **51**, 1090.
17 Ashmore, P. G. and Levitt, B. P. *Advances in Catalysis and Related Subjects*, 1957, **9**, 367.
18 Ashmore, P. G. and Levitt, B. P. *Trans. Faraday Soc.* 1956, **52**, 835.
19 Ashmore, P. G. and Levitt, B. P. *Seventh Symposium on Combustion*, p. 45. Butterworths, London, 1958.
20 Tyler, B. J. Thesis, University of Cambridge, 1961.
21 Tyler, B. J. and Ashmore, P. G. *Ninth Symposium on Combustion*, p. 201. Academic Press, New York, 1963.
22 Ashmore, P. G. and Tyler, B. J. (*a*) *J. Catalysis*, 1962, **1**, 39; (*b*) *Trans. Faraday Soc.* 1962, **58**, 1108.
23 Baldwin, R. R. *Fuel*, 1952, **31**, 312.
24 Baldwin, R. R., Corney, N. S., Doran, P., Mayor, L. and Walker, R. W. *Ninth Symposium on Combustion*, p. 184. Academic Press, New York, 1963.
25 Baldwin, R. R., Jackson, D., Walker, R. W. and Webster, S. J. *Tenth Symposium on Combustion*, p. 423. Combustion Institute, 1965, Pittsburgh.
26 Blackmore, D. R., O'Donnell, G. and Simmons, R. F. *Tenth Symposium on Combustion*. Combustion Institute, 1965, Pittsburgh.
27 Norrish, R. G. W. and Husain, D. *Proc. Roy. Soc.* A, 1963, **273**, 145.
28 Gray, P. and Lee, J. C. *Trans. Faraday Soc.* 1954, **50**, 719.
29 Norrish, R. G. W. and Foord, S. G. *Proc. Roy. Soc.* A, 1936, **157**, 503.
30 Norrish, R. G. W. and Patnaik, D. *Nature, Lond.*, 1949, **163**, 883.
31 Enikolopyan, N. S. *Seventh Symposium on Combustion*, p. 157. Butterworths, 1959.
32 (*a*) Bone, W. A. and Allum, R. E. *Proc. Roy. Soc.* A, 1932, **134**, 582.
32 (*b*) Karmilova, L. V., Enikolopyan, N. S. and Nalbandyan, A. B. *Zh. fiz. Khim.* 1957, **31**, 851.
33 Ashmore, P. G. Thesis, University of Cambridge, 1950.
34 Preston, K. F. Thesis, University of Cambridge, 1963.
35 Norrish, R. G. W., Porter, G. and Thrush, B. A. *Proc. Roy. Soc.* A, 1955, **227**, 423.
36 Christie, M., Collins, J. M. and Voisey, M. A. *Trans. Faraday Soc.* 1965, **61**, 462.
37 Buckler, E. J. and Norrish, R. G. W. *Proc. Roy. Soc.* A, 1938, **167**, 292, 318; 1939, **172**, 1.
38 Dixon-Lewis, G. and Linnett, J. W. *Trans. Faraday Soc.* 1953, **49**, 756, 766.
39 Hoare, D. E. and Walsh, A. D. *Trans. Faraday Soc.* 1954, **50**, 37.
40 Crist, R. H. and Roehling, O. C. *J. Am. Chem. Soc.* 1935, **57**, 2196.
41 Ashmore, P. G. and Norrish, R. G. W. *Nature, Lond.*, 1951, **167**, 390.
42 Norrish, R. G. W. (*a*) *Trans. Faraday Soc.* 1925, **21**, 63; (*b*) *Z. phys. Chem.* 1926, **120**, 205.
43 Norrish, R. G. W. and Ritchie, M. (*a*) *Nature, Lond.*, 1932, **129**, 243; (*b*) *Proc. Roy. Soc.* A, 1933, **140**, 99, 112, 713.
44 Norrish, R. G. W. and Griffiths, J. G. A. (*a*) *Trans. Faraday Soc.* 1931, **27**, 451; (*b*) *Proc. Roy. Soc.* A, 1931, **130**, 592; 1932, **135**, 69.

45 Norrish, R. G. W. and Griffiths, J. G. A. *Proc. Roy. Soc.* A, 1934, **147**, 140.
46 Norrish, R. G. W. and Ashmore, P. G. *Proc. Roy. Soc.* A, 1950, **203**, 472; 1950, **204**, 34.
47 Ashmore, P. G. *Fifth Symposium on Combustion*, p. 700. Reinhold, 1955.
48 Ashmore, P. G. and Chanmugan, J. *Trans. Faraday Soc.* 1953, **49**, 254.
49 Ashmore, P. G. and Wesley, T. A. B. *Tenth Symposium on Combustion*, p. 217. Combustion Institute, 1965, Pittsburgh.
50 (a) Polanyi, J. C. *Applied Optics, Supplement on Chemical Lasers*, pp. 115, 125, 1965.
50 (b) Airey, J. R., Polanyi, J. C. and Snelling, D. R. *Tenth Symposium on Combustion*. Combustion Institute, 1965, Pittsburgh.
51 Ashmore, P. G. *Nature, Lond.*, 1953, **172**, 449.
52 Apin, A. J. *Acta Physicochim U.R.S.S.* 1940, **12**, 406.
53 Norrish, R. G. W. and Briggs, A. G. *Proc. Roy. Soc.* A, 1964, **278**, 27.

12

THE PYROLYSES OF PARAFFINS

J. H. PURNELL AND C. P. QUINN

1. Introduction

The problems of paraffin pyrolysis have exercised the chemist's imagination for almost exactly a century but the first serious kinetic studies date only from about 1930. [1,2] Most of our modern attitudes towards the reactions derive directly from the work of Rice and his colleagues, who, applying Paneth techniques, first showed that radicals were present during the pyrolysis of paraffins, [3] and that unbranched chains, propagated by alkyl radicals, could give rise to the observed product distributions [4] and overall kinetic characteristics. [5] Since that time advances have been associated almost entirely with the development of new techniques, particularly mass spectrometry and gas chromatography. The mass spectrometer permitted valuable isotopic distribution studies while gas chromatography has brought about a revolution in experimental methods. Before its advent analytical studies of the reactions were too tedious to be more than subsidiary to investigations based on the manometric method. By 1960, however, detailed analysis of reaction products could be carried out with such speed and accuracy by gas chromatography that it became possible to follow the course of a reaction by repeated sampling from a single reaction mixture. [6] The availability of high sensitivity detectors allowed the technique to be applied to the estimation of the trace products arising from chain initiation and termination processes, [7] or, alternatively, to the earliest stages of the reactions when secondary processes are least important [8] and when vessel surfaces have not been modified by the deposition of carbon. [9] The investigator who uses this approach not only provides himself with a means to collect a greater volume of information of a more fundamental nature but also gives himself a more subtle advantage. His attention becomes focused upon the elemental processes by which individual products are formed, rather than on the kinetics of the more complex processes, reactant removal and pressure rise. The simplicity of the methods of gas chromatography and the advantages of detailed analysis as a method of following chemical reactions are, we believe, so great that the manometric method of kinetic study should now be regarded as redundant.

2. Inhibition of paraffin pyrolyses

The observation that paraffin pyrolyses are inhibited by nitric oxide and by propylene provided further early evidence for the participation of free radicals, but at the same time posed a new problem. Staveley[10] demonstrated that no matter how much nitric oxide was added, the initial rate of pyrolysis of ethane, measured manometrically, never fell below a certain limit. For the last twenty-five years the nature of this 'residual reaction' has been a major point of controversy.

The most important aspect of that controversy is undoubtedly the development, and ultimately the refutation, of the 'molecular hypothesis' advanced by Stubbs and Hinshelwood.[11] They observed that the residual rate for n-pentane pyrolysis was the same whether nitric oxide or propylene was used as inhibitor and so proposed that the residual reaction was a true molecular process. They then studied the kinetics of several inhibited paraffin pyrolyses,[12] in some cases over an extremely wide pressure range,[13] and as a result suggested modifications of the theories of unimolecular processes.[13,14] What was not generally recognized was that the identity of residual rate was observed only in a single set of experimental conditions and, indeed, attempts to demonstrate this coincidence of rate with three inhibitors over a wide range of conditions[15] were largely unconvincing. Indeed Norrish and Pratt[16] reviewing the evidence conclude: 'We think that the equality of limiting rates has not been established as exact and general. The explanation of this equality, therefore, is not a necessary requirement of any mechanism postulated for the (residual) reaction.'

Finally, the shapes of the progress curves with different inhibitors are so unlike that the molecular hypothesis must be suspect on this account.[17]

The evidence *against* the molecular hypothesis is by now very strong and derives from three main sources.

Eltenton, [18] using a specially designed mass spectrometer, measured directly the methyl radical concentration in ethane pyrolysing in a fast flow reactor. He observed that the introduction of nitric oxide to even 50% by volume influenced the methyl radical concentration only trivially.

Wall and Moore [19] carried out isotope 'scrambling' experiments and found that the isotopic composition of hydrogen and methane formed from a mixture of C_2H_6 and C_2D_6 could only be explained on a radical basis and was unaffected by addition of up to 2·5% of nitric oxide. Rice and Varnerin, [20] similarly, studied pyrolysis of mixtures of CH_4 and C_2D_6 and also found evidence of much scrambling and that its extent was independent of the nitric oxide concentration. Both groups of workers were thus led to suppose that the nitric oxide-inhibited reaction involved the agency of free radicals. Later studies with mixtures of $C_3H_8 + D_2$ [21] and $C_4H_{10} + C_4D_{10}$ [22] led to the same conclusions.

Further evidence that the residual reaction is not a molecular process derives from detailed product analyses of the pyrolysis of *n*-butane. [23] In the absence of an inhibitor, the initial product distribution is a function of the reactant pressure, and the variations may be explained in terms of the competition of the reactions,

$$C_2H_5 \cdot \rightarrow C_2H_4 + H \cdot$$

and $$C_2H_5 \cdot + C_4H_{10} \rightarrow C_2H_6 + C_4H_9 \cdot$$

which are of different total order. In the reaction with added propylene the change of product distribution with inhibitor pressure is such that if molecular processes take place then they must mirror, *in every detail*, the kinetic behaviour of the radical scheme, including the competition between the processes consuming the ethyl radical. Since it has not been possible to devise such a molecular scheme it is clear that we must assume that the residual reaction in the presence of propylene is a modification of the free radical mechanism which accounts for the whole of the reaction taking place in the absence of inhibitors. It is also worth noting that the initial distribution of products is *more* complicated in the presence of inhibitors than in their absence. [17]

In summary, it now seems established beyond doubt that paraffin pyrolysis is entirely a free radical process and, as a corollary, that failure to inhibit a reaction is worthless as a test for molecular processes. Indeed inhibited pyrolyses now appear as complex two-component reaction systems which may well turn out to be of only limited interest.

2.1. The mechanism of inhibition. The 'molecular hypothesis' could not be regarded as completely discounted until a free radical mechanism had been devised to give a satisfactory account of the residual reaction. The first of a number of recent attempts to devise such a mechanism was due to Wojciechowski and Laidler. [24] For ethane, for example, they suggest, [25] at complete inhibition, the mechanism

$$C_2H_6 + NO \to C_2H_5\cdot + HNO \tag{1}$$

$$C_2H_5\cdot \to C_2H_4 + H\cdot \tag{2}$$

$$H\cdot + C_2H_6 \to C_2H_5\cdot + H_2 \tag{3}$$

$$H\cdot + NO \rightleftharpoons HNO \tag{4}$$

$$C_2H_5\cdot + HNO \to C_2H_6 + NO \tag{-1}$$

This scheme yields an overall rate expression for the steady state:

$$\frac{d}{dt}[C_2H_4] = \left(\frac{k_1 k_2 k_3 k_{-4}}{k_{-1} k_4}\right)^{\frac{1}{2}} [C_2H_6]$$

The rate constants of reactions involving the inhibitor occur as the ratio $(k_1 k_{-4}/k_{-1} k_4)$, which is the equilibrium constant of

$$C_2H_6 \rightleftharpoons C_2H_5\cdot + H\cdot$$

Thus, the overall rate expression is not only independent of [NO] but of the *nature* of the inhibitor provided that it can participate in initiation and termination in a fashion analogous to (1) and (4). For propylene, this demands that

$$C_2H_5\cdot + C_3H_6 \to C_2H_6 + C_3H_5\cdot$$

be regarded as a termination.

This approach clearly establishes one important thing; *if it is necessary* to write inhibition mechanisms such that residual rates are independent of inhibitor identity, *this can be done*. This removes the last intellectual difficulty involved in the rejection of the 'molecular hypothesis'. On the other hand, the mechanism suggested is unrealistic in that it is far too simple. In particular, in contrast with accepted evidence, the reactions

$$R \cdot + NO \rightarrow RNO \rightarrow oxime$$

and their consequences, are not included, while for olefinic inhibitors it is known that product distributions are highly complex and are hardly to be explained by such a simple scheme. Finally, the mechanism cannot accommodate the well-established fact that at inhibitor pressures above those required for minimum rate, an acceleration is observed. From the practical point of view, recent work on the reaction of ethyl radicals and NO by Pratt and Purnell [26] flatly contradicts the proposed inhibition mechanism, while the detailed studies [27, 28] of pentane and hexane inhibition by NO also reveal a complex state of affairs with heterogeneous processes probably participating. Thus, attempts to extend the mechanism into wider fields, especially the elucidation of un-inhibited decompositions, are seen to be not only questionable but liable to confuse the issue further.

A more realistic attitude has been adopted by Norrish and Pratt [16] and by Quinn, [17] the former considering NO inhibition, and the latter propylene inhibition, as complex two-component processes.

The reaction of alkyl radicals with NO is known to produce oximes, the pyrolyses of which [6, 29] occur rapidly at temperatures well below those of paraffin decomposition. Indeed, formaldoxime reacts explosively with NO. [30] Pratt and Purnell [6, 29] elucidated the main features of the mechanism of acetaldoxime pyrolysis and showed that in the NO-accelerated reaction the processes

$$NO + CH_3CHNOH \rightarrow (CH_3CHN) \cdot + HNO_2$$

$$(CH_3CHN) \cdot \rightarrow CH_3 \cdot + HCN$$

$$HNO_2 \rightarrow OH \cdot + NO \cdot$$

were very important and could lead to chain branching in systems involving ethyl radicals and NO. Steady-state treatment of a general mechanism involving these reactions was shown by

Norrish and Pratt[16] to lead to a rate equation which correctly predicts the shape of the NO inhibition curves over the whole range of NO pressure. This mechanism, moreover, leads to reasonable values for the velocity constants of certain elementary reactions, and correctly indicates a trivial rate of loss of NO at partial inhibition.

This view of inhibition is clearly much more in line with experimental results, and offers the further attraction of describing the three regions of rate in inhibition experiments in terms of a single scheme. Even so, as Norrish and Pratt point out, [16] this may still be an over-simplification, since numerous other reactions of NO in hydrocarbon systems may be envisaged. Further, it is now well established that the reaction of NO with alkyl radicals depends very much on temperature. In room-temperature photolyses, for example, the R/NO adduct is stable and the ratio [NO]/[R] in the adduct may exceed unity. [31] At intermediate temperatures, *ca.* 250 °C, acetaldoxime derived from $C_2H_5 \cdot + NO$ via tetraethyl lead pyrolysis has been shown to decompose very rapidly at the vessel wall. [26] The subject of NO inhibition is thus seen to be fraught with difficulties and the utility of its application is correspondingly questionable.

Quinn[17] has treated the decomposition of *n*-butane in the presence of propylene as a co-pyrolysis involving mutual sensitization. Taking the simplest possible case, he adds to the *n*-butane scheme, [8] outlined on page 339, the extra reactions

$$C_2H_5 \cdot + C_3H_5 \cdot \rightarrow C_5H_{10} \text{ (or disproportionation)}$$

$$C_3H_5 \cdot + C_3H_5 \cdot \rightarrow C_6H_{10} \text{ (or disproportionation)}$$

$$C_2H_5 \cdot + C_3H_6 \rightarrow C_2H_6 + C_3H_5 \cdot$$

$$C_3H_5 \cdot + C_4H_{10} \rightarrow C_3H_6 + C_4H_9 \cdot$$

$$C_3H_5 \cdot + C_3H_6 \rightarrow C_6H_{11} \cdot \rightarrow \text{products}$$

Making reasonable assumptions a steady-state solution is possible, and Quinn showed that the calculated 'inhibition curve' shows a flat minimum followed by an increase in rate at higher propylene pressures—entirely in accord with experiment. Significantly, in this approach the minimum rate is shown to be of no fundamental significance and, further, as with the mechanism of Norrish and Pratt, it is quite unnecessary to postulate chain initiation by the inhibitor.

We see, therefore, that inhibition processes may be very complex and that the operative mechanism may depend not only upon the identity of reactants but also upon the physical conditions. There need be no simple, unique mechanistic interpretation to be sought and the assumption that one exists may only lead back to an impasse comparable with that which developed over the rôle of molecular processes. Obviously, such a situation is to be avoided.

3. The uninhibited pyrolyses

3.1. *Surface effects.* Although few workers would now support the views expressed there, Voevodsky's paper [21] of 1959 has strongly influenced present-day views of the rôle of surfaces in uninhibited paraffin pyrolyses. The paper drew attention to the need to carry reactions only to very small fractions of completion if the state of the vessel surface was not to be modified or 'conditioned'. Experiments were described which seemed to indicate that the rate of pyrolysis of propane was a function of the surface–volume ratio in a clean quartz vessel, and could be further affected by coating the vessel surface with magnesium perchlorate, or treating it with a mixture of nitric oxide and hydrogen sulphide. Voevodsky interpreted these effects as evidence that the radical chains responsible for the reaction were propagated in the gas phase but initiated and terminated at the vessel wall. A similar mechanism had earlier been suggested in a most interesting paper by Rice and Herzfeld. [32] It now seems well established, however, that the effects described by Voedovsky were artifacts caused by trace quantities of oxygen which were either adventitious [33] or produced by the slow decomposition of the magnesium perchlorate coatings. [9, 33]

An analytical study [9] of the pyrolysis of *n*-butane at 527 °C showed that for pressures in the range 10–25 mmHg the reaction rate and product distribution were unaffected by increasing the surface–volume ratio of clean reaction vessels from 1·2 to 4·4 cm⁻¹, by coating them with potassium chloride, or by treating them with hydrofluoric acid. The reaction under such conditions is consequently believed to be homogeneous. In a similar way, a manometric study of propane pyrolysis [34] showed clearly that in the absence of traces of oxygen, the pyrolysis of propane at the pressures employed is unaffected by surface in clean Pyrex vessels or in vessels coated with potassium chloride or lead oxide.

When, in the n-butane study, [9] the vessel was 'conditioned' by carrying previous pyrolyses to large extents of reaction, reaction rates were depressed by as much as 40%, although the product distributions were unchanged. Similar effects were found when the vessel was coated with magnesium oxide, which is known to be particularly effective in removing hydrogen atoms. Thus it seems that the commonly employed conditioning of reaction vessels achieves reproducibility in depositing on the walls a coating of carbonaceous material which is unmodified by further deposition, but which actually *introduces* a heterogeneous chain termination reaction absent in clean vessels. This effect could be the cause of the heterogeneity reported by Sagert and Laidler, [35] and vitiates the conclusions drawn from much early work on these reactions.

The pyrolyses of isobutane and ethane, on the other hand, are affected quite dramatically by the nature, extent and previous history of the vessel surface. Fusy et al. [36] noted depressions of the initial rate of pressure rise of up to 40% when the clean Pyrex vessel used for isobutane pyrolysis was packed to increase its surface/volume ratio from 0·8 cm^{-1} to 10 cm^{-1}. Coating the packed vessel with lead oxide depressed the rate even further. Quinn [7] and Gordon [37] have shown that in ethane pyrolysis, the initiation of reaction chains (as reflected by primary methane formation) is not affected by surface in unconditioned quartz vessels, while on the other hand the *overall* rate of pyrolysis is markedly dependent on the nature and extent of the vessel surface. In packed quartz vessels, Quinn [7] observed irreproducibility in the length of induction periods. Fusy et al., [26] on the other hand, using clean Pyrex vessels found no evidence of irreproducibility, but did observe that the overall pyrolysis of ethane was slower in packed than in unpacked vessels, and that lead oxide coating of the packed vessel produced further depressions of the rates.

Sharp transitions in the order of ethane pyrolysis from 1·0 at higher pressures to 1·5 at lower pressures have been reported by Laidler and Wojciechowski, [25] while Kudryavtseva, Pavlov and Vedeneev [38] have shown that the order may rise even higher at lower pressures and Fusy *et al.* [36] have confirmed the general shape of the order plot. Laidler and Wojciechowski [25] originally explained this transition as resulting from a change in *homogeneous* chain termination process from

$$C_2H_5\cdot + C_2H_5\cdot \rightarrow C_4H_{10} \text{ (or } C_2H_4 + C_2H_6)$$

to

$$H\cdot + C_2H_5\cdot \rightarrow C_2H_6 \text{ (or } C_2H_4 + H_2)$$

as the pressure (and hence the ratio $[C_2H_5\cdot]/[H\cdot]$) is reduced. The changes seem too sharp to be compatible with this explanation, however, and Gordon and Niclause have suggested that the change in order results from the termination of chains by the removal of hydrogen atoms at the walls. Marshall and Quinn [39] have analysed the rate data of Kudryavtseva *et al.* and find them compatible with this interpretation if the efficiency of the wall removal process is $10^{-2·5}$, a value well in line with other measurements.

It seems, then, that the experimental data at present available indicate that in clean quartz or Pyrex vessels uninhibited paraffin pyrolyses are entirely homogeneous as long as hydrogen atoms are not important chain carriers. When they are, they may be removed sufficiently rapidly at the vessel walls for this process to contribute significantly to chain termination. When this happens, reaction rates become a function of the nature and extent of the vessel surface, and reaction orders may change rapidly with total pressure.

3.2. *Oxygen effects*. Niclause and his co-workers have studied the effects of trace quantities of oxygen on the uninhibited pyrolyses of paraffins. In Pyrex vessels of low surface/volume ratio the pyrolyses of propane, *n*-butane, isobutane and neopentane are accelerated by traces of oxygen, but the effect is limited to the initial stages. [40] For vessels of high surface/volume ratio it has been shown that for propane [41,42] and isopentane, [42] at least, oxygen may also exert an inhibiting effect which depends strongly on the nature and extent of the vessel surface. Indeed, in a vessel of surface/volume ratio 4·6 cm^{-1} coated with PbO, the initial rate of pyrolysis of propane may be reduced almost to zero in the presence of less than 2 % of oxygen. This experiment is probably the most unequivocal demonstration yet made of the absence of molecular processes. Niclause and his co-workers [42] explain the effects of traces of oxygen in terms of the interaction of two radical chain systems, one characteristic of oxidation, the other of pyrolysis. The oxygen introduces an additional chain initiation process, and the oxygenated radicals of the oxidation chain are assumed to be removed more efficiently at surfaces than the radicals propagating the pyrolysis chain. It is interesting that though these investigations have included analytical studies of the reaction products, no products uniquely characteristic of an oxidation process have ever been reported.

4. Mechanisms of uninhibited pyrolyses

As we have shown, there is now a good deal of evidence to the effect that, under suitably chosen experimental conditions, paraffin pyrolyses may be regarded as essentially homogeneous processes, in the absence of inhibitors. Not all of the earlier investigations of the uninhibited processes were carried out under such conditions. We will therefore limit discussion to work published since 1959. The results of these investigations are a remarkable testimony to the foresight shown in Rice's early work; with the exception of a few minor details we find that, *in the initial stages*, Rice's mechanisms are in excellent agreement with modern work on the reactions.

4.1. *n-Butane pyrolysis.* The paraffin pyrolysis of which the mechanism is now least controversial is undoubtedly that of *n*-butane. Under conditions in which no indications of heterogeneity are observed [9] the kinetics of formation of all the products in the initial stages may be explained quantitatively by the following mechanism [8] in the considerable temperature range 420–530 °C.

$$C_4H_{10} \rightarrow C_2H_5 \cdot + C_2H_5 \cdot \qquad (5a)$$

$$C_4H_{10} \rightarrow C_3H_7 \cdot + CH_3 \cdot \qquad (5b)$$

$$C_4H_9 \cdot \rightarrow C_2H_4 + C_2H_5 \cdot \qquad (6)$$

$$C_4H_9 \cdot \rightarrow C_3H_6 + CH_3 \cdot \quad (7)$$

$$C_2H_5 \cdot + C_4H_{10} \rightarrow C_2H_6 + C_4H_9 \cdot \quad (8)$$

$$CH_3 \cdot + C_4H_{10} \rightarrow CH_4 + C_4H_9 \cdot \quad (9)$$

$$H \cdot + C_4H_{10} \rightarrow H_2 + C_4H_9 \cdot \quad (10)$$

$$C_2H_5 \cdot \rightarrow C_2H_4 + H \cdot \quad (2)$$

$$C_2H_5 \cdot + C_2H_5 \cdot \rightarrow C_4H_{10} \quad (-5a)$$

$$C_2H_5 \cdot + C_2H_5 \cdot \rightarrow C_2H_4 + C_2H_6 \quad (11)$$

Reactions $(-5a)$ and (11) were identified as the main chain termination processes by observing that, of all the chain propagation products, ethane alone was formed with an activation energy which was independent of reactant pressure. Thus it appears that ethyl is at a much greater steady-state concentration than all the other radicals involved, presumably because it is less reactive. Measurements of ethylene/ethane ratios over a wide range of conditions led to the conclusion that the unimolecular process (2) was in its pressure-dependent region (though not second-order) at reactant pressures of 20–120 mmHg, and Kassel parameters for the process were determined by curve fitting. Although butenes are formed late in the reaction, they were shown not to be primary products and so the several reactions

$$C_4H_9 \cdot \rightarrow C_4H_8 + H \cdot$$

are clearly too slow to compete effectively with (6) and (7). A later manometric investigation [35] was in agreement with the majority of the conclusions of Purnell and Quinn, although some evidence of heterogeneous effects was found in a vessel of unspecified surface treatment. The manometric method proved incapable of detecting the pressure dependence of reaction (2).

4.2. *Ethane pyrolysis.* The packing of the reaction vessel, or the modification of its surface, can have an appreciable effect on the rate of ethane pyrolysis, particularly at low temperatures and pressures. [7,25,36,37] These effects may be explained quantitatively by the introduction of a first-order wall-removal process of hydrogen atoms. [39] Under conditions in which this heterogeneous termination is unimportant, the reaction *in the initial stages* is well described by the following mechanism. [7,37,43]

$$C_2H_6 \rightarrow CH_3 \cdot + CH_3 \cdot \tag{12}$$

$$CH_3 \cdot + C_2H_6 \rightarrow CH_4 + C_2H_5 \cdot \tag{13}$$

$$H \cdot + C_2H_6 \rightarrow H_2 + C_2H_5 \cdot \tag{3}$$

$$C_2H_5 \cdot \rightarrow C_2H_4 + H \cdot \tag{2}$$

$$C_2H_5 \cdot + C_2H_5 \cdot \rightarrow C_4H_{10} \tag{-5a}$$

$$C_2H_5 \cdot + C_2H_5 \cdot \rightarrow C_2H_4 + C_2H_6 \tag{11}$$

The steady-state treatment of this mechanism for long chains yields the rate equation

$$\frac{d}{dt}[C_2H_4] = \frac{d}{dt}[H_2] = k_2 \left(\frac{k_{12}[C_2H_6]}{k_{-5a} + k_{11}} \right)^{\frac{1}{2}}$$

while the overall order of reaction has been found to be close to unity for most conditions in which wall effects are unimportant. This order and the overall rate equation may be made compatible in two ways. Kuchler and Theile, [44] and later Laidler and Wojciechowski, [25] suggested that of the two unimolecular reactions involved (12) shows *second*-order and (2) *first*-order behaviour. On the other hand Quinn [7] has suggested that (12) is normally in its *first*-order region, and that (2) shows a pressure dependence similar to that observed in *n*-butane pyrolysis. [8] Quinn's interpretation is supported by the results of his analytical investigation of methane formation by reactions (12) and (13). The methane yields were found to be exactly first order in ethane, showing unequivocally that (12) is normally in its first-order region. Quinn then used this fact and the Kassel parameters for reaction (2), to give a quantitative description of ethane pyrolysis. [7] In the same study, trace quantities of n-butane were found among the initial products, supporting the choice of reactions ($-5a$) and (11) as the main chain termination. This work has since been repeated by three groups, one [37] obtaining results and drawing conclusions in complete agreement with Quinn. The other two studies [43,45] gave results in general agreement but which suggested that the ethane dissociation, (12), is of slightly greater order than unity at the lower end of the pressure range used. Lin and Back [43] have also used gas-chromatographic analysis of the products of the initial stages to compare yields of methane and n-butane. Their results suggest that small discrepancies between the yields may indicate the participation of another termination process which becomes more important at low pressures. It may be that this process is the wall removal of H atoms discussed by Marshall and Quinn. [39]

The pyrolysis of ethane shows self inhibition, but the effect is much less marked than in other paraffin pyrolyses. Laidler and Wojciechowski [25] attempted to give an account of the time course of ethane pyrolysis by including the processes

$$H\cdot + C_2H_4 \rightarrow C_2H_5\cdot \qquad (-2)$$

and

$$C_2H_5\cdot + H_2 \rightarrow C_2H_6 + H\cdot \qquad (-3)$$

in their Kuchler–Theile mechanism which we now know to be incorrect.

In his study of the formation of methane in the ethane pyrolysis, [7] Quinn noted that methane was formed not only by a primary mechanism attributable to reactions (12) and (13), but also by a secondary process accelerated considerably by the addition of ethylene. In a later paper [46] the kinetics of this secondary process were investigated by the analytical method, and shown to be capable of description by the rate equation

$$\left(\frac{d}{dt}[CH_4]\right)_{\text{secy.}} = k[C_2H_4]^2[C_2H_6]^{\frac{1}{2}}$$

The half-order dependence upon ethane supports Quinn's contention that reaction (1) is first order under the conditions used. The second-order dependence on ethylene is explained in terms of the mechanism

$$C_2H_5{\cdot} + C_2H_4 \rightleftharpoons CH_3CH_2CH_2CH_2{\cdot}$$

$$CH_3(CH_2)_2\dot{C}H_2 + C_2H_4 \rightleftharpoons CH_3(CH_2)_4\dot{C}H_2$$

$$CH_3(CH_2)_4\dot{C}H_2 \rightleftharpoons \underset{\substack{CH_2\ CH_2 \\ CH\ CH_2 \\ CH_3\ H}}{CH_2} \rightleftharpoons CH_3\dot{C}H(CH_2)_3CH_3$$

$$CH_3\dot{C}H(CH_2)_3CH_3 \rightarrow CH_3CH{=}CH_2 + CH_3CH_2\dot{C}H_2$$

$$CH_3CH_2\dot{C}H_2 \rightarrow \dot{C}H_3 + C_2H_4$$

The observation of second-order behaviour indicates that under these commonly used experimental conditions, the n-butyl radical isomerizes more rapidly by way of addition of ethylene, and subsequent isomerization, than by any internal hydrogen transfer of the type:

$$CH_3CH_2CH_2\dot{C}H_2 \rightleftharpoons CH_3\underset{\substack{\\ H}}{CH}\overset{\substack{CH_2 \\ }}{\diagdown}CH_2 \rightarrow CH_3\dot{C}HCH_2CH_3$$

Experiments which have been interpreted as providing evidence of such internal isomerization of small alkyl radicals [47-49] have invariably been carried to large extents of reaction when product olefins would facilitate an isomerization mechanism of the type projected by Quinn. Such a mechanism for radical isomerization was first suggested on theoretical grounds by Rice and Kossiakoff, [50] is supported by work from other fields, [51-53] and is compatible with the results of the interesting experiments of Brodskii *et al*. [54] The occurrence of the secondary mechanism for methane formation explains the difficulties which earlier workers [55] encountered when they used insensitive techniques to measure methane yields, and identified processes (12) and (13) as the sole mechanism for methane formation.

4.3. *Propane pyrolysis.* Only two papers on the kinetics of the uninhibited pyrolysis of propane have been published in the past five years. That of Laidler, Sagert and Wojciechowski [56] is based almost entirely on manometric experiments; the only analytical data relate to the products of one reaction carried to 14% conversion. The authors reported an order of unity at low temperatures and high pressures with an activation energy of 67·1 kcal mole^{-1}; at low pressures and high temperatures an order of 1·5 was reported. The nature of the reaction vessel surface was not described, but packing the vessel produced decreases in rate, with attendant changes in order and activation energy. The mechanism suggested for that part of the reaction believed to be homogeneous was of the Rice–Herzfeld type. In earlier work on ethane pyrolysis [25] Laidler and Wojciechowski had assumed—as we now know, incorrectly—that methyl radical recombination was in its third-order region under pyrolysis conditions. In order to accommodate this assumption with the orders which they measured for propane decomposition, Laidler *et al.* [56] were forced into the difficult position of postulating second-order behaviour for the unimolecular reactions

$$C_3H_8 \rightarrow CH_3\cdot + C_2H_5\cdot$$

and

$$C_4H_{10} \rightarrow CH_3\cdot + C_3H_7\cdot$$

but first-order kinetics [25,35] for the similar process,

$$C_4H_{10} \rightarrow C_2H_5\cdot + C_2H_5\cdot$$

These difficulties of interpretation do not arise if one accepts the experimental evidence that methyl recombination is in its second-order region under the conditions used for the experiments.

A much more thorough study of the reaction was carried out by Martin, Dzierzynski and Niclause. [34] This work made good use of chromatographic analysis of the products although relying on manometric determination of rates. Martin *et al.* showed that contrary to earlier reports, [21] no heterogeneous effects could be detected in clean Pyrex or silica vessels at 546 °C with pressures of 25 or 50 mmHg of propane. Detailed investigations [42] of the effects of trace quantities of oxygen provided an explanation of Voevodsky's observations [21] while it seems likely that the vessel used by Laidler et al. [56] had been 'aged' and its surface thereby rendered active. Martin et al. [34] developed an extrapolation technique to avoid the measurement of initial rates, and for conditions in which the reaction was homogeneous found an order of 1.25 ± 0.05 and an activation energy of 67 kcal mole^{-1}.

These observations suggest to us that there may be competition between homogeneous chain termination processes involving at least two radicals at comparable concentrations.

Much the most detailed study of propane pyrolysis is that carried out by Leathard, Purnell and Quinn, [57] as yet unpublished. Detailed analytical studies have been carried out in the range of 0–2% decomposition over the wide-temperature range, 445–545 °C, in conditions in which the contribution of oxygen effects and of heterogeneity can be discounted. Initial rates of C_2H_4 formation (i.e. < 0.2% decomposition) were found to be expressible by the equation

$$\frac{d}{dt}[C_2H_4] = \frac{k_a[C_3H_8]^{\frac{3}{2}}}{1 + k_b[C_3H_8]}$$

over the whole range of initial pressures and temperatures studied. For temperatures in the region of 540 °C and pressures close to 100 mmHg the measured orders and activation energies are in good agreement with the results of earlier manometric studies. An interesting feature never previously observed is the fall of the order at 50 mmHg towards 0.5 when the temperature is reduced to 440 °C. The characteristics of the pyrolysis at higher extents of decomposition originate not only in the complexity of the initial reaction scheme but from the early intervention of radical–olefin polymerization and decomposition reactions of the type observed by Quinn in ethane pyrolysis. [46] In the propane system, such reactions can be far more complex and at least thirty processes can be visualized. A detailed account must, therefore, await further study. However, it is already clear that attempts by earlier workers to correlate initial manometric rates with late analytical data are unlikely to do more than confuse interpretation.

5. The derivation of fundamental rate constants

The use of powerful analytical methods has brought about a remarkable advance in our understanding of paraffin pyrolysis mechanisms over the last eight years. Indeed, it is fair to say that the mechanisms of ethane and of n-butane pyrolysis are now understood about as well as is the photolysis of acetone, which is commonly regarded as a 'clean' radical source. This understanding enables us to measure velocity constants for some of the elemental processes involved in the radical chains of the uninhibited pyrolysis. The values we obtain, however, are only as valid as the mechanisms we assume for their derivation, and it is on this score that we must set aside the work of Blackmore and Hinshelwood. [58]

The determinations fall into two classes, relative, and absolute. The latter consist of the three measurements [7,37,43] of Arrhenius parameters for the dissociation

$$C_2H_6 \rightarrow CH_3\cdot + CH_3\cdot$$

As is commonly the case, the velocity constants measured by the three sets of investigators are in much better agreement than are the Arrhenius parameters derived from them. It seems likely that errors in temperature scales may be responsible for the differences in the activation energies.

From a study of the kinetics of formation of ethylene and ethane during the initial stages of the pyrolysis of n-butane, the velocity constants of the reactions

$$C_2H_5\cdot \rightarrow C_2H_4 + H\cdot$$

and
$$C_2H_5\cdot + C_4H_{10} \rightarrow C_2H_6 + C_4H_9\cdot$$

have been determined. [8] A steady-state treatment of the mechanism of page 339 of this chapter indicates that, in the initial stages, and with long reaction chains,

$$\frac{[C_2H_4]}{[C_2H_6]} = 1 + \frac{2k_2}{k_8[C_4H_{10}]}$$

so that from observations of this product ratio, the ratio of the velocity constants of reactions (8) and (2) may be calculated. The same reaction scheme also yields the initial rate expression,

$$\frac{d}{dt}[C_2H_6] = k_8 \sqrt{\frac{k_{5a}+k_{5b}}{k_{-5a}+k_{11}}} [C_4H_{10}]^{\frac{3}{2}}$$

when the reaction chains are long. The ratio

$$(k_{5a}+k_{5b})/(k_{-5a}+k_{11})$$

may be related [8] to the statistical calculation of the equilibrium constant, k_{5a}/k_{-5a}, for n-butane dissociation to ethyl radicals. Thus, observations of the rate of ethane formation allow the calculation of k_8 and, hence, k_2.

The velocity constant for the dissociation of the ethyl radical has also been measured by Lin and Back [43] in their study of ethane pyrolysis. They compared the rate of formation of hydrogen by the chain propagation processes

$$C_2H_5 \cdot \rightarrow C_2H_4 + H \cdot$$

$$H \cdot + C_2H_6 \rightarrow H_2 + C_2H_5 \cdot$$

with the rate of formation of n-butane by the main chain termination

$$C_2H_5 \cdot + C_2H_5 \cdot \rightarrow C_4H_{10}$$

Thus they deduce the velocity constant for ethyl dissociation in terms of that for ethyl recombination. Although this method is more satisfactory than that of Purnell and Quinn, [8] it is not beyond criticism. Ethane pyrolysis, unlike that of n-butane, is known to involve a surface termination [7,36,37,39,43] under certain conditions, and then radical concentrations are not homogeneous within the vessel. Under these conditions it is not necessarily valid to assume, as Lin and Back have done, that the mean square concentration of radicals is equal to the square of their mean concentration.

Pyrolysis of propane [57] is yielding measurements of fundamental rate constants, in particular of the reactions

$$CH_3 \cdot + C_3H_8 \rightarrow CH_4 + C_3H_7 \cdot$$

and
$$s\text{-}C_3H_7 \cdot \rightarrow C_3H_6 + H \cdot$$

As more is known of processes occurring later in the reaction, other rate constants may in turn be measurable. Indeed, this is true of all paraffin pyrolyses and it is not improbable that these systems may, in the future, provide reliable and simple methods for the study of radical reaction rates. Such measurements are urgently needed since few published data can be accepted unequivocally.

6. Conclusion

It has been the thesis of this chapter that the use of modern analytical methods has helped to bring about a considerable clarification of our understanding of paraffin pyrolyses. The new picture which is emerging is essentially simpler than the old. The uninhibited reactions seem to take place *entirely* by radical chains which differ only in points of detail from those which Rice and Herzfeld suggested thirty years ago. In contrast, the inhibited reactions appear to be relatively complex modifications of these same radical chain schemes.

The uninhibited pyrolyses are in general entirely homogeneous, but wall effects do occur when hydrogen atoms are important chain carriers, that is, when molecular hydrogen is a major product of the reaction.

The work on ethane indicates quite clearly that the unimolecular dissociations, and bimolecular radical associations, which initiate and terminate the chains of the uninhibited pyrolyses, do not involve third bodies to any great extent under normal reaction conditions.

Our own work on *n*-butane pyrolysis shows that there ethyl radical recombination dominates all other chain terminations, the presumption being that ethyl is at a much greater steady-state concentration than the other centres involved. The order for propane pyrolysis, however, suggests that for that reaction at least two radicals (presumably methyl and propyl) are at comparable concentrations. This contrast of behaviour has led us to form the opinion that, of all the chain carriers involved in paraffin pyrolyses, ethyl is the least reactive. Thus, for a paraffin pyrolysis in which ethane is an appreciable primary product, we expect the main chain termination to be the mutual destruction of ethyl radicals.

Obviously there is still much experimental work to be done, and this discussion points clearly to the fact that such future work must concentrate first on the study of product formation in the very earliest stages of reaction, and only then on the measurement of overall orders and energies. Far more emphasis should be placed on the study of heterogeneous effects, and efforts should be made to study the reactions only under conditions for which wall effects are calculable. If these attitudes become generally accepted, it seems that it should soon be possible to form a general theory of paraffin pyrolysis, and thereby provide a useful method for the resolution of some of the problems of radical reactions in the gas phase.

REFERENCES

1   Pease, R. N. *J. Am. Chem. Soc.* 1928, **50**, 1779.
2   Pease, R. N. and Durgan, E. S. *J. Am. Chem. Soc.* 1930, **52**, 1262.
3   Rice, F. O., Johnston, W. R. and Evering, B. L. *J. Am. Chem. Soc.* 1932, **54**, 3529.
4   Rice, F. O. *J. Am. Chem. Soc.* 1934, **55**, 3035.
5   Rice, F. O. and Herzfeld, K. F. *J. Am. Chem. Soc.* 1934, **56**, 284.
6   Pratt, G. L. and Purnell, J. H. *Proc. Roy. Soc.* A, 1961, **260**, 317.
7   Quinn, C. P. *Proc. Roy. Soc.* A, 1963, **275**, 190.
8   Purnell, J. H. and Quinn, C. P. *Proc. Roy. Soc.* A, 1962, **270**, 267.
9   Purnell, J. H. and Quinn, C. P. *J. chem. Soc.* 1961, p. 4128.
10 Staveley, L. A. K. *Proc. Roy. Soc.* A, 1937, **162**, 557.
11 Stubbs, F. J. and Hinshelwood, Sir C. *Proc. Roy. Soc.* A, 1950, **200**, 458.
12 Stubbs, F. J. and Hinshelwood, Sir C. *Proc. Roy. Soc.* A, 1950, **201**, 18.
13 Stubbs, F. J., Ingold, K. U., Spall, B. C., Danby, C. J. and Hinshelwood, Sir C. *Proc. Roy. Soc.* A, 1952, **214**, 20.
14 Ingold, K. U., Stubbs, F. J. and Hinshelwood, Sir C. *Proc. Roy. Soc.* A, 1950, **203**, 486.
15 Jach, J., Stubbs, F. J. and Hinshelwood, Sir C. *Proc. Roy. Soc.* A, 1954, **224**, 283.
16 Norrish, R. G. W. and Pratt, G. L. *Nature, Lond.*, 1963, **197**, 143.
17 Quinn, C. P. *Symposium Kinetics of Pyrolytic Reactions.* Chemical Institute of Canada, Ottawa, 1964.
18 Eltenton, G. C. *J. chem. Phys.* 1947, **15**, 455.
19 Wall, L. A. and Moore, W. J. *J. Am. Chem. Soc.* 1951, **73**, 2840.
20 Rice, F. O. and Varnerin, R. E. *J. Am. Chem. Soc.* 1954, **76**, 324.
21 Voevodsky, V. V. *Trans. Faraday Soc.* 1959, **55**, 65.
22 Kuppermann, A. and Larson, J. G. *J. chem. Phys.* 1960, **33**, 1264.

23 Purnell, J. H. and Quinn, C. P. *Nature, Lond.*, 1961, **189**, 656.
24 Wojciechowski, B. W. and Laidler, K. J. *Can. J. Chem.* 1960, **38**, 1027.
25 Laidler, K. J. and Wojciechowski, B. W. *Proc. Roy. Soc.* A, 1961, **260**, 91, 103.
26 Pratt, G. L. and Purnell, J. H. *Trans. Faraday Soc.* 1964, **60**, 371.
27 Bryce, W. A. and Chrysochoos, J. *Trans. Faraday Soc.* 1963, **59**, 1842.
28 Chrysochoos, J. and Bryce, W. A. *Trans. Faraday Soc.* 1965, **61**, 2447.
29 Pratt, G. L. and Purnell, J. H. *Trans. Faraday Soc.* 1962, **58**, 692.
30 Taylor, H. A. and Bender, H. *J. chem. Phys.* 1941, **9**, 761.
31 Christie, M. I. *Proc. Roy. Soc.* A, 1958, **249**, 258.
32 Rice, F. O. and Herzfeld, K. F. *J. phys. Chem.* 1951, **55**, 975.
33 Martin, R. and Niclause, M. *J. chim. phys.* 1964, **61**, 802.
34 Martin, R., Dzierzynski, M. and Niclause, M. *J. chim. phys.* 1964, **61**, 286.
35 Sagert, N. H. and Laidler, K. J. *Can. J. Chem.* 1963, **41**, 838.
36 Fusy, J., Sacchi, G., Martin, R., Combes, A. and Niclause, M. *C. r. hebd. Séanc. Acad. Sci., Paris*, 1965, **261**, 2223.
37 Gordon, A. S. *Symposium Kinetics of Pyrolytic Reactions.* Chemical Institute of Canada, Ottawa, 1964.
38 Kudryavtseva, Yu. I., Pavlov, B. V. and Vedeneev, V. I. *Zh. fiz. Khim.* 1964, **38**, 978.
39 Marshall, R. M. and Quinn, C. P. *Trans. Faraday Soc.* 1965, **61**, 2671.
40 Engel, J., Combe, A., Letort, M. and Niclause, M. *Revue Inst. fr. Pétrole*, 1957, **12**, 627.
41 Martin, R., Dzierzynski, M. and Niclause, M. *J. chim. phys.* 1964, **61**, 790.
42 Niclause, M., Martin, R., Combes, A. and Dzierzynski, M. *Can. J. Chem.* 1965, **43**, 1120.
43 Lin, M. C. and Back, M. H. *Symposium Kinetics of Pyrolytic Reactions.* Chemical Institute of Canada, Ottawa, 1964 (and by private communication).
44 Küchler, L. and Theile, H. *Z. phys. Chem.* B, 1939, **42**, 359.
45 Dexter, R. W. and Trenwith, A. B. *Proc. chem. Soc.* 1965, p. 394.
46 Quinn, C. P. *Trans. Faraday Soc.* 1963, **59**, 2543.
47 Kerr, J. A. and Trotman-Dickenson, A. F. *J. chem. Soc.* 1960, p. 1602.
48 Kerr, J. A. and Trotman-Dickenson, A. F. *Trans. Faraday Soc.* 1959, **55**, 921.
49 Heller, C. A. and Gordon, A. S. *J. phys. Chem.* 1958, **62**, 709.
50 Kossiakoff, A. and Rice, F. O. *J. Am. Chem. Soc.* 1943, **65**, 590.
51 Frey, F. E. and Hepp, H. J. *J. Am. Chem. Soc.* 1933, **55**, 3357.
52 Sefton, V. B. and Leroy, D. J. *Can. J. Chem.* 1956, **34**, 41.
53 McNesby, J. R., Drew, C. M. and Gordon, A. S. *J. chem. Phys.* 1956, **24**, 1260.

54 Brodskii, A. M., Kalinenko, R. A., Lavrovskii, K. P. and Shevel'-kova, L. V. *Kinet. Katal.* 1964, **5**, 49.
55 Stepukhovich, A. D. and Derevenskikh, L. V. *Zh. fiz. Khim.* 1960, **34**, 2315.
56 Laidler, K. J., Sagert, N. H. and Wojciechowski, B. W. *Proc. Roy. Soc.* A, 1962, **270**, 242.
57 Leathard, D., Purnell, J. H. and Quinn, C. P. To be published.
58 Blackmore, D. R. and Hinshelwood, Sir C. *Proc. Roy. Soc.* A, 1961, **268**, 36.
59 Purnell, J. H. and Quinn, C. P. *J. chem. Soc.* 1964, p. 4049.

SUBJECT INDEX

absorption spectra, *see* spectra

acetaldehyde
as branching agent, 244, 281
formed in oxidation of propane, 261;
in photolysis of nitroethane, 11
oxidation of, 233, 259
photolysis of, 14, 121
polymerization of, 14, 223-4

acetaldehyde peroxide, chemiluminescence from, 53

acetaldoxime, 334, 335

acetamide, photolysis of, 49

acetic acid
effect of, on polymerization of
acetaldehyde, 223
elimination of, in cross-linking of
polyvinyl acetate, 219
inhibits chain reaction of ozone
with H_2O_2, 35
photovoltaic effect in aqueous solution of, 30

acetone
from oxidations of hydrocarbons,
272, 273, 274, 280
photolysis of, 6-7; flash, 13, 14, 104,
121

acetyl radical, 7

acetylene
deactivation (vibrational) of C_2D_2
and, 164
explosive decomposition of, 288
oxidation of, 126, 233
from photolysis of ethylene, 11

acetylhydroperoxide radical, 244

acidity constants, for triplet states,
106

acids, carboxylic, photolysis of, 44-5

acrylonitrile
graft polymers containing, 217
polymerization of, 202, 207, 208

actinometry, 64, 65

activation energies, 65, 73
for endothermic reactions, 230, 241,
277
prediction of, 90
for propagation and termination of
polymerization, 207
in propane pyrolysis 344-5
for reactions in oxidation of hydrocarbons, 230, 231, 232
23

for reactions in solvents, 27
for surface recombination processes,
128
see also Arrhenius parameters

after-glows, 129, 184, 189

'ageing' of reaction vessels, 66, 337

alcohols, from cool-flame oxidation of
a kanes, 256

aldehydes
as branching agents, 244, 250-1,
254-5, 268, 272, 281, 283
oxidation of, 246, 254.
in oxidation of hydrocarbons, 12,
229-31, 256, 258
photolysis of, 4, 7-8, 42-7, 201;
flash, 121

alkali metals
reaction of, with alkyl halides, 192
spin-orbit relaxation rates for, 172

alkanes
oxidation of, 251, 272-5
photochlorination of, 74, 75, 85-7
pyrolyses of, 230-51
see also hydrocarbons

alkenes, *see* olefins

alkoxy radicals
dissociation of, 202
as initiators of polymerization, 202
in oxidation of hydrocarbons, 230,
231

alkyl halides
photochlorination of, 73
reactions of, with alkali metals, 192;
with chlorine in mixtures with alkanes, 86; with lithium ethyl, 231

alkyl nitrates, explosive decomposition
of, 288

alkyl nitrites, as photosensitizers for
combustion of hydrocarbons, 126,
128

alkyl radicals
decomposition of, *see* cracking of
hydrocarbons
in oxidation of hydrocarbons, 230,
269-71, 276
reaction of, with NO, 334, 335

allyl alcohol, photolysis of H_2O_2 in
presence of, 33-4

allyl radical, 125
as initiator of polymerization, 202

[353]

APA

aluminium trimethyl, as polymerization catalyst, 226
amides, photolysis of, 15, 48–9
amines, photolysis of, 15, 48
amino acids (aromatic), flash photolysis of, 107
amino (NH$_2$) radical, 48, 129
spectrum of, 122, 124–5
ammonia
deactivation (vibrational) of ND$_3$ and, 164
inhibition of H$_2$/Cl$_2$ reaction by, 3, 67, 287
oxidation of, 126, 129, 310
photolysis of, 15, 47
sensitization of H$_2$/O$_2$ ignition by, 310
amyl nitrite
as photosensitizer for oxidation of hydrocarbons, 128
anthracene
co-polymerization of, with styrene, 218
oxidation of, 204
as photosensitizer for polymerization of styrene, 203
triplet-triplet annihilation reaction in, 103
anti-knock, mechanism of, 25, 128
argon
deactivation of (rotational), 148
deactivations by (electronic), 170, 173; (rotational), 148; (vibrational), 142, 155, 159, 160, 161, 164
inhibition of ignitions by, 295, 304, 305
as third body in recombination of iodine atoms, 115
aromatic free radicals, 101, 105
aromatics
flash photolysis of, 118
oxidation of, 126, 251
Arrhenius parameters
for abstractions of hydrogen, 304
for dissociation of ethane, 346
for photochlorinations, 77, 78, 85, 86, 87
in thermal ignitions, 290–2.
arsenite, photolysis of H$_2$O$_2$ in presence of, 34
arsine, deactivation (vibrational) of AsD$_3$ and, 164
atomic line reversal, measurement of vibrational temperature by, 149

auto-acceleration
and ignitions, 236–7, 292
in oxidation of hydrocarbons, 235, 236, 251, 253
in polymerizations, 204–9, 215
azo(bis)isobutyronitrile, as initiator, for autoxidation of peroxides, 54; for polymerization, 56, 201
azomethane, ignition boundary for, 288, 291, 292
benzene
deactivation (electronic) by, 187
oxidation of, 233
photolysis of carbonyl compounds in, 47
triplet absorption spectrum of, 102
benzene hydroperoxide, chemiluminescence from, 53
benzophenone
flash photolysis of, by laser pulse, 109
as sensitizer for photolysis of carbonyl compounds, 47
benzoyl peroxide
chemiluminescence from, 53
as initiator, of oxidation of tetralin, 55; of polymerization, 56, 200, 201, 205, 208, 210
photolysis of, 52
variation with pressure of rate of decomposition of, 199–200, 210–11
benzoyloxy radicals, 200, 202, 217
benzyl formic acid, photolysis of, 45
benzyl radical, 101, 200
identification of, 119, 125
bi-t-butyldiperoxyoxalate, as initiator for autoxidation of peroxides, 54
biradicals, 45, 203, 233, 243
block co-polymers, 217
BO$_2$ radical, 122, 123
Boltzmann equation, 145
Boltzmann equilibria, 94–5, 105
bond energies, 5, 73, 90, 277, 281–2
and bond lengths, 125
and spectra, 5
boron trifluoride, promotion of polymerizations by, 222, 225
branching, kinetic
aldehydes as agents of, see aldehydes
degenerate (delayed), in ignitions, 289, 311, 315; in oxidation of hydrocarbons, 24, 129, 229, 235, 236–7, 241, 243, 245, 250, 251

branching, kinetic (*cont.*)
in ignitions, 241, 289–90, 293, 295, 300, 301, 312, 317, 326
in oxidation of hydrocarbons, 229–41 *passim*, 270, 276, 281
in polymerization of formaldehyde, 221
possible participation of excited species in, 241–6
in pyrolyses, 334
quadratic, 295, 307, 309
branching, structural, of polymer molecules, 217, 221, 226
bromine
reaction of, with hydrogen, 68
recombination of atoms of, 115–16
bromine oxide (BrO) radical, 117
i-butane
inhibition of H_2/O_2 ignition by, 303
oxidation of, 270, 272, 273, 274–5
pyrolysis of, 337, 338
n-butane
inhibition of H_2/O_2 ignition by, 303
formation of, in photolysis of ethylene, 11, and of methyl butyl ketone, 7; in pyrolysis of ethane, 342
pyrolysis of, 332, 335, 336–7, 339–40, 348
butanone, cool-flame oxidation of, 258
i-butene
formed in oxidation of *i*-butane, 274; of neopentane, 280
isomerization of, 84
oxidation of, 271, 272, 273
polymerization of, 225
t-butoxy radical 52
t-butyl hydroperoxide, photolysis of, 14, 50, 201
n-butyl radical, from pyrolysis of butane, 343
i-butyraldehyde, from neopentane, 280
butyramide, photolysis of, 48, 49
n-butyric acid, photolysis of, 44

C_2 radical, 127
C_3 radical, 119, 122, 125, 127
cadmium atoms, deactivation of (electronic), 187
'cage effect'
in gas phase, 42
pressure and, 201
in solution, 27–8, 41, 44, 105
carbene, as biradical, 243

carbon
formation of, in flash photolysis, 13, 104, 105, 127
heat of sublimation of, 5
carbon-carbon double bond, effect of substituents on reactivity of, 213
carbon dioxide
deactivation of (rotational), 147; (vibrational), 154
deactivations by, (electronic), 172, 173, 174, 186; (rotational), 148; (vibrational), 158, 160
formation of, in decomposition of benzoyl peroxide, 200; in oxidations of CO, 38, 317, 318, and of hydrocarbons, 241, 273
inhibition of ignitions by, 295, 304, 305
spectra of, 122, 123
carbon diselenide, flash photolysis of, 155, 157, 170
carbon disulphide
flash photolysis of, 118, 155
oxidation of, 126, 246, 289
reaction of, with NO, 287
carbon monoxide
deactivation of (vibrational), 162, 164
deactivations by, (electronic), 167, 168, 169, 170, 175, 186; (vibrational), 150–1, 159
exchange of vibrational energy between NO and, 157, 159, 162
formation of, in oxidation of hydrocarbons, 261, 272, 273; in photolyses of amides, 48, 49, of carbonyl compounds, 4, 6, 8, 43, 44, 45, 47, and of cyclic ketones, 9
liquid, photolysis of ozone in, 38
oxidation of, 24, 34, 241, 289, 316–20
photochlorination of, 72, 157
reaction of, with hydroxy radicals, 318
thermal stability of, 127
vibrational excitation of, by transfer of electronic energy, 176, 177
carbon oxysulphide, flash photolysis of, 118
carbon tetrachloride
photolyses of peroxides in, 40–1, 50, 51–2, 200, 210
in vinyl polymerization, 57–8, 59
carbonyl compounds
from oxidation of alkenes, 272

23-2

carbonyl compounds (*cont.*)
photochemistry of, 3–4, 42–7, 121
three types of decomposition of,
7–9, 43–5, 47
triplet states of, 105
see also aldehydes, ketones
carbonyl radicals, in oxidation of
hydrocarbons, 245
CCl_3 radical, as initiator of vinyl polymerization, 57, 59
cellulose, photo-oxidation of, 54
cerous ions, initiation of polymerization by, 202
CF_2 radical, 124
CH radical, 127
CH_2 and CH_3 radicals, *see* methylene
and methyl radicals
chain reactions
in decomposition of benzoyl peroxide, 199–200
in ignitions, 241, 244, 289, 292–
327
in oxidation of aldehydes, 11; of CO,
34; of hydrides, 129; of hydrocarbons, 25, 127, 229–47, 276–
81
in photochlorinations, 1, 10, 23,
66–7, 70–1
in photolyses, 43, 50, 52
in pyrolyses, 339–41
see also initiation, propagation, and
termination of chain reactions,
and branching, kinetic
chemiluminescence, 38, 53, 125, 129
see also cool-flame oxidation
chloric acid, from photolysis of ClO_2
in water, 39
chlorine
deactivation of (vibrational), 150
dissociation of molecule of, 65
oxidation of, 67, 116–17
photochemical reactions of, *see*
photochlorinations
reactions sensitized by, 3, 9, 24, 68,
294, 312
recombination of atoms of, 79–81,
116
see also hydrogen/chlorine reaction
chlorine dioxide, photolysis of, 15–16,
39–41, 104, 120; flash, 117
chlorine monoxide, flash photolysis of,
117
chlorine oxide (ClO) radical, 15, 16,
116–17
chloroalkyl radicals, excited, 65, 81–4

chlorobenzene, photolysis of tetralin
hydroperoxide in, 52
chloroform, photolysis of benzoyl
peroxide in, 53
chlorophyll, flash photolysis of, 107–8,
109
chloropicrin
'lighthouse effect' produced in
CO/O_2 mixtures by, 320
oxidations sensitized by, 24, 294,
297, 312, 320
chloroplasts, flash photolysis of, 107–8
CHO (formyl) radical, 5, 121
spectrum of, 3, 122
chromium carbonyl, as initiator of
polymerization, 59
chrysene, photosensitization of styrene
polymerization by, 218
circular dichroism, possible application of, to flash photolysis, 110
Cl_3 radical, 69, 75, 80, 81
ClO (chlorine oxide) radical, 15, 16,
116–17
CN radical, *see* cyanogen radical
co-catalysts, for polymerization, 208,
225
colloidal smokes, anti-knock effect of,
128
combustion
contributions of Norrish to knowledge of, 22–5
ignitions in, 287–329
low-temperature, 250–86
study of, by flash photolysis, 105,
126–9
see also cool-flame oxidations, ignitions, oxidations
computer of average transients (CAT),
97
computer, use of, for calculations
relating to
combustion problems, 22
$H_2/Cl_2/NO/NOCl$ reaction, 322–3
$H_2/O_2/NO$ reaction, 300
inhibitions by HBr, 309
oxidation of hydrocarbons, 239–
40
cool-flame oxidation, of hydrocarbons,
25, 234, 241, 244–6, 250–86, 289,
314
co-polymerization, 207, 212–19
co-pyrolysis, 335
cracking of hydrocarbons, 229, 232,
257, 269
oxygen and, 237

cross-linking of polymers, 201, 217–18, 218–19
CS radical, 118
  deactivation of (vibrational), 156–7, 174
CSe radical, deactivation of (vibrational), 157, 174
cumene, autoxidation of, 53–4
cumene hydroperoxide
  chemiluminescence from, 53
  photochemistry of, 15, 50, 51–2, 201
cumyloxy radical, 52
cyanogen
  photolysis of, 112, 118, 155, 192
  reaction of NO and, 3
cyanogen (CN) radical
  from cyanogen, 112, 118, 155, 192
  from hydrocarbon/O₂/NO₂ reaction, 316
  reactions of, 119–20
cyclic ketones, photolysis of, 9, 43, 45–7
cycloheptatriene, spectrum of, 119
cycloheptatrienyl radical, 125
cyclohexane
  cool-flame oxidation of, 258
  photolyses in, 47, 48
cyclopentadienyl radical, 119, 125
cyclopentanone, photolysis of, 45–6

dehydrochlorination, 88
dehydrogenation, in oxidation of hydrocarbons, 237–41
deprotonation, rate constants for, 106
deuterides, deactivation of (vibrational), 152–3
deuterium
  deactivations by (electronic), 173; (rotational), 148; (vibrational), 149, 151
  deactivations of compounds of hydrogen and of, 162, 164–5
  for determining number of hydrogen atoms in radicals, 118
  reaction of, with chlorine, 87
  for study of pyrolyses of paraffins, 332
deuterium oxide
  deactivations by (electronic), 167, 169, 186
  deactivations (vibrational) by water and by, 157, 162
diacetyl, photolysis of, 14, 121
diazirine, photolysis of, 41

diazomethane
  photolysis of, 41, 42
  products of flash photolysis of, 119, 124
1,1-dichloroethylene, photochlorination of, 78
cis-1,2-dichloroethylene
  isomerization of, 79, 81, 82–4
  photochlorination of, 77–9, 81, 82
dichloromethane, deactivation of (vibrational), 154
dicumyl peroxide, photochemistry of, 15, 50, 51, 201
dielectric constants of solvents, and rate of cross-linking of polymers, 220
diethyl ether, cool-flame oxidation of, 259, 261
diethyl ketone, photolysis of, 47
dihydroanthracene, effect of, on styrene polymerization, 204
dihydroperoxides, in cool-flame oxidation, 282–3
dilatometers, 208, 225
dioxan, photolyses in, 45, 50–1
2,5-diphenyloxazole, sensitizes photolysis of benzoyl peroxide, 52
di-n-propyl ketone, photolysis of, 43, 44
divinyl compounds, co-polymerizations of styrene and methyl methacrylate with, 212, 219
drying agents, acting as retarders of reaction, 67
dyes, photosensitizations by, 54–7

egg albumin, flash photolysis of, 107
electron spin resonance, 13
electronic energy, 135, 155
  increased by absorption of light, 98
  interconversion of, with translational and vibrational energy, by crossing of potential curves, 175–80; by resonance process, 165–74
  transfer of, between atoms and complex molecules, 185; between atoms of different elements, 180–5
electrons
  hydrated, 29
  radiolysis by pulse of, 108
  solvated, 108
  transfer of, in initiation of polymerization, 203
  transfer rates for, 106

emission spectra, *see* spectra
energy
 absorption and dissipation of, in
  photochemistry, 26
 chains of, 16, 36, 325
 transfer of, in molecular collisions,
  133–97; from solvent to molecules
  undergoing photolysis, 53
 see also electronic, rotational, translational, *and* vibrational energies
entropies of activation (rotational,
  translational, vibrational), 89
epoxides, from oxidation of olefines,
  272, 281
equilibrium
 displacement of, by light flash, 98–9
 protonic, 106–7
esters, carbonyl group of, in photolysis,
  44–5
ethane
 formation of, in photolysis of
  ethylene, 111; in pyrolysis of *n*-
  butane, 340, 346
 inhibition of H_2/O_2 ignition by, 303
 oxidation of, 127, 270, 272, 278
 pyrolysis of, 332, 333, 337, 338,
  340–4, 346, 347, 348
ethanol, as radical scavenger, 31
ethers
 cool-flame oxidation of, 246
 ring, as products of cool-flame
  oxidation of alkanes, 256
ethyl acetate
 photolysis of, 45
 and photolysis of water, 29
ethyl iodide, deactivation (electronic)
  by, 173
ethyl radicals, in pyrolyses of paraffins,
  340, 347, 348
ethylene
 crystalline polymer of, 226
 deactivation of (vibrational), 154
 deactivations by (electronic), 186,
  187
 formation of, in oxidation of propane, 261, 269; in photolyses of
  amides, 48, of carboxylic acids and
  esters, 44–5, and of ketones, 45, 46,
  47; in pyrolyses of paraffins, 345
 oxidation of, 127, 254–5, 271–2, 273
 photochlorination of, 3, 70, 78
 reaction of fluorine with, 230, 231
ethylene oxide, from oxidation of
  ethylene, 271
ethylmethyl amine, photolysis of, 48

explosions
 conditions for, 68, 236, 237, 264,
  290–8
 homogeneous, 126
 inhomogeneous, 128
 spontaneous, 288

ferric chloride, cross-linking of polymers by, 219–20
ferrocene, as anti-knock compound,
  128
flame propagation, of cool flames, 246
flash discharge lamps, 13, 97–8, 99
flash photolysis, 12–18, 93–11, 250
 and steady photolysis compared, 121
 in study of combustion, 126–9; of
  free radicals in gas phase, 112–32
 vibrational excitation by, 155–6
flash spectroscopy, 95–9, 99, 103
fluorenone, sensitizes photolysis of
  tetralin hydroperoxide, 52
fluor cence
 of cumene hydroperoxide, 52
 of electronically excited indium, 183;
  mercury, 184; sodium 181–2;
  thallium, 183
 of irradiated cerous salts, 202
 in study of deactivation, (electronic),
  166, 167; (rotational), 146–7,
  147–8; (vibrational), 151, 160
fluorine, kinetics of reactions of, 230,
  231, 243
fluorite region, photochemical reactions in, 5, 11
formaldehyde
 as branching agent, 233, 250, 252,
  254–5, 271, 310–11, 315
 electronically excited, as light emitter in cool-flame oxidations, 25,
  259–60
 formation of, in oxidation of hydrocarbons, 261, 269–70, 272; in
  photolyses of dioxan, 51, and of
  nitromethane, 11
 and induction period of cool-flame
  oxidations, 25, 245
 inhibitions by, of H_2/O_2 ignition,
  303, 305; of low-temperature
  oxidations, 273
 oxidation of, 11, 254
 photolysis of, 5, 43
 polymerization of, 14, 220–2, 224
formaldoxime, 334
formic acid, promotes polymerization
  of formaldehyde, 221

formyl (CHO) radical
  from photolysis of aldehydes and
    ketones, 5, 121
  spectrum of, 3, 122
free radicals
  acidity and protonation constants
    for, 106
  aromatic, 101
  flash-photolysis studies of, in gas
    phase, 112–32
  initiation of polymerization by, 201,
    204
  interreactions of, in flash and steady
    photolysis, 121
  in oxidation of hydrocarbons, 256
  in photochemical and thermal
    reactions, 4
  produced by dissociation, 105, 199;
    by pulse radiolysis, 108
  in pyrolyses of paraffins, 332–3
  reactivity of 227
  spectroscopy of, 95, 100–1, 121–6
  trapping of, 100, 207
  see also individual radicals
Friedel-Crafts catalysts, cross-linking
    of polymers by, 219–20

gas chromatography, 13, 64, 73–4, 250
  for analysis of products of low-
    temperature combustions, 256,
    269, 273–4; of products of
    pyrolyses of paraffins, 330, 331
gas phase
  'cage effect' in, 42
  comparison of photochemistry in
    liquid phase and in, 26–8
  flash-photolysis studies of free
    radicals in, 112–32
  kinetics of flash photolysis in, 102–5
  oxidation of hydrocarbons in, 229–
    49, 251
  photochlorination in, 64–92
  polymerization of ethylene in, 226
glyoxal
  photolysis of, 3
  polymerization of, 224
graft co-polymers, 215–17

haemoglobin/oxygen reaction, rate
    constants for, 107
halides, and initiation of vinyl poly-
    merization, 57–8
halogenated compounds
  flash photolysis of (methanes), 118
  and ignitions, 308–9

halogens, kinetic studies of free
    radicals of, 114–18
HCCl radical, 122, 124, 125
HCF radical, 122, 124, 125
heat
  production of, in cool flames, 262
  production of, in sensitized igni-
    tions, calculated by computer,
    222, 327
  production and removal of, in chain
    reactions, as a function of
    temperature, 263–6
helium
  deactivation of (electronic), 183–4,
    185
  deactivations by, (electronic), 173;
    (rotational), 148; (vibrational), 151
  inhibition of ignitions by, 295, 304,
    305
  as third body in recombination of
    iodine atoms, 103
helium-neon laser, 134
n-hexane
  cool-flame oxidation of, 256
  photolyses of peroxides in, 47, 50,
    51–2
n-hexene, oxidation of, 271–2
HNCN radical, 119, 125
HNO (nitroxyl) radical, 122, 124, 125,
    129, 310
HO radical, see hydroxyl radical
HO₂ (perhydroxyl) radical, 100, 293
  in H₂/Cl₂ reaction, 70
  in ignitions, 293, 299, 301
  in oxidation of hydrocarbons, 229,
    233, 234, 238, 252, 269, 277–9,
    281
  in photolysis of H₂O₂, 35; of water,
    30
HS₂ radical, 101
HSiBr radical, 122, 124
HSiCl radical, 122, 124
hydrazine
  inhibition of H₂/O₂ ignition by, 310
  oxidation of, 126, 129, 288, 310
hydrides
  deactivation of (vibrational), 143,
    145, 150, 154, 192
  deactivations by (electronic), 174;
    (vibrational), 162–3
  exchange of vibrational energy
    between, 158, 163
  oxidation of, 129
  spectra of, 122
  see also individual hydrides

hydrocarbons
  exchange of vibrational energy
    between oxygen and, 164
  inhibition of ignitions by, 302–6
  oxidation of, 12, 126, 127, 229–49,
    308; slow, 24–5, 269–81
  photochlorination of, 74–84
  photosensitization of carbonyl com-
    pounds by, 47
  pyrolyses of, 125, 330–51
  unsaturation produced in, by free
    radicals, 8, 15, 48, 49
  see also alkanes, olefines
hydrochloric acid
  deactivation by (electronic), 187
  deactivation of (vibrational), 191
  inhibits reaction of ozone and H₂O₂,
    35
  from photolysis of ClO₂ in water, 39
  photovoltaic effect in 31
hydrogen
  atoms of, formed in photolysis of
    water, 29
  deactivation of (rotational), 147;
    (vibrational), 155
  deactivations by, (electronic), 170,
    174, 175, 186; (rotational), 148,
    149; (vibrational), 151, 164
  formation of, in oxidation of pro-
    pane, 261; in photolyses of
    amides, 48, of amines, 15, 48, of
    ammonia, 47–8, of aqueous solu-
    tions of cerous salts, 202, and of
    water, 29
  initiation of polymerization by, 202–
    3
  ortho and para, 73, 164
  reactions of, with bromine, 68; with
    chlorine, see hydrogen/chlorine
    reaction; with hydroxyl radicals,
    126; with nitrogen dioxide, 297,
    298, 300, 301; with oxygen, see
    hydrogen/oxygen reaction; with
    sulphur, 3, 23
  sensitization of CO/O₂ reaction by,
    24, 316, 317
hydrogen azide, flash photolysis of,
  118
hydrogen bromide, suppression of
  H₂/O₂ ignitions by, 309
hydrogen chloride
  catalysis of polymerization of
    formaldehyde by, 222
  elimination of, in cross-linking of
    chlorinated polymers, 219

inhibitions by, of H₂/Cl₂ reaction,
  10, 68, 70; of H₂/O₂ ignition, 309
  see also hydrogen/chlorine reaction
hydrogen/chlorine reaction, 1, 9–10,
  66–70, 87, 242
  effects on, of oxygen 10, 23–4, 67,
    69, 70; of water, 2, 3, 67
  inhibition of, by HCl, 10, 68, 70; by
    NCl₃, 3, 18, 67, 320; by NH₃, 3, 67,
    287
  sensitized ignitions of, 320–6
  thermal, in presence of NO and
    NOCl, 72–3
hydrogen iodide
  deactivation by (electronic), 173
  deactivation of (vibrational), 150,
    155
hydrogen ions, in photovoltaic effects,
  21
hydrogen/oxygen reaction
  ignition in, 241, 242, 246, 289
  inhibition of, by halogenated com-
    pounds, 308–9; by hydrazine,
    310; by hydrocarbons, 302–6;
    NO₂ by, 10, 24, 196, 294, 297
    301; by water, 306–8
  sensitization of ignition of, 3, 9, 24,
    68, 294–302, 310, 320
  study of, by flash photolysis, 14, 126
hydrogen peroxide
  chemiluminescence from reactions
    of, 38
  formation of, in H₂/O₂ reaction, 9,
    67, 293; in oxidation of hydro-
    carbons, 292, 233, 258, 269, 276,
    278, 281; in photolyses of ClO₂,
    39–40, and of water, 29–30
  reaction of, with NO, 300
  sensitization of oxidations by, 24
hydrogen sulphide
  ignition boundary for, 292
  oxidation of, 14, 126, 129
hydrogenations
  mercury-sensitized, 47
  in photolysis of carbonyl compounds
    44
hydroperoxides, in oxidation of hydrocarbons, 231, 232, 234–5, 244,
  258–9, 282–3
hydroxyacetic acid, from photolysis of
  monochloracetic acid, 26–7
hydroxyl radicals
  combination of, 67
  in ignitions, 293, 299, 300, 301, 318
  initiation of polymerization by, 57

hydroxyl radicals (*cont.*)
    in oxidations of aldehydes, 252, 255;
        of CO, 34; of hydrocarbons 127,
        277, 278, 281, 283; of hydrogen,
        126, 229, 233, 238, 252, 255–6
    in photolyses, of ClO_2, 40; of H_2O_2,
        33; of ozone, 36, 38; of water, 29,
        30
    in photovoltaic phenomena, 31
    reaction of, with molecular hydro-
        gen, 126
    vibrationally excited, 17
    in water vapour, 112
hypochlorite, chemiluminescence from
    reaction of, with H_2O_2, 38

ignition boundaries, 259, 288–94
    for CO/O_2, 316–17
    for H_2/Cl_2, 320–2
    for $H_2/O_2/NO_2$, 295, 296, 299, 301
    for hydrazine/O_2, 310
    for methane/O_2, 311, 314
    quantitative predictions of, 292, 299,
        327
ignitions, 265, 287–8, 326–9
    branched chain thermal theory of,
        294–310
    and cool flames, 241, 244, 246
    isothermal, 290, 292–4, 301, 306
    sensitized, of CO/O_2, 316–20; of
        H_2/Cl_2, 294, 320–6; of H_2/O_2,
        294; of methane/O_2, 294, 310–16
    thermal, 290–2, 301, 306
    transition between isothermal and
        thermal, 306
    two-stage, 257
imino (NH) radical, 118, 129, 316
impurities
    catalysis of vibrational energy trans-
        fer by, 157
    in $H_2/\ l_2$ reaction, 10
    in photochlorinations, 66
    in polymerizations, 207–8
    in solvents, 106
indium, transfer of electronic energy
    from mercury to, 183, 185
induction periods
    in cool-flame oxidations, 262
    in oxidation of hydrocarbons, 25,
        243–4, 254
    in polymerizations, 207
    in sensitized ignitions, 128, 295, 297,
        301, 311
inert gases
    deactivations by, (electronic), 166,

175, 178; (rotational and vibra-
    tional), 16, 148
dilution with, in flash photolysis,
    103, 105, 105, 115, 116, 119
inhibition of ignitions by, 295, 304,
    305
in lamps for flash photolysis, 13
see also individual gases
infra-red spectrometry, 96, 109
    for analysis of energy transfer, 174,
        189, 190, 191
inhibitions, *see individual reactions and
    inhibitors*
initiation of chain reactions, 10
    in ignitions, 312
    in oxidation of hydrocarbons, 229,
        230
    in polymerizations, 199–204, 221
insertion reactions, 41–2
iodine
    as anti-knock substance, 128
    deactivation of, (electronic), 173;
        (rotational and vibrational), 148,
        150, 165
    recombination of atoms of, 14, 15,
        103, 114–15
iodine oxide (IO) radical, 117
ionization potentials
    correlation of, with quenching cross-
        section, 165, 175
    of methylene, methyl, and tropyl
        radicals, 125–6
    of third body, 115
iron carbonyl, flash photolysis of, 170
iron pentacarbonyl, as anti-knock
    compound, 128
isobutane, etc., *see i*-butane, etc.
isomerization
    of butene, 84
    of *cis*-1,2-dichlorethylene, 79, 81,
        82–4
    of peroxide radicals in oxidation of
        hydrocarbons, 231, 232
    in photolysis of cyclic ketones, 46
    in pyrolysis of ethane, 343
    of vinyl compounds, 106
isotopes, in study of
    branching of polymers, 217
    initiation of polymerization, 57
    oxidation of methane, 241
    peroxy radicals, 54
    photolysis of benzoyl peroxide, 200–
        1; of ClO_2, 40; of ozone, 35
    pyrolyses of paraffins, 332
    see also deuterium

ketene
  flash photolysis of, 13, 14, 104
  photolysis of, 12, 41, 43
  photolysis of mixture of methyl
    chloride and, 42
α-keto acids and esters, photolysis
  of, 45
ketones
  photolysis of, 3–4, 6–9, 42–7, 201;
    flash, 121
  from photolysis of organic peroxides,
    50, 51, 52
ketyl radicals, 105
kinetic spectrophotometry, 96–8
kinetic spectroscopy, 118
krypton, deactivation by, 148

Lambert-Salter line diagrams
  for spin-orbit relaxation, 171, 173
  for vibration-translation energy
    transfer, 153, 161
lasers
  energy transfer in, (helium/neon),
    134; (neon/oxygen), 188
  for flash photolysis, 109, 157
lead oxide (HgO), as anti-knock compound, 128
lead tetra-ethyl
  as anti-knock compound, 128
  pyrolysis of, 335
liquid phase, photochemistry in, 26–
  63; see also solution
lithium ethyl, reactions of, with alkyl
  halides, 231
lumiflavin, flash photolysis of, 107

magnetic resonance, possible detection of transient species by, 109
manganese carbonyl, as initiator of
  vinyl polymerization, 57–8, 59
mass spectrometry, 64, 74, 88, 109, 330
  agreement between ionization
    potentials and results of, 125–6
  measurement of methyl radicals by,
    332
matrix isolation technique, 100
menthone, photolysis of, 9
mercury
  deactivation of (electronic), 166–8,
    176–7, 181–3, 186–8
  electronically excited, 184
  photosensitization by, 17, 47, 232
  seal of, for greased stopcocks, 222
metal carbonyls, as initiators of vinyl
  polymerization, 57

methacrylic acid, initiation of polymerization of, 202
methane
  deactivation by (electronic), 186,
    187
  deactivation of (vibrational), 162;
    compared with that of CD_4, 164
  formation of, in photolysis of
    acetaldehyde, 121; in pyrolysis of
    ethane, 341, 342
  ignition of oxygen and, 294, 310–16
  inhibition of ignition of $H_2/O_2/NO_2$
    by, 302, 304–5
  oxidation of, 10, 25, 127, 232, 233–4,
    241, 251–2
  and oxidation of CO, 319
methanol
  effect of, on photolysis of water, 29
  oxidation of, 260
  from oxidation of methane, 311,
    315; of propane, 201
methyl acrylate, initiation of polymerization of, 202
N-methyl-β-alanine methyl ester,
  photolysis of, 45
methyl butyl ketone, photolysis of, 6,
  7, 43
methyl butyrate, photolysis of, 44
methyl chloride
  deactivation by (rotational), 148
  photolysis of mixture of ketene and,
    42
methyl ethyl ketone, photolysis of,
  43
methyl iodide
  photolysis of, 255
  reaction of, with fluorine, 243
methyl isopropyl ketone, photolysis of,
  44
methyl methacrylate
  co-polymerization of, with divinyl
    compounds, 213, 219; with sty-
    rene, 213–14
  graft polymers containing, 217
  polymerization of, 204–6, 218
methyl nitrate, ignition boundary for,
  291, 292
methyl pent-2-ene, oxidation of, 271–2
methyl propyl ketone, photolysis of,
  43
methyl (CH_3) radical, 100
  formation of, in oxidations of
    methane, 233, 252, and of
    propane, 269–70; in photolyses
    of acetone, 6, 7, of aldehydes and

methyl (CH$_3$) radical (*cont.*)
  ketones, 121, of diazomethane, 119, of ketene, 12, and of methyl iodide, 255; in pyrolyses of paraffins, 332, 344
  inhibition of H$_2$/O$_2$ ignition by, 305
  ionization potential of, 125–6
  oxidation of, 255
  recombination of, 121
methyl vinyl ketone polymers, photolysis of, 14
methylene chloride, reaction of, with fluorine, 243
methylene (CH$_2$) radical, 12, 14, 122
  formed in photolysis of amides, 49; of ketene, 12; of ketones, 7, 8
  insertion reactions of, 41–2
  ionization potential of, 125–6
  triplet and singlet states of, 100, 124
molecules
  study of rotational relaxation by crossed beam of, 147, 189
  triatomic, (linear), bending of, 123–4; prediction of ground and excited states of, 122–3
molybdenum carbonyl, as initiator of vinyl polymerization, 59
monochloracetic acid, photolysis of aqueous solution of, 26–7

N$_3$ radical, 122, 125
naphthalene, triplet-triplet annihilation reaction in, 103
NCl (nitrogen monochloride) radical, 18, 68, 118
NCl$_2$ (nitrogen dichloride) radical, 18, 68, 118, 326
NCN radical, 119, 122, 125
NCO radical, 122, 123, 125
NCS radical, 122, 125
neon
  deactivation by, 148, 151
  energy transfer (electronic) from helium to, 183–4, 185
neon/oxygen laser, energy transfer in, 188
neopentane
  inhibition of ignitions by, 304–5
  products from oxidation of, 280–1
  pyrolysis of, effect of oxygen on, 338
Nernst chain reaction, H$_2$/Cl$_2$ reaction as example of, 10, 66
NH (imino) radical, 118, 129, 316

NH$_2$ (amino) radical, 48, 129
  spectrum of, 122, 124–5
nitric oxide
  deactivation of, (electronic), 17; (rotational), 147–8; (vibrational), 142, 151–2, 157, 159–60, 162
  deactivations by (electronic), 166, 170, 175, 186, 187
  exchange of energy between nitrogen and, (electronic), 189–9; (vibrational), 159–61, 162
  formation of, by dissociation of NO$_2$, 3, 10; in oxidation of NH$_3$, 310; in photolysis of NOCl, 17, 120, 155, 192
  inhibition of pyrolyses of paraffins by, 125, 332, 333, 334
  potential diagram of, 152
  reactions of, with alkyl radicals, 334, 335; with CS$_2$, 287; with CN radical, 119; with cyanogen, 3; with H$_2$O$_2$, 300
  sensitizations by, of H$_2$/Cl$_2$ ignition, 320–2; of H$_2$/O$_2$ ignition, 24, 294–302, 310; of methane/O$_2$ ignition, 312–16
  and thermal H$_2$/Cl$_2$ reaction, 72–3
  as third body in recombination of iodine atoms, 115
  vibrationally excited, 17, 120
nitroethane, photolysis of, 11
nitrogen
  deactivation of, (rotational), 147; (vibrational), 150, 158
  deactivations by, (electronic), 167, 168, 169, 170, 173–4, 175, 186; (rotational), 148; (vibrational), 16, 159, 160, 161
  exchange of energy between NO and, (electronic), 188–9; (vibrational), 159–61, 162
  formation of, in oxidation of hydrazine, 310; of methane in presence of NO$_2$ and NO, 315
  inhibition of ignitions by, 295, 304, 305
  liquid, photolysis of ozone in, 37
  vibrationally excited, 176
nitrogen dichloride radical, 18, 68, 118
nitrogen dioxide
  and CO/O$_2$ reaction, 319
  and H$_2$/O$_2$ reaction, 10, 24, 126, 294, 297, 301
  photolysis of, 3, 10; flash, 14, 118, 120, 155

nitrogen dioxide (*cont.*)
    reactions of, with hydrogen, 297, 298,
        300, 301; with oxygen, 242, 300
    sensitizations by, of acetylene/O_2
        reaction, 126; of hydrocarbon/O_2
        reaction, 127; of methane/O_2
        reaction, 10, 311–16
    spectrum of, 122
nitrogen monochloride radical, 18, 68,
    118
nitrogen trichloride
    chlorine-photosensitized decompo-
        sition of, 9, 18, 68, 118
    explosive decomposition of, 288
    ignition of H_2/Cl_2 by thermal
        decomposition of, 326
    inhibition of photochemical H_2/Cl_2
        reaction by, 3, 18, 67, 320
nitromethane, photolysis of, 11
nitrosyl bromide, photolysis of, 120
nitrosyl chloride
    inhibition of H_2/Cl_2 ignition by,
        320–5
    photolysis of, 10, 17, 192; flash, 120,
        155
    reaction of, with chlorine atoms, 72,
        80
    sensitizations by, of H_2/O_2 reaction,
        24, 294, 297; of methane/O_2
        reaction, 312
    and thermal H_2/Cl_2 reaction, 72–3
nitrosyl halides, flash photolysis of, 118
nitrosyl iodide, 115
nitrous oxide
    deactivation of (rotational), 147
    deactivations by, (electronic), 170,
        186, 187; (rotational), 148; (vibra-
        tional), 158
    explosive decomposition of, 288
    formation of, in ultra-violet photo-
        lysis of ozone in nitrogen, 37
    in photovoltaic reactions, 31
    sensitization of H_2/O_2 ignition by,
        310
nitroxyl (HNO) radical, 122, 124, 125,
    129, 301
NOH (oximino) radical, 11
nuclear magnetic resonance, 13

olefins
    as branching agents, 273, 281
    formation of, in oxidation of hydro-
        carbons, 229, 237, 240, 256, 270,
        277; in photolysis of carbonyl
        compounds, 9, 44

inhibition of pyrolyses of paraffins
    by, 331, 332, 333, 334
oxidation of, 251, 271–2, 276, 281
photochlorination of, 70–1, 74–
    84
polymerization of, 226–7
reactions of, with fluorine, 243
vibrationally excited radicals from,
    84
optical masers, 183
optical rotatory dispersion, possible
    application of, to flash-photolysis
    experiments, 110
oxidations
    of hydrocarbons in gas phase, 229–
        49
    sensitized, 10
    see also cool-flame oxidations, com-
        bustion, ignitions, *and individual
        compounds*
oximes, from reaction of alkyl radicals
    with NO, 334–5
oximino (NOH) radical, 11
oxygen
    in chemiluminescent reactions, 38,
        53, 129
    deactivation of, (rotational), 147;
        (vibrational), 150, 164
    deactivations by, (electronic), 170,
        172, 174, 175, 186; (vibrational),
        150, 164
    from dissociation of NO_2, 3, 10
    effect of, on cracking of hydro-
        carbons, 237; on H_2/Cl_2 reaction,
        10, 23–4, 67, 69, 70; on H_2/S
        reaction, 3; on pyrolyses of
        paraffins, 336, 338–9, 345
    as impurity in solvents, 106
    insertion reactions of, 41
    photolysis of, in liquid nitrogen, 37
    in photolyses, of ClO_2, 39; of H_2O_2,
        35; of ozone, 36; of water, 29, 30
    in photosensitization of polymeriza-
        tion, 57
    quenching of triplets in gas phase
        by, 103
    reactions of, with chlorine, 67, 116–
        17; with hydrogen, *see* hydrogen/
        oxygen reaction; with vibration-
        ally excited sulphur, 189–90
    vibrationally excited, 15–16, 17, 37,
        39, 104, 120, 155, 157, 164, 242
oxygen-oxygen bond in peroxides,
    fission of, on photolysis, 14, 33,
        49–50, 202

ozone
explosive decomposition of, 288
photolysis of, 9, 16–17, 35–8; flash,
17, 36, 118, 155
vibrationally excited oxygen molecules from, 37, 120, 155, 242

P_2 radical, vibrationally excited, 190
paraffinoid solutions, photolyses in, 8,
43–4
paraffins, pyrolyses of, 330–51; see also
alkanes, hydrocarbons
pentachloroethane, dehydrochlorination of, 88
pentane
oxidation of, 252, 256, 258
pyrolysis of, 331
peracid, in cool-flame ignition, 244
perchloric acid, photovoltaic effect in
aqueous solution of, 30
perhydroxyl radical, see HO_2
peroxides
as branching agents, 281
cool flames from decomposition of,
268
organic, photolysis of, 49–56
in oxidation of hydrocarbons, 11,
231, 232, 252–3, 257–8, 261, 280,
282
perturbation methods, 94
pH
and chain carriers in decomposition
of H_2O_2, 34–5
choice of, for observation of protonic and ground-state equilibria,
107
PH radical, 129
phenols, polyhydric, chemiluminescence from, 38
phenoxyl radical, 107
phenyl radical, 101, 125
from benzene and mono-halo benzenes, 119
from benzoyl peroxide, 200, 217
phosgene from reaction of chlorine
and CO, 72
phosphine
deactivations by (electronic), 187
deactivation of (vibrational), compared with that of PD_3, 164
flash photolysis of, 118
oxidation of, 126, 129
phosphorus, oxidation of, 241, 242,
289
photobiology, 107–8

photochemical equivalence, law of,
1
photochemistry
foundations of, 2, 19
spectroscopy and, 5–6
photochlorination, 1
absolute rate constants for, 71–4
competitive, 84–8
in gas phase, 64–92
of hydrocarbons, mechanisms for,
74–84
theoretical interpretation of reaction
rates of, 88–90
photodecarboxylation, 45
photogelation, 57
photolysis, see individual substances,
and flash photolysis
photolytic cell, 31
photometry
continuous, analysis by, 68–9
plate, 103
photomultiplier cells, 114, 127–8
for recording distortion of shock
fronts, 146
photovoltaic effects
use of flash photolysis to study, 18
in water, 30–3
pic d'arête, in slow oxidations, 260
PN radical, vibrationally excited, 190
polarizability, correlation of, with
quenching cross-section, 175
polarography, 13
for investigation of photovoltaic
phenomena, 30–1
polyacetaldehyde, 223, 224
polyethylene, 226
polyformaldehyde, 270–2, 224
polymers, 198–228
cross-linking in, 201, 217–18, 219–20
macromolecular radicals of, 216
precipitation of, and rate of polymerization, 207, 208, 214
polymerization
of aldehydes, 14, 220–4
auto-acceleration during, 204–9,
215
cationic and Ziegler–Natta, 224–7
flash photolysis, and study of, 121
at high pressures, 210–12
photosensitization of, 56–9
initiation of, 199–204
polymethylvinylketone
graft co-polymers with, 219
photolysis of, 14, 45, 215–17, 220
polyoxymethylene, 223

polystyrenes, chlorinated, cross-linking of, 219–20
polyvinyl acetate, cross-linking of, by Friedel-Crafts catalysts, 219
polyvinyl chloride, cross-linking of, by Friedel-Crafts catalysts, 219–20
porphyrins, flash photolysis of, 107
potassium salts, in photovoltaic reactions, 31–2
predissociation, 2, 26, 40, 41, 65, 148
  in spectra of free radicals, 100, 101, 125
pressure
  and decomposition of benzoyl peroxide, 199–201
  and induction period of cool-flame oxidation, 262
  and oxidation of methane, 251
  and polymerizations, 199, 210–12
  pulses of, in cool flame oxidation, 261, 262; at lower sensitizer ignition limit, 301
  and recombination of radicals, 201
  relations of ignition boundaries to temperature and, 260, 288–9, 293
propagation of chain reactions
  in ignitions, 293, 299, 312, 317
  in oxidation of hydrocarbons, 229, 321, 233, 276–81; of hydrogen, 293
  in photochlorinations, 84, 85
  in polymerizations, 221
  reversal of, in H_2/Cl_2 reaction, 70
propane
  cracking/oxidation ratio for, at different temperatures, 269
  deactivations by (electronic), 173, 186, 187
  inhibition of H_2/O_2 ignition by, 303
  oxidation of, 239, 271; cool-flame, 256, 260, 261–2
  from photolysis of methyl butyl ketone, 7
  pyrolysis of, 336, 337, 338, 344–8
propionamide, photolysis of, 48, 49
i-propyl benzene, hydrogenation of ethyl radicals by, 47
propylene (propene)
  effect of, on oxidation of propane, 271
  inhibition of pyrolyses of paraffins by, 331, 332, 333
  oxidation of, 271–2; oxidation of propane catalysed by, 270

from oxidation of propane, 261, 269, 270
from photolysis of methane, 9
proteins, flash photolysis of, 15, 107
protonation, rate constants for, 106
pulse radiolysis, 108
pulsed heating methods, 94
pyrene, photosensitization of styrene polymerization by, 218
pyrogallol, chemiluminescence from, 38
pyrolyses
  of hydrocarbons in presence of NO, 125
  of paraffins, 269, 330–51; inhibition of, 331–6; mechanisms of, 339–45
pyruvic acid, photolysis of, 45

quantum theory, 1, 2
quantum yields
  in flash photolysis, 99
  in H_2/Cl_2 reaction, 66, 69
  in initiation of polymerization, 57–8
  in photochlorinations, 66, 71
  in photolyses, 29, 33–4, 35, 36, 37, 39, 40, 41, 43, 50–2
  reduced by cage effect, 105
quenching coefficients (relative) for dyes, 56
quinones, flash photolysis of, 107

radicals, see free radicals
radiolysis
  and photolysis, of H_2O_2, 34; of water, 29
  by pulse of electrons, 108
rate constants
  for abstractions of hydrogen, 302
  for co-polymerizations, 214, 217
  for deactivations (electronic), 17, 170
  for decomposition of benzoyl peroxide, 199
  for flash photolyses, 99–100, 104, 107
  measurement of, 71–4
  for propagation and termination of polymerization, 206, 207, 211–12
  for protonation and deprotonation, 106
  for reactions in ignitions, 300; in oxidation of CN, 120; in oxidation of hydrocarbons, 231, 282; in pyrolyses of paraffins, 333, 335, 346–8; in recombination of atoms, 103, 115, 120–1

rate equations, 87–8, 335
reaction rates
of photochlorinations, 88–90
of pyrolyses, 348
reaction vessel surfaces
catalytic reaction at, 240
in H_2/Cl_2 reaction, 70, 77, 79, 80
in pyrolyses of paraffins, 336–8
propagation and termination of
chain reactions on, in oxidation of
hydrocarbons, 233, 234, 276
removal of heat at, 114, 263–6, 290–2
removal of hydrogen atoms at, 337,
338, 341, 342, 348
removal of HO_2 radicals at, 293
recombination of chlorine atoms on,
79
and yields of minor products in
oxidation of hydrocarbons, 274–5
280
reaction vessels
'ageing' of, 66
conditioning of, 337
limiting diameter of, for oxidations,
254
shape of, and ignitions, 292
size of, and ignitions, 301
recombination of atoms and radicals
germinate (cage effect), 28, 201
of halogens, 103, 115–16
in liquid and gas phases, 28
in oxidation of hydrocarbons, 229–
30, 236, 260
in photolysis of aldehydes and
ketones, 201
primary and secondary, 27–8
relaxation techniques, 94, 95, 106
relaxation times, electronic and vibrational, compared, 177, 179
resonance interaction, long-range,
185
resonance quenching, 12, 177
rhenium carbonyl, as initiator of vinyl
polymerization, 57, 59
riboflavin, flash photolysis of, 107
ring contraction, in photolysis of
cyclic ketones, 46
ring ethers, as products of cool-flame
oxidation of hydrocarbons, 256
rotating sector technique, 56, 59, 71–2
rotational energy, 26, 125, 135
transfer of, to translational energy,
145–9
rotational entropy of activation, 89
rubber, graft polymers on, 217

Rydberg series, determination of
ionisation potentials of free
radicals from, 125, 126

scavengers, 4, 31
schlieren photography, for computing
vibrational relaxation time, 149
Schüler discharge, 113
Schumann-Runge region of spectrum,
12, 16, 104
selenium
deactivation of (electronic), 169–72
deactivation (vibrational) of CSe
radical by, 157
semiquinone radicals, 105, 107
sensitizers
and ignition boundaries, 290
and thermal ignitions, 294–302
SH (sulphydryl) radical, 129
shock-tube techniques, 94, 116, 149
shock waves, deduction of rotational
relaxation rates from, 146, 147
silicon hydride, deactivation of (vibrational), compared with that of
SiD_4, 164
SO radical, 118, 129
chemiluminescent reaction of oxygen with, 129
vibrationally excited, 190
sodium
deactivation of (electronic), 12, 166,
175, 176
electronically excited, 133
transfer of electronic energy from
mercury to, 181, 182, 185
solution
'cage effect' on reactions in, 27–8,
105
kinetics of flash photolysis in, 105–7
photochemistry in, 26–63
pulse radiolysis in, 108
study of radicals in, 101
solvents
deactivations by, 26
dielectric constants of, and rate of
cross-linking of polymers, 220
hydrocarbon, unsaturation produced in, by free radicals, 8, 15,
48, 49
interaction of, with carbonyl compounds, 43–4, 46; with ClO_2, 41
nature of, and rate of polymerization, 206–7
as third bodies in recombinations,
27

spectra, absorption, 5
  of excited states, 13, 101, 102, 124, 149
  of free radicals, 13, 100–1, 121–6
  of fuels, 129
  of transient species in flash photolysis, 13, 95–8, 109–10
spectra, emission, 5
  of excited states, 101–2, 149
  of free radicals, 100, 112–13
spectrometers, rapid-scanning, 96
spectroscopy
  flash, 95, 96
  kinetic, 96–8, 99, 103, 109, 118, 190
  and photochemistry, 5–6
  vacuum, 113
spin-orbit relaxation, 166, 170, 172
stannic chloride
  cross-linking of polymers by, 219–20
  promotes polymerization of formaldehyde, 222; of isobutene, 225
steric factors, in oxidation of hydrocarbons, 230
Stern-Volmer diagrams, for vibrational deactivation of NO, 160, 161
stopped-flow technique, 94, 107
styrene
  co-polymerization of, 213–14, 218–19
  effect of pressure on polymerization of, 199, 206, 210–12
  photosensitization of polymerization of, 57, 203–4
styrene-anthracyl radical, 219
sulphur
  deactivation (vibrational) of CS radical by, 156
  oxidation of, 189–90, 241
  reaction of, with hydrogen, 3, 23
sulphur dioxide
  afterglow emission of, 129
  deactivation by (rotational), 148
  deactivation by (vibrational), 154
  molecule of, 124
  photolysis of, 16
  spectrum of, 122
sulphur hexafluoride, deactivation (electronic) by, 173
sulphur oxides, flash photolysis of, 118
sulphur trioxide, photolysis of, 16, 190; flash, 155
sulphuric acid, in photovoltaic effects, 30, 33

sulphydryl (SH) radical, 129
surface polarity, and chlorinations, 3
surfaces of reaction vessels, see reaction vessel surfaces

tellurium dimethyl, as anti-knock compound, 128
temperature
  ceiling, for polymerization, 221–2
  of cool-flame oxidation, 261
  electronic and translational, 176, 179, 180
  during flash photolysis, 114–15
  and ratio of cracking to oxidation products of hydrocarbons, 232, 257, 269
  relation of ignition boundaries to pressure and, 260, 288–9, 293
  rotational and vibrational, 17
  and vibration-translation energy transfer, 149, 150
temperature coefficients
  classification of reactions by, 2
  negative, for oxidation of higher hydrocarbons, 234, 245, 257–9; and cool flames, 264; interpretation of, 281–3
  of recombination of iodine atoms, 103, 115
temperature-jump technique, 94
termination of chain reactions
  in ignitions, 293, 294, 305, 307, 312, 317
  in oxidation of hydrocarbons, 229, 234, 240, 253, 276, 281
  in polymerizations, 221
  in pyrolyses, 333, 347, 348
  quadratic, 294, 295, 301, 302, 307, 309
tetrachloroethylene, photochlorination of, 78, 80
tetralin, photosensitized oxidation of, 54, 55–6
tetralin hydroperoxide, photolysis of, sensitized by fluorenone, 52
textiles, 'tendering' of, 54
thallium, transfer of electronic energy from mercury to, 183, 185
third bodies in recombination of atoms and radicals, 27, 103, 114, 115
tin tetraethyl, as anti-knock compound, 128
titanium tetrachloride, promotes polymerization of isobutene, 225

toluene, sensitizes photolysis of benzoyl peroxide, 52
translational energy, 134, 135
  transfer to, of rotational energy, 145–9; of vibrational energy, 135–45
translational entropy of activation, 89
trichloroethylene, photochlorination of, 78, 88
trioxan, preparation of polymers from, 224
triplet states
  absorption spectra of, 14, 101–2
  annihilation reaction of, 103
  of carbonyl compounds, 105
  decay of, in solution, 106
  of methylene radical, 124
  of photosensitizers of polymerization, 203
  in photolyses of carbonyl compounds, 44
  quenching of, by oxygen, 103
tropyl radical, ionization potential of, 125–6
tungsten carbonyl, as initiator of vinyl polymerization, 59

ultrasonic dispersion, in study of
  transfer of rotational to translational energy, 145
  transfer of translational energy, 135
  vibrational deactivation, 133, 152, 163
unsaturation
  produced in hydrocarbon solvents by radicals, 8, 15, 48, 49
  produced in polyvinylchloride by Friedel-Crafts catalysts, 220
uranyl oxalate
  in actinometer, 13
  in photolytic cell, 31–2
i-valeraldehyde, photolysis of, 43
valeramide, photolysis of, 48, 49
vapour pressure, used to follow copolymerization, 214–15
velocity constants, see rate constants
vibrational energy, 134, 135
  of complex molecules, 107
  heteromolecular transfer of, 157–64
  interconversion of, with electronic energy, 165–80; with rotational energy, 164–5; with translational energy, 16, 133, 149–57

in nascent bonds, 190, 325
from photochlorinations, 181–4
from photolyses, 38, 46, 120, 155, 192
theory of transfer of 135–45
vibrational entropy of activation, 89
vinoxy radical, 125
vinyl acetate, graft polymers containing, 217
vinyl chloride
  co-polymerization of, 214–15
  photochemical initiators for polymerization of, 56–9
  polymerization of, 207–9
vinyl compounds, isomerization of, 106
vinylidene chloride
  co-polymerization of, 214–15
  polymerization of, 207
viscosity
  of solution, and rate of polymerization, 204, 206, 207, 212
  of solvent, and encounters of particles, 27

walls of reaction vessels, see reaction vessel surfaces
water
  and CO/O₂ ignition, 287, 316
  as co-catalyst in polymerizations, 225
  deactivations by, (electronic), 167, 169, 186; (vibrational), 157, 162, 191
  exchange of vibrational energy between oxygen and, 157
  and H₂/Cl₂ reaction, 2, 3, 67
  inhibitions by, of H₂/O₂ ignition, 306–8; of polymerization of acetaldehyde, 223
  from oxidation of alkanes, 241, 269
  photolysis of, 28–30, 118
  and photolysis of ozone, 16
  photovoltaic phenomena in, 30–3

xenon, deactivations by, (electronic), 173, 174; (rotational), 148

Ziegler-Natta polymerizations of α-olefins, 226–7
zinc oxide, sensitization of photolysis of water by, 30

AUTHOR INDEX

Agnew, W. G. and J. T. Agnew, 259, 261

Aivazov, B. V. and M. B. Neiman, 257

Airey, J. R., R. R. Getty, J. C. Polanyi and D. R. Snelling, 242, 243

— J. C. Polanyi and D. R. Snelling, 326

Akeroyd, E. I. and R. G. W. Norrish, 43

Alfrey, T. and C. C. Price, 213

Allen, A. C., and R. A. Holroyd, 29

Allen, P. E. M. and C. R. Patrick, 207

Andersen, V. S. and R. G. W. Norrish, 203, 218

Anson, P. C., P. S. Fredericks and J. M. Tedder, 86

Antokol'skii, V. L. and V. Ya. Shtern, 245

Apin, A. J. 321

Arnold, J. S., R. J. Browne and E. A. Ogryzlo, 39, 53

Ashmore, P. G., 297, 320, 326

— and J. Chanmugan, 72, 322

— F. S. Dainton and R. G. W. Norrish, 296

— and B. P. Levitt, 297

— and R. G. W. Norrish, 319

— and B. J. Tyler, 298, 300

— and T. A. B. Wesley, 322

Ausloos, P., 45

Axford, D. W. E. and R. G. W. Norrish, 250, 252, 254

Ayscough, P. B., A. J. Cocker and F. S. Dainton, 78, 82

— — and S. Hirst, 75, 78, 88

— — — and M. Weston, 78, 82

— F. S. Dainton and B. Fleischfresser, 78, 84

Bader, L. W. and E. A. Ogryzlo, 79, 80

Bailey, H. C. and G. W. Godin, 51

— and R. G. W. Norrish, 250, 251, 256, 259

Baldwin, R. R. 302

— and D. Bratten, 278

— N. S. Corney, P. Dovan, L. Mayor and R. W. Walker, 303, 306, 319

— D. Jackson, R. W. Walker and S. J. Webster, 304, 318

— and L. Mayor, 278, 279

Bamford, C. H. 48

— W. G. Bart, A. D. Jenkins and P. F. Onyon, 207, 214

— S. Brumby and R. P. Wayne, 59

— J. C. Casson and R. P. Wayne, 42

— P. A. Crowe, J. Hobbs and R. P. Wayne, 57, 59

— — and R. P. Wayne, 57

— and M. J. S. Dewar, 54, 56, 57

— G. C. Eastmond and V. J. Robinson, 57

— J. Hobbs and R. P. Wayne, 59

— A. D. Jenkins and R. Johnston, 213

— and R. G. W. Norrish, 7, 8, 11, 43, 45, 47

Bancroft, J. D., M. Hollas and D. A. Ramsay, 122, 124, 125

Bardwell, J. and C. N. Hinshelwood, 253, 257, 258

Barrett, J. and J. H. Baxendale, 29

Bartlett, P. D. and J. G. Traylor, 53

Barton, D. H. R. and K. E. Howlett, 79

— and P. F. Onyon, 79

Basco, N., A. B. Callear and R. G. W. Norrish, 17, 120, 151, 157, 158, 162, 192

— J. E. Nicholas, R. G. W. Norrish and W. H. J. Vickers, 118, 119, 120, 155

— and R. G. W. Norrish, 17, 37, 118, 119, 120, 155, 192, 242

Bass, A. N. and D. Garvin, 190

Bates, D. R., 172

Bauer, H. J., H. O. Kneser and E. Sittig, 151, 166

Baxendale, J. H. and J. A. Wilson, 29, 33

Beckett, A., A. D. Osborne and G. Porter, 105, 111

Bell, E. R., J. H. Raley, F. F. Rust, F. H. Senbold and W. E. Vaughan, 50

Ben-aim, R. and M. Lucquin, 260

Bengough, W. I. and H. W. Melville, 207

— and R. G. W. Norrish, 207, 208, 209, 214

Bennett, W. R., Jr, 181
— W. L. Faust, R. A. McFarlane and
   C. K. M. Patch, 188
Benson, S. W., 258, 281
— and A. M. North, 207
Benton, E. E., F. A. Matson, E. E.
   Ferguson and W. W. Roberts, 184
Berman, J. D., J. H. Stanley, W. V.
   Sherman and S. G. Cohen, 47
Berthelot, D. and H. C. Gaudechon, 6
Berthoud, A. and H. Bellerot, 71
Beutler, B. and H. Josephy, 181
Bevington, J. C., 200, 202, 214, 224
— G. M. Guzman and H. W. Melville,
   217
— H. W. Melville and R. P. Taylor,
   218
— and R. G. W. Norrish, 219, 222,
   223
Blackmore, D. R., G. O'Donnell and
   R. F. Simmons, 309
— and C. N. Hinshelwood, 346
Blat, E. I., M. I. Gerber and M. B.
   Neiman, 259
Bloch, B. M. and R. G. W. Norrish,
   43
Blundell, A. and G. Skirrow, 271
Blythe, A. R., 147
Bodenstein, M. and W. Z. Dux, 68, 69
— P. Harteck and E. Padelt, 40
— and I. Unger, 68, 69
Bolland, J. L. and G. Gee, 55
Bone, W. A. and R. E. Allum, 311
Bonner, B. H. and C. F. H. Tipper,
   258, 259, 272, 281
Booth, G. H. and R. G. Norrish, 15, 48
Borkowski, R. P. and P. Ausloos, 47
Borrell, P., 47
— and R. G. W. Norrish, 44, 47
Bowen, E. J., 39, 40, 53
— and W. M. Cheung, 26, 39, 41
— and R. A. Lloyd, 38, 39, 53
Bradley, J. N., 149
— G. A. Jones, G. Skirrow, and
   C. F. H. Tipper, 261
— and A. Ledwith, 41
Brady, W. T. and H. R. O'Neal, 224
Brennen, W. R. and G. B. Kistia-
   kowsky, 188
Bridge, N. K. and G. Porter, 105
Briggs, A. G. and R. G. Norrish, 18,
   68, 116
Brodskii, A. M., R. A. Kalinenko,
   K. P. Lavrovskii and L. V. Shevel'-
   kova, 343

Broida, H. P. and T. Carrington, 147
Brown, R. L. and W. Klemperer, 148,
   165
Bryce, W. A. and J. Chrysochoos, 334
Buckler, E. J. and R. G. W. Norrish,
   316, 327
Bueso-Sanllehi, F., 123
Bull, T. H. and P. B. Moon, 192
Burgess, R. H. and J. C. Robb, 232
Burgoyne, J. H., 257, 259
Burns, G. and D. F. Hornig, 116
— and R. G. W. Norrish, 117
Burns, W. G. and F. S. Dainton, 72
Buxton, G. and W. K. Wilmarth,
   34
Bykhovskii, V. K. and E. E. Nikitin,
   172, 188

Callear, A. B., 151, 155, 158, 162, 165,
   166, 175, 190
— and R. J. Cvetanović, 187
— J. A. Green and G. J. Williams,
   184
— and R. G. W. Norrish, 17, 118, 128,
   155, 167
— and R. J. Oldman, 170
— and I. W. M. Smith, 151, 159, 162,
   188
— and W. J. R. Tyerman, 155, 169,
   170
— and G. J. Williams, 162, 167, 186
— and J. F. Wilson, 174
Callomon, H. H. and D. A. Ramsay,
   118
Cario, G. and J. Franck, 17
Carrington, T., 176
Carruthers, J. E. and R. G. W. Norrish, 11, 14, 221, 222, 256
Cartlidge, J. and C. F. H. Tipper, 259,
   281
Cashion, J. K. and J. C. Polanyi, 242,
   243
Castellano, E. and H. J. Schumacher,
   36
Chapman, D. L., F. Briers and B.
   Walter, 71
— and F. B. Gibbs, 68, 69
Chapman, M. C. C., 68, 69
Charters, P. E., B. R. Khare and
   J. C. Polanyi, 242, 243
— and J. C. Polanyi, 191, 242, 243
Chernyak, N. Ya., V. L. Antonovskii,
   A. F. Revzin and V. Ya. Shtern,
   231, 239
— and V. Ya. Shtern, 257

Chiltz, G., P. Goldfinger, G. Huy-brechts, G. Martens and G. Ver-becke, 71, 78, 88
Chow, C. C. and E. F. Greene, 145, 150, 155
Christie, M. I., 335
— J. M. Collins and M. A. Voisey, 316
— A. J. Harrison, R. G. W. Norrish and G. Porter, 103, 114, 115
— R. G. W. Norrish and G. Porter, 14, 15, 114, 115
Chrysochoos, J. and W. A. Bryce, 334
Ciamician, G., 6
Clouston, J. G., Gaydon, A. G. and I. I. Glass, 176
Clyne, M. A. A. and B. A. Thrush, 125
Cohen, A. and G. Jung, 67
Copeland, A. W., O. D. Black and A. B. Garrett, 33
Cottrell, T. L., 277
— R. C. Dobbie, J. McClain and A. W. Read, 164
— and J. C. McCoubrey, 135, 146, 147
— and A. J. Matheson, 164
Cowan, G. C. and D. F. Hornig, 146
Crist, R. H. and O. C. Roehling, 319
Crone, H. G. and R. G. W. Norrish, 6
Cullis, C. F., A. Fish and D. W. Turner, 271, 276
— and C. N. Hinshelwood, 229, 253 256
— and E. J. Newitt, 276
Currie, C. L. and F. S. Dainton, 34, 35
— and D. A. Ramsay, 125
Cvetanović, R. J., 84, 186

Daen, J. and P. C. T. de Boer, 154
Dainton, F. S., 41, 71, 294
— and K. J. Ivin, 221
— K. J. Ivin and G. A. Creak, 90
— and D. G. L. James, 203
— D. A. Lomax and M. Weston, 75, 77, 78, 88
— and R. G. W. Norrish, 10
Dalby, F. W., 122, 124
Damon, G. H. and F. Daniels, 6
Daniels, M., 34
Darwent, B. de B. and F. G. Hurtu-bise, 167
Datz, S., and E. H. Taylor, 192
Davies, W., Jr. and W. A. Noyes, Jr. 44

Davy, H., 259
Day, R. A. and R. N. Pease, 259
Decius, J. C., 154
DeMore, W. B. and S. W. Benson, 42
— and O. F. Raper, 37
Dexter, R. W. and A. B. Trenwith, 342
Dickens, P. G., J. W. Linnett and O. Sovers, 177
Dingledy, D. P. and J. G. Calvert, 247, 270
Dixon, R. N., 122, 123, 125
— and D. A. Ramsay, 122, 125
Doering, W. E., R. G. Buttery, R. G. Laughlin and N. Chaudhuri, 41
Donovan, R. J. and D. Husain, 115, 173, 186
Douglas, A. E. and W. J. Jones, 122, 125
Dressler, K. and D. A. Ramsay, 122, 125
Drysdale, D. and R. G. W. Norrish, 250, 280
Dubovitskii, F. I., 246
Durie, R. A. and D. A. Ramsay, 116, 117
Dusoleil, S., P. Goldfinger, G. Mar-tens, A. M. Mahieu-Van der Auwera and D. Van der Auwera, 88
Dyachokovskii, F. S., N. N. Bubnov and A. E. Shilov, 230
— and A. E. Shilov, 230

Edgecombe, F. H. C. and R. G. W. Norrish, 202
— R. G. W. Norrish, and B. A. Thrush, 14, 16
Eigen, M., 94
Einstein, A., 1
Eltenton, G. C., 332
Emeleus, H. J., 259
Engel, J., A. Combe, M. Letort and M. Niclause, 338
Enikolopyan, N. S., 311
Erhard, K. H. L. and R. G. W. Norrish, 128
Evans, M. G., and M. Polanyi, 27, 90

Falconer, J. W. and J. H. Knox, 269
Falconer, W. E., J. H. Knox and A. F. Trotman-Dickenson, 277
Farkas, L., 47
Fettis, G. C. and J. H. Knox, 73, 88, 90

Fettis, G. C., J. H. Knox and A. F. Trotman-Dickenson, 89
Findlay, F. D. and J. C. Polanyi, 191
Fish, A., 280
Fitzsimmons, R. V. and E. J. Bair, 155
Foner, S. N. and R. L. Hudson, 277
Foord, S. G. and R. G. W. Norrish, 250, 251, 254, 294, 295, 302
Franck, J., 2
— and E. Rabinovitch, 27
Frank-Kamenetskii, D. A., 266, 291
Frey, F. E. and H. J. Hepp, 343
Frey, H. M., 84, 124
— and G. B. Kistiakowsky, 42
Fricke, H. and E. J. Hart, 29, 35
Frish, S. E. and O. P. Bachhova, 181
Frosch, R. P. and G. W. Robinson, 188
Fujimoh, M. and D. G. E. Ingram, 33
Fusy, J., G. Sacchi, R. Martin, A. Combes, and M. Niclause, 337, 338, 341, 347

Gausset, L., G. Herzberg, A. Lagerquist and B. Rosen, 122, 125
Gaydon, A. G. and I. R. Hurle, 149, 176, 178
— and H. G. Wolfhard, 260
Gibson, Q. H., 107
Givens, W. G. and J. E. Willard, 116
Godfrey, T. S. and G. Porter, 102
— G. Porter and P. Suppan, 107
Goldfinger, P., 75, 77, 88, 270
— G. Huybrechts and G. Verbeke, 88
— and H. S. Johnston, 89
Goodeve, C. F. and N. O. Stein, 29
Gordon, A. S., 337, 338, 341, 342, 346, 347
Gray, J. A., 232
Gray, P. and J. C. Lee, 310
Greene, E. F. and J. P. Toennies, 149
Griffiths, J. G. A. and R. G. W. Norrish, 3, 9, 18, 67, 118, 294, 295, 297
Grossweiner, L. I. and W. A. Mulac, 107
Guillet, J. E. and R. G. W. Norrish, 45, 204, 215, 217, 219, 220

Hammond, G. S., C. E. Boozer, C. E. Hamilton and J. N. Sen, 54
Harding, A. J. and R. G. W. Norrish, 250, 251, 254, 271
Harrison, A. G., L. R. Honnen, H. J. Dauben and F. P. Lossing, 126

Hartridge, H. and F. J. W. Roughton, 94
Hay, J., J. H. Knox and J. M. C. Turner, 270, 272, 274, 280, 281
Heidt, L. J., 33, 36
— and G. S. Forber, 36
— and V. R. Landi, 35
Heller, C. A. and A. S. Gordon, 343
Henri, V. and M. C. Teves, 2
Herr, K. C. and G. C. Pimental, 94, 96, 109
Herschbach, D. R., 192
Herzberg, G., 29, 100, 112, 121, 122, 124, 125, 180, 182
— and J. W. C. Johns, 122, 124
— and D. A. Ramsay, 121, 122, 124
— and D. N. Travis, 119, 122
— and R. D. Vernon, 122
— and P. A. Warsop, 125
Herzfeld, K. F., 154
— and T. A. Livovitz, 135, 144, 147, 150, 157, 162
Heyrovsky, M. and R. G. W. Norrish, 30
Hilpern, J. W., G. Porter and L. J. Stief, 106
Hirschfelder, J. O., C. F. Curtis, and R. B. Bird, 135, 144, 146
Hirschlaff, E. and Norrish, R. G. W., 11
Hoare, D. E. and A. D. Walsh, 319
Hochanadel, C. J., 34
Hoey, G. R. and K. O. Kutschke, 321
Holmes, R., G. R. Jones and R. Lawrence, 147
— — N. Pusat and W. Tempest, 148
Hooker, W. J. and R. C. Millikan, 154
Hsieh, M. S. and D. T. A. Townend, 246
Hunt, J. P. and H. Taube, 33
Hurle, I. R., 180
Husain, D. and R. G. W. Norrish, 129, 203, 310
Hutton, E., 80
— and M. Wright, 80
Huybrechts, G., L. Myers and G. Verbeke, 88

Ingold, K. O., F. J. Stubbs and C. N. Hinshelwood, 331
Irving, G. W. and J. H. Knox, 277

Jack, J., F. J. Stubbs and C. N. Hinshelwood, 331
Jackson, G. and R. Livingston, 106

Jackson, G. and G Porter, 106
Jackson, J. M. and N. F. Mott, 135,
140
Jarvie, J. M. and A. H. Laufer, 47
Javan, A., W. R. Bennett, Jr and D. R.
Herriott, 184
Jennings, K. R., 113
Jeunehomme, M. and A. B. F.
Duncan, 160
Johns, J. W. C., 123, 124
— S. H. Priddle and D. A. Ramsay,
121, 122, 124

Kane, C. P., E. A. C. Chamberlain
and D. T. A. Townend, 257, 259
Kane, R. A., J. J. McGarvey and
W. D. McGrath, 190
Kapralova, G. A. and A. E. Shilov, 230
— E. A. Trofimova, L. Yu. Rusin,
A. M. Ghaikin and A. E. Shilov,
243
Karl, G. and J. C. Polanyi, 176
Karmilova, L. V., N. S. Enikolopyan
and A. B. Nalbandyan, 241, 311,
312
— N. S. Enikolopyan, A. B. Nalban-
dyan and V. I. Il'in, 233
Kaufman, F., 133
Kerr, J. A. and A. F. Trotman-
Dickenson, 270, 343
Khan, A. V. and M. Kasha, 38, 39, 53
Khan, M. A., R. G. W. Norrish and
G. Porter, 14, 121
Kirk, A. D. and J. H. Knox, 278, 282
Kirkbride, F. W. and R. G. W.
Norrish, 4, 5, 43
Kistiakowsky, G. B., 36
— and R. Spence, 241
Kleimenov, N. A. and A. B. Nal-
bandyan, 232
Knowles, A. and E. M. F. Roe, 107
Knox, J. H., 86, 237, 251, 270, 278,
281
— and R. L. Nelson, 86
— and R. G. W. Norrish, 231, 232,
250, 251, 256, 257, 258, 259, 261
— R. G. W. Norrish and G. Porter,
14, 105
— and J. Riddick, 78, 84
— R. F. Smith and A. F. Trotman-
Dickenson, 277
— and A. F. Trotman-Dickenson, 89
— and J. M. C. Turner, 237, 277
— and C. H. J. Wells, 237, 270, 271,
276

Kondratiev, V. N., 112, 259
Kossiakoff, A. and F. O. Rice, 343
Kowalskii, A. A., 293
— P. Ya. Sadovnikov and N. M.
Chirkov, 235
Küchler, L. and H. Theile, 341
Kudryavtseva, Yu. I., B. V. Pavlov
and V. I. Vedeneev, 338
Kuppermann, A. and J. G. Larson,
332
Kyriacos, G, H. R. Menapace and
C. E. Boord, 256

Laidler, K. J., 178, 186
— N. H. Sagert, and B. W. Wojcie-
chowski, 344, 345
— and B. W. Wojciechowski, 333,
338, 341, 342, 344
Lambert, J. D., 161
— A. J. Edwards, D. Pemberton and
J. L. Stretton, 163
— D. G. Parks-Smith and J. L.
Stretton, 163
— and R. Salter, 152-3, 154, 161
Land, E. J. and G. Porter, 101
Landau, L. and E. Teller, 133, 135
Langer, A., J. A. Hipple and D. P.
Stevenson, 126
Leathard, D., J. H. Purnell and C. P.
Quinn, 345, 347
Leermakers, P. A. and G. F. Versley,
45
Legay-Sommaire, N., L. Henry and
F. Legay, 159
Letort, M., 223
— and A. -J. Richard, 223
Levitt, B. P. and D. F. Hornig, 146
Lewis, B., 237
— and G. von Elbe, 241, 256, 293
Lewis, G. N. and M. Kasha, 101
Lin, M. C. and M. H. Bade, 341, 342,
346, 347
Linnett, J. W. and M. H. Booth, 79
Linschitz, H. and K. Sarkanan, 107
— C. Steel, and J. A. Bell, 106
Lipman, R. D. A. and R. G. W.
Norrish, 226, 227
Lipscomb, F. J., R. G. W. Norrish
and G. Porter, 67, 104
— — and B. A. Thrush, 15, 118, 120,
155, 190
Livingston, R., G. Porter and M.
Windsor, 107
— and V. Ryan, 107
Lloyd, W. G. and C. E. Lauge, 54

Losev, S. A., and A. I. Osipov, 150
Lossing, F. P., K. U. Ingold and I. H. S. Henderson, 126
Lucquin, M., 260
Lyon, R. K., 42

McCormac, M., and D. T. A. Townend, 249
McCoubrey, J. C., R. C. Milward and A. R. Ubbelohde, 154
Macrae, K. and R. G. W. Norrish, 250
McDonald, R. D. and R. G. W. Norrish, 11
McDowell, C. H. and J. R. Thomas, 244
McGowan, I. R. and C. F. H. Tipper, 257
McGrath, W. D. and R. G. W. Norrish, 16, 36, 37, 118, 120, 155, 242, 243
McKellar, J. F. and R. G. W. Norrish, 231, 245, 250, 251, 255, 271
McKinley, J. D., D. Garvin and M. J. Boudart, 242
McMillan, G. R., J. G. Calvert and J. N. Pitts, 44
McNesby, J. R., C. M. Drew and A.S. Gordon, 343
Mains, G. J., J. L. Roebber and G. K Rollefson, 121
Malherbe, F. E. and A. D. Walsh, 262
Mann, D. F. and B. A. Thrush, 118
Markham, M. C. and K. J. Laidler, 30, 33
Marshall, J. G. and P. V. Rutledge, 35
Marshall, R. and N. Davidson, 103, 114
Marshall, R. M. and C. P. Quinn, 338, 342, 347
Martens, G., 270
Martin, J. T. and R. G. W. Norrish, 14, 50, 201
Martin, R., M. Dzierzynski and M. Niclause, 337, 338, 344, 345
— and M. Niclause, 336
Massey, H. S. W. and E. H. S. Burhop, 180, 185
Matheson, M. S. and L. M. Dorfman, 108
Matland, C. G., 168
Merer, A. J. and D. N. Travis, 122
Merrett, F. M., 217
— and R. G. W. Norrish, 206, 210
Meyer, R. T., 109

Mies, F. H., 142, 144
Millikan, R. C., 150, 151, 160, 162
— and C. A. Osburg, 164
— and D. R. White, 149, 164
Minkoff, G. J. and C. F. H. Tipper, 251, 252, 257, 261, 269
Mitchell, A. C. G. and M. W. Zemansky, 119, 133, 166, 167, 168
Montashyan, A. A. R. I. Moskina and A. B. Nalbandyan, 232
Montashyan, A. A., and A. B. Nalbandyan, 232
Montroll, E. W. and K. E. Shuler, 154
Morgan, J. E. and H. I. Schiff, 157–8
Mott, N. F. and H. S. W. Massey, 180
Mulcahy, M. F. R., 255
Mullen, J. D. and G. Skirrow, 271, 272
Müller, A., B. Rumberg and H. T. Witt, 108
Müller, K. and H. J. Schumacher, 70, 71

Nagai, Y. and C. F. Goodeve, 40
Nalbandyan, A. B., 246
— and N. V. Fok, 232
Neiman, M. B. and L. Egorov, 241
— and A. I. Serbinov, 235
Nelson, L S and J. L. Lundbergh, 105
Newitt, D. M. „ L. S. Thornes, 239, 258, 259
Nicholson, A. E. and R. G. W. Norrish, 199, 206, 210
Niclause, M., R. Martin, A. Combes and M. Dzierzynski, 338, 339, 344
Nikitin, E. E., 142, 151, 152, 172
Norman, I. and G. Porter, 100
Norrish, R. G. W., 2, 3, 6, 7, 9, 10, 12, 16, 67, 68, 93, 104, 114, 129, 135, 201, 233, 242, 243, 250, 251, 252, 256, 257, 270, 272, 283, 294, 320
Norrish, R. G. W. and M. E. S. Appleyard, 6, 7, 43
— and P. G. Ashmore, 294, 320
— and C. H. Bamford, 7, 8, 9, 43
— and A. G. Briggs, 326
— and E. F. Brookman, 204, 213
— H. G. Crone, and D. D. Saltmarsh, 6, 43
— and F. S. Dainton, 294, 295, 296
— and S. G. Foord, 233, 310
— and J. G. A. Griffiths, 3, 10, 224, 320
— and D. Hussain, 310

Norrish, R. G. W. and F. W. Kirkbride, 5
— and L. H. Long, 5
— and G. H. Neville, 9
— and W. A. Noyes, Jr, 5
— and G. A. Oldershaw, 16, 118, 129,
155, 190
— and S. Paszyc, 18, 31
— and D. Patnaik, 12, 311
— and G. Porter, 12, 14, 93, 104, 105,
113, 126, 250, 276, 294
— — and B. A. Thrush, 105, 126, 127,
128, 316
— and G. L. Pratt, 331, 334, 335
— and J. D. Reagh, 248, 254
— and E. K. Rideal, 3, 31
— and M. Ritchie, 69, 320
— and K. E. Russell, 225
— and M. H. Searby, 15, 50, 51, 201
— and J. P. Simons, 203, 204, 218
— and F. F. P. Smith, 3
— and R. R. Smith, 205, 206, 207, 215
— and W. MacF. Smith, 12
— and G. W. Taylor, 250, 251
— and B. A. Thrush, 113
— and J. Wallace, 233, 250, 251, 294,
312
— and R. P. Wayne, 36, 37, 39, 47
— and A. P. Zeelenb- 14, 129
North, A. M. ? .. Reed, 207
Noyes, R. . .., 20
Noyes, W. A., Jr. and P. A. Leighton, 186

Oldenberg, O., 133
— and F. F. Riecke, 112
Oldershaw, G. A. and R. J. Cvetanović, 84
Oster, G., 57

Padley, P. J. and T. M. Sugden, 176
Palmer, H. B. and D. F. Hornig, 116
Paneth, F. A. and W. Hofeditz, 4
Pariser, R., 102
Patnaik, D. and R. G. W. Norrish,
250, 255
Paul, D. E. and F. W. Dalby, 119
Pease, R. N., 330
— and E. S. Durgan, 330
— and W. R. Munro, 257
Perkin, W. H., 259
Phelps, A. V., 188
— and S. C. Brown, 183, 184
Plesch, P. H., 225
Polanyi, J. C., 177, 178, 189, 190, 191,
192, 326

— and S. D. Rosner, 242, 243
Polyak, S. S., N. S. Enikolopyan, and
V. Ya. Shtern, 241
— and V. Ya. Shtern, 231
Poroikova, A. I., and A. B. Nalbandyan, 232
— V. V. Voevodskii and A. B. Nalbandyan, 232
Porter, G., 12–13, 93, 97, 99, 101, 103,
105, 113
— and G. Black, 118
— and J. A. Smith, 103, 115
— and G. Strauss, 108
— Z. G. Szabo, and M. G. Townsend
103, 115
— and B. Ward, 101, 119, 125
— and P. West, 103
— and F. Wilkinson, 105, 106
— and M. Windsor, 102
— and M. R. Wright, 101, 106, 116,
118, 119, 125
Poutsma, M. L. and R. L. Hinman, 84
Pouyet, B., 48
Pratt, G. L. and J. H. Purnell, 330,
334, 335
Preston, K. F., 313
Prettre, M., P. Dumanois and P. C.
Laffitte, 259
Pringsheim, P., 133, 175, 181, 183
Pritchard, H. O., J. B. Pyke and A. F.
Trotman-Dickenson, 73, 86
Purnell, J. H. and C. P. Quinn, 330,
332, 336, 337, 339, 341, 342, 346,
347

Quinn, C. P., 330, 331, 332, 334, 335,
337, 341, 342, 345, 346, 347

Rabinowitch, E. and W. C. Wood, 27,
114
Ramsay, D. A., 122
Raper, O. F. and W. B. DeMore, 37
Rapp, D., 162
— and P. Englander-Golden, 162
Rautian, S. G. and I. I. Sobelman, 181
Renner, R., 123
— and E. Teller, 123
Repa, L. A. and V. Ya, Shtern, 231
Revzin, A. F., G. B. Sergeev and V.
Ya. Shtern, 232
Rice, F. O., 330, 339
— and K. F. Herzfeld, 330, 336, 348
— W. F. Johnston and B. L. Evening,
330, 341
— and R. E. Varnerin, 332

Rideal, E. K. and R. G. W. Norrish, 1
Ritchie, M., and R. G. W. Norrish, 3, 9, 68, 69
Rodebush, W. H. and W. C. Klingelhoefer, 73
Rudberg, E., 27
Rüppel, von H., V. Bültemann, and H. T. Witt, 97
Rusin, L.Yu, A. M. Chaikin and A. E. Shiov, 243
Russell, K. E., and J. Simons, 103, 114, 115

Sadovnikov, P. Ya., 235
Sagert, N. H., and K. J. Laidler, 337, 340
— and B. A. Thrush, 188
Saltmarsh, O. D. and R. G. W. Norrish, 6, 9, 43, 45
Sampson, R. J., 278
Sandler, S. and J. A. Beech, 256
Satterfield, C. N. and R. C. Reid, 232, 269
Scheer, M. D., 192
— and J. Fine, 168
Schmitz, R. and J. H. Schumacher, 70, 71
Schüler, H., 113
Schultz, G. V. and F. Blaschke, 205
Schumacher, R., 70, 71
Schwartz, R. N. and K. F. Herzfeld, 144
— Z. I. Slawsky, and K. F. Herzfeld, 133, 140, 144, 150, 152, 154, 155
Seakins, M. and C. N. Hinshelwood, 234–6, 253, 254, 258
Sefton, V. B., and D. J. Leroy, 343
Semenov, N. N., 112, 229, 230, 231, 233, 242, 243, 250, 251, 263, 269, 290, 291, 292
— and A. E. Shilov, 243
Sergeev, G. B. and V. Ya. Shtern, 232
Setser, D. W., R. Littrell and J. C. Hassler, 42
— B. S. Rabinowitch and D. W. Plazek 187
Sette, D., A. Busala and H. C. Hubbard, 154
Shilman, A. and R. A. Marcus, 121
Shtern V. Ya., 241, 251, 252, 257, 261, 262, 269, 270, 271, 272, 281
— and S. S. Polyak, 245
Shu, N. W. and J. Bardwell, 257, 259
Sigal, P., 45
Simons, J. P. and A. J. Yarwood, 118

Skirrow, G., 271
— and A. Williams, 271, 276
Sokdova, N. A., A. M. Markevitch and A. B. Nalbandyan, 230
Spence, K. and D. T. A. Townend, 246
Spence, R., 235, 241
Srinivasan, R., 44, 45, 47
Standinger, H., 223
Staveley, L. A. K., 331
Steacie, E. W. R., 112, 187
Steiner, H. and E. K. Rideal, 73
Steinfeld, J. I. and W. Klemperer, 148, 165
Stepp, E. E. and R. A. Anderson, 184
Stepukovich, A. D. and L. V. Derevenskikh, 343
Strausz, O. P. and H. E. Gunning, 186, 187
Stretton, J. L., 154, 164, 165
Stubbs, F. J. and C. N. Hinshelwood, 331
— K. U. Ingold, B. C. Spall, C. J. Danby and C. N. Hinshelwood, 331

Takayanagi, K., 144
Tanczos, F., 154
Taube, H., 35, 41
Taylor, G. W., 273
Taylor, H. A., and H. Bender, 30
Thomas, J. K. and E. J. Hart, 29
Thomas, J. M. and R. G. W. Norrish, 230
Thorson, W. R., 172
— and J. W. Moskowitz, 172
Thrush, B. A., 113, 118, 119, 121, 122, 125, 128
— and J. J. Zwolenik, 118, 119, 125
Thynne, J. C. J. and P. Gray, 230
Tipper, C. F. H., 244
Titlestadt, N., 31
Topps, J. E. C. and D. T. A. Townend, 260
Townend, D. T. A., 257
— and E. A. C. Chamberlain, 246
— and M. R. Mandlekar, 259
Townes, C. H. and A. L. Schawlow, 135
Travers, M. W., 223
Traylor, T. G., 54
Trotman-Dickenson, A. F., 73
Tuesday, C. S. and M. Boudart, 157
Turner, J. M. C., 271, 272, 280
Tyler, B. J., 298, 300
— and P. G. Ashmore, 298, 300

Ubbelohde, A. R., 252, 256, 258
Ueberreiter, K. and W. Bruns, 52

Vanderslice, J. T., E. A. Mason and
    W. G. Maisch, 188
Van't Hoff, J. H., 209
Vasil'ev, I. N. and V. A. Kronganz, 52
Vedeneev, V. I., A. M. Chaikin and
    A. E. Shilov, 249
— G. N. Gerasimov and A. P. Purnal,
    35
Voevodskii, V. V., 332, 336, 344,
    345
— and N. I. Vedeneev, 241
Volman, D. H., 187
— and J. C. Chen, 33
— and R. Seed, 33
Voronkov, V. G., and N. N. Semenov,
    246

Wall, L. A. and W. J. Moore, 332
Walsh, A. D., 122, 123, 253, 256, 257,
    258, 266, 273

Walters, W. D., 230
Watson, J. S. and B. M. B. Darwent,
    232
White, A. B. and E. I. Gordon, 184
White, A. G., 246
White, D. R., 155, 164
White, J. U., 112
Wilkinson, F., 188
Winter, T. G., 154
Witteman, W. J., 154
Wojciechowski, B. W. and K. J.
    Laidler, 333
Wray, W. L., 151, 152
Wright, F. J., 118

Yamazaki, H. and R. J. Cvetanović,
    174

Zahra, A. and W. A. Noyes, 44
Zeelenberg, A. P., 280
— and A. F. Bickel, 270, 274, 280
Zener, C., 133
Ziegler, K. and G. Natta, 226

For EU product safety concerns, contact us at Calle de José Abascal, 56–1°,
28003 Madrid, Spain or eugpsr@cambridge.org.

www.ingramcontent.com/pod-product-compliance
Ingram Content Group UK Ltd.
Pitfield, Milton Keynes, MK11 3LW, UK
UKHW010852090126
466816UK00011B/180